Strategies for Engineering Communication

Susan Stevenson
Simon Fraser University

Steve Whitmore
Simon Fraser University

With a Chapter by Margaret Hope

John Wiley & Sons, Inc.
New York / Chichester / Weinheim / Brisbane / Singapore / Toronto

Acquisitions Editor   *Joseph Hayton*
Editorial Assistant   *Steven Peterson*
Marketing Manager   *Katherine Hepburn*
Senior Production Editor   *Norine M. Pigliucci*
Senior Designer   *Dawn Stanley*
Production Management Services   *Suzanne Ingrao*
Cover Image   © *Corbis Digital Stock*

This book was set in *Times Roman* by *UG / GGS Information Services, Inc.* and printed and bound
by *Hamilton Printing*
The cover was printed by *Phoenix Color Corp.*

This book is printed on acid-free paper.

Library of Congress Cataloging in Publication Data:
Stevenson, Susan.
    Strategies for engineering communication / Susan Stevenson, Steve Whitmore; with a
    chapter by Margaret Hope.
        p. cm.
    Includes bibliographical references.
    ISBN 0-471-12817-1 (pbk. : alk. paper)
        1. Communication in engineering.   I. Whitmore, Steve   II. Title.

    TA158.5 .S74 2001
    620'.001'4—dc21   2001033008

Printed in the United States of America

10 9 8 7 6 5 4 3 2 1

Preface

The main goal of this text is to help you prepare for the communicative challenges of the engineering workplace. We do so by outlining the importance and relevance of rhetorical principles and by offering an array of strategies you can apply to improve your efficiency, effectiveness, and flexibility in writing a wide range of documents. We also aim to help you increase your confidence and effectiveness when presenting, when speaking less formally, and when working on a team.

Rhetorical effectiveness and *strategies* are key concepts underpinning our approach. We want you to primarily focus on strategies for achieving your goals and not merely on a set of rules (which may or may not be applicable or effective in any given situation). We also want you to think critically about the various elements of a rhetorical situation and to use those elements to determine for yourselves the appropriate content, organization, style, and format for a document. These concepts help strengthen confidence and reduce the need to copy models or to use boilerplate text.

We consider the focus on rhetorical principles and on strategies to be major strengths of this text. Another strength is that it directly addresses you, as engineering students, and your communicative needs. This text is engineering-centered in terms of being based on real-life experience, geared to practical applications, and supported by many examples written by student and practicing engineers.

CONTENT

The text can be divided into three main emphases. The first emphasis (Chapters 1 and 2) acknowledges that writing is an individual and idiosyncratic process and offers a range of strategies for planning, inventing, organizing, drafting, revising, and editing. We also offer a range of strategies for dealing with procrastination and overcoming writer's block. Three key features of this emphasis on writing processes are our approach to creativity, our focus on drafting, and our attention to the health issues relating to writing.

The second emphasis (Chapters 3, 4, and 5) draws attention to the importance of rhetorical situation, explores the features of persuasive and informative writing, and provides numerous strategies for improving the efficiency and effectiveness of both types of writing. From the perspective of our rhetorical, strategy-based approach, we present writing not as a solitary activity, but rather as highly interactive and interpersonal, requiring constant attention to the needs and responses of readers. We also address the growing complexity of engineering teams as they increasingly become not only cross-functional and international, but also virtual. Further, we include an important section on listening skills, address strategies for team writing, provide suggestions for making the most of routine workplace communications, and offer approaches for effectively chairing meetings. The chapter on oral presentations addresses the essential skill of preparing, practicing, and delivering both individual and team presentations.

The third emphasis (Chapters 6, 7, and 8) identifies a range of stylistic habits that affects readability and suggests strategies for improving order and emphasis, for clarifying the connections among ideas, and for increasing clarity and conciseness. In these chapters, we make an exception to our emphasis on strategies rather than rules by providing a punctuation guide organized in terms of rules for punctuating common sentence structures. We also focus on the rhetorical and dynamic nature of form, and offer general principles for understanding and employing the elements of format. In addition, we supply a general guide for referencing information sources. Finally, we discuss some specific documents engineers routinely write, from informal e-mail messages to formal reports and proposals. We provide sample documents, but also point to the limitations of models and to the need to follow the strategies developed throughout this text in order to adapt to specific contexts and to create new genres when the need arises.

FEATURES

Beyond its content and approach, other features of this text include checklists, heuristics, and exercises. Every chapter but the last includes checklists that you can use while creating or revising documents and while preparing or practicing for individual or team presentations. We also provide a number of heuristics, or lists of questions, you can use to analyze your writing processes, audiences, writing styles, and formats, to help guide your team writing, and to improve your delivery of oral presentations. Exercises at the end of chapters allow you to explore various aspects of communication and offer opportunities for practicing the strategies outlined in those chapters. As we explain in the following section, we envision these exercises as supplementing the writing assignments and oral presentations required for many engineering courses. If such integration is not possible, then the exercises we offer can help fill that void and can be used in conjunction with other relevant assignments that simulate real-world communication.

HOW TO USE THIS TEXT

One way to use this text is the way we use it—for communication courses that are integrated with core courses so that students write documents and prepare presentations that become assignments for both courses. It could also be used for engineering courses that incorporate writing assignments and presentations. In the series of one-credit courses we offer at Simon Fraser University, we use selected elements of the text. For example, our first course is corequisite with a course on engineering and society in which students must write persuasive research papers and so we concentrate on the writing process, rhetorical situation, persuasion, and research strategies. In the next course, students are writing reports, and we focus on informative writing, format, style, and genre. In an upper-level course, students are working on team projects, so we concentrate on project documents such as proposals and specifications, and deal with issues such as team writing, group dynamics, and listening skills.

In the same vein, we focus on oral presentations when students are preparing to give technical presentations and on workplace issues, résumés, and cover letters when students are applying for work terms. In other courses, various sections are used for review or to address specific problems, challenges, or weaknesses.

Of course, integration is not always possible, and this text also provides support for a wide variety of assignments and activities that could be required in stand-alone communication courses.

While we have arranged the topics in this text in a way that makes sense to us, we have also tried to organize and present information to maximize flexibility. We imagine instructors addressing issues in whatever order makes sense to them and having students move backward and forward through the text to suit their approach. For example, rhetorical situation and persuasion could be addressed first while students write proposals, and then students could be asked to critique their writing processes after their documents are drafted. Discussions and examples of various genres can be introduced at any point, and various sections can be highlighted or skipped over to accommodate the focus and goals of a particular course. Material not specifically covered in courses may nevertheless be useful reference material for engineering students at some other time, when they are working in teams, making technical presentations, applying for summer jobs, on work terms, and so on.

SUPPLEMENTS

Faculty who adopt this text have access to a website that contains teaching aids such as notes, figures, and tables that can be duplicated as overhead transparencies or pasted into a PowerPoint presentation. Additional sample documents are provided as .pdf files, and examples for revision exercises are also provided. This on-line teachers' manual is a work-in-progress that we will update as new examples become available, as we develop new techniques in the classroom, and in response to input we receive from those who adopt or use this text. For access to the manual, contact your local Wiley representative or visit the Wiley website at www.wiley.com/college, and search on 'Stevenson'.

HOW TO CONTACT US

We would appreciate hearing from those of you who use this textbook. We want to know about your successes, frustrations, suggestions, and questions. Our e-mail addresses are stevenso@sfu.ca and whitmore@cs.sfu.ca. To ensure a timely response, we recommend addressing your e-mail to both addresses.

ACKNOWLEDGMENTS

This text reflects an equal partnership and joint effort on the parts of Steve Whitmore and Susan Stevenson. We have worked together for many years and first developed much of the material in this text for handbooks used for the courses we teach in the engineering communication program that we jointly coordinate. A few sections were planned and drafted collaboratively, but most were drafted separately and then augmented and revised as drafts were passed back and forth until we achieved agreement with respect to content, arrangement, and style.

This collaborative effort involves many other people as well. Aside from Margaret Hope who drafted an entire chapter and parts of another, we received advice and examples from our colleagues in engineering, from engineering undergraduate students, both present and long-since graduated, and from the practicing engineers and other professionals who, over the years, have attended our continuing studies and in-house courses. In a very real sense, all students who have taken our courses and all faculty members who have taught co-requisite courses have played a role in creating this text.

We are especially grateful to Margaret Hope for authoring the chapter on oral presentations and for her help with the material on workplace communication. Without her contribution, this text would lack key information we believe to be essential to your success as engineers.

We also thank our past and present colleagues in the School of Engineering Science at Simon Fraser University for providing advice, examples, and support for the text, most notably John Bird, Jim Cavers, Chao Cheng, Tim Collings, John Dill, Doug Girling, John Jones, Ron Marteniuk, Andrew Rawicz, and Jacques Vaisey.

We thank all the engineering students, participants in the Management Skills for Advanced Technology program, and other practicing engineers who have contributed to this text, including those who agreed to let us use their documents: Caroline Dayyani, Frederick Ghahramani, Eric Hennessey, May Huang, Yann Le Du, Dewey Liu (whose award-winning report was unfortunately cut from the text because of its length), Glenn Mahoney, and Shirley Wong. We simply cannot begin to name all the people and companies who provided us with examples for the chapter on style, but we are grateful for their contributions.

Another group we cannot hope to name in full are the scholars, researchers, and teachers who have influenced our thinking and our teaching, most directly Richard Coe, Andrea Lunsford, and Don Wilson. Further, this text would likely never have been written without Robert Hendricks, Director of Materials Science and Engineering at Virginia Polytechnic Institute and State University, who brought our work to the attention of John Wiley and Sons. At John Wiley and Sons, we received much help and support from the following people: Joseph Hayton, Norine Pigliucci, Charity Robey, and Sharon Smith. We are also grateful to the outside services of Suzanne Ingrao, Ingrao Associates. We thank the reviewers of the various drafts of this text: Valerie Arms, Drexel University; Dianne Atkinson, Purdue University; Craig James Gunn, Michigan State University; Rollie Jenison, Iowa State University; Sharon M. Jones, University of California-Berkeley; and Anne Parker, University of Manitoba. Their thoughtful comments and suggestions have led to many improvements.

Last, but certainly not least, we thank Patricia and Minh, our long-suffering partners, who have supported us on the home front. Without their support, encouragement, and patience we would never have completed the text.

No doubt, we have inadvertently omitted the names of others who have had a hand in the genesis of the text. To them we offer our sincere apologies and profound thanks. And, as always, much credit goes to others; any faults are our own.

Contents

Chapter 1

Planning and Inventing Strategies

1.1 SUMMARY

Writing, like any engineering task, requires that you devote time to careful planning in order to meet your goals. This chapter introduces you to the processes of planning technical documents and thinking creatively in order to generate new insights. In addition, we suggest a range of strategies that you can use to enhance your creativity while writing. Careful attention to the writing process can increase your efficiency as well as your comfort while writing.

This chapter begins with a general discussion of the writing process, focusing upon strategies rather than rules in order to help you develop flexibility as a writer. We then outline issues relating to your writing environment and address concerns relating to using computers for writing, as follows:

1. **Strategies for Planning**

   To meet any goal that you may set for yourself, you must carefully plan *how* and *when* you intend to undertake the work necessary to achieve that goal. Careful attention to time management, audience and purpose, and the preliminary organization of your content will save you time and thus ensure you meet deadlines.

   a. **Time Management and Procrastination**

   Time management is critical to your success. If you fail to carefully structure your time when writing, you risk failing to meet deadlines, which can result in lower grades in school and lost opportunities for advancement at work. While most of us are aware that we must meet certain deadlines, we all too often procrastinate with tasks we dislike or with which we are unfamiliar. The material we provide about procrastination considers *why* we procrastinate and suggests some strategies for dealing with the problem. An important distinction is also made between *procrastination* and *incubation* (i.e., between putting off a task and allowing time to think about it).

b. Rhetorical Issues

In the long run, you can increase your efficiency while writing by carefully considering the issues of audience and purpose while planning the writing task. Clearly understanding the purpose for writing helps ensure that you do not inadvertently discuss interesting, but irrelevant, topics. Similarly, having a clear sense of *who* you are writing for allows you to determine what information and arguments will be most likely to achieve your goals. Carefully considering your audience and purpose also increases the likelihood that you will adopt an appropriate perspective and follow the forms and conventions your audience expects. Failing to attend to audience and purpose while planning increases the likelihood that you will be making time-consuming and frustrating revisions later on.

2. Strategies for Inventing

By the term *inventing*, we are referring to the many ways of generating new insights into topics and, through research, to incorporating ideas and information that others have already generated. We provide a wide range of strategies that you can experiment with in order to begin the process of writing.

a. Creativity

All too often, creativity is thought of as something that is mystical, innate, or—at least in relation to writing—largely restricted to the Humanities. Instead, we present creativity as something that everyone possesses, irrespective of discipline or predisposition. Creativity is something that you can develop through practice. A well-written proposal may differ from a good short story in terms of its structure and its metaphors, but both forms can demonstrate high levels of creativity. We encourage you to experiment with some of the strategies we suggest.

b. Problem-Solving Strategies

Engineers are in the business of solving problems. In order to effectively solve problems with writing tasks, engineers must not only learn to think creatively in relatively unstructured ways, but must also learn to apply structured formulas, sometimes called *heuristics*, in order to generate new ideas and new questions. We suggest several traditional methods for solving problems that you may find of use.

c. Collaborative Strategies

Ideas do not occur in a vacuum. We collaborate in many ways with others in order to generate documents. Some of these approaches may involve things as simple as brainstorming or discussions with friends and colleagues.

d. Research Strategies

Research, of course, is one of the most important strategies for inventing that engineers employ. We outline a range of strategies related to researching topics using library resources, databases, and the World Wide Web. We also provide strategies for evaluating web sites.

3. Organizational Strategies

All writing must be organized in some fashion if your audience is to easily comprehend what you are trying to communicate. In order to deal with this issue while planning, many of us develop an initial outline for the document. But outlining is not the only way to organize our thoughts, and for some people, outlines can interfere with the creative elements of writing. We suggest some alternative forms for outlines, and we also suggest other ways to structure the writing task that involve using graphics and writing introductory paragraphs.

1.2 INTRODUCTION TO THE WRITING PROCESS

Those who work in engineering and other technical fields too often view writing as an unpleasant—although necessary—task. This all-too-common dislike for writing may originate in part with past unpleasant experiences, perhaps when a teacher or reviewer returned an assignment covered with so much red ink that it appeared to be bleeding. Although the comments would have been well intentioned, the recipient might well have experienced a sense of failure and developed a range of problems including procrastination, anxiety, and writer's block. In our experience, many engineering students would benefit from a more positive attitude concerning their abilities as writers. Better understanding 'the writing process is one way to develop a more realistic assessment of your strengths as a writer.

Another potential impediment writers face is the common belief that writing follows rules. Those who hold this view contend that if they could simply learn all the rules, they would become successful writers. Unfortunately, this *rule-bound* perspective not only deflects attention from more productive concerns, but it also limits a writer's choices in unfortunate ways, as we explain below.

Typically, these rules are expressed using an absolute term such as *never*:

- Never end a sentence with a preposition.
- Never use the passive voice.
- Never start a sentence with *because*.

This last rule is a case in point: those who follow it may well have been cautioned at an early age not to start sentences with *because*. In context, this advice would have made good sense. For example, it could have been made in response to a rash of incomplete sentences written in response to test questions:

Test Question: Why did Spot chase Fluff in the story?

Answer: Because Fluff is a cat and Spot is a dog.

The advice to avoid starting sentences with *because* would have been meant to encourage students to write complete sentence such as "Spot chased Fluff in the story because Fluff is a cat and Spot is a dog." Unfortunately, those who did not understand (or who forgot) the specific reason for not starting a sentence with *because* might very well have generalized this advice into a rule that was carried into adulthood.

In reality, communicating ideas clearly sometimes requires that sentences begin with the word *because*. Many writers avoid doing so by using *since* or *as*. Unfortunately, these alternatives risk confusing readers as illustrated in the following sentence:

As the truck was in gear, the load of bricks slid off the tailgate.

The reader could interpret the sentence in two different ways:

While the truck was in gear, the load of bricks slid off the tailgate.

or

Because the truck was in gear, the load of bricks slid off the tailgate.

In other words, following the rule to never start a sentence with *because* risks confusing readers and leaving them uncertain whether the *timing* or the *cause* of the event is most important. Starting sentences with *because* is perfectly all right if your intent is to emphasize cause and effect.

Rather than simply focusing on *rules* for writing, you can more productively focus on *strategies* for writing. If you achieve your goal with a given text, then it is successful, even if you have broken various stylistic rules. However, keep in mind that readers expect you to follow certain conventions in terms of style and form. If you consistently violate readers' expectations, you risk confusing or irritating them; or worse, readers may question your competence and professionalism.

Therefore, you should violate the expected conventions of writing only when you have a clear reason for doing so. For example, you might choose to emphasize a particularly important point by writing it as a sentence fragment. Used sparingly, sentence fragments prove an effective strategy to emphasize a point (in part, because the fragment will generally cause the reader to pause for a moment). In other words, while a *rule-based approach* to writing may suggest that you never use a sentence fragment, a *strategy-based approach* to writing suggests that using the occasional sentence fragment is acceptable if it helps you achieve your goals.

This need to focus on strategies rather than rules is perhaps nowhere more evident than when considering the writing process. In many ways, the writing process is idiosyncratic, differing from individual to individual. For example, some people may work most productively in a noisy and disorganized open office, while others may be happiest with absolute silence and a carefully organized private space.

As you read this chapter and the next one, consider your writing process in terms of the strategies you employ and how effectively they work for you. The efficiency of your composing process and the effectiveness of your completed documents are largely determined by the strategies you employ while writing. We provide a range of potential strategies, but you must determine which are most appropriate for you in particular situations.

Also note that actual planning, drafting, revising, and editing activities rarely fall into the neat 1-2-3 sequence we must inevitably use to describe them. Figure 1.1 portrays the writing process as if it were a linear sequence.

In reality, you will often find it necessary to circle back to earlier strategies and to work through certain stages in the process more than once. Studies of how professional writers compose suggest that experienced writers constantly circle back and reexamine what they have previously thought or written in the light of what comes to mind as they move forward. As Figure 1.2 suggests, writing is actually an iterative, or recursive, process.

The following strategies and advice for various aspects of the writing process address problems we commonly encounter in the writing of both engineers and engineering students. These strategies are designed to help you capitalize on the strengths and overcome the weaknesses in your present composing process so that you can write more effectively and efficiently. Keep in mind, however, that if one of the strategies we suggest interferes with your preferred approach, you should try a different strategy.

Our goal is to help you write a first draft as quickly as possible and to write one that requires as little revision as possible. We realize that engineers and engineering students must often write to strict deadlines and under severe time constraints. However, we are also aware that engineers must be good revisers, often being called upon by their supervisors or project managers to make major changes in their reports.

Further, because engineering is a professional field, employers and clients expect (and often demand) that engineers present themselves as professionals in all aspects of

Figure 1.1 Writing Viewed as a Linear Process

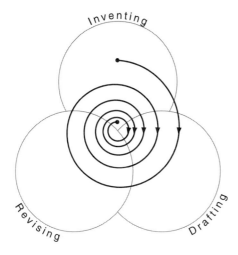

Figure 1.2 Writing Viewed as an Iterative (Recursive) Process

their work, not least of all in their documentation. Similarly, instructors expect you to adopt a professional attitude toward your assignments and may provide you with opportunities to revise documents after they have read and commented on them.

1.2.1   Writing as Idiosyncratic

Everyone has idiosyncrasies when it comes to writing. For example, you may prefer particular environments, specific software packages, and certain types of audiences. In general, these idiosyncrasies fall into two general categories: your usual or preferred practices and your attitudes toward writing.

Your usual practices will fall somewhere between extremes. For example, consider the level of noise that works for you. Some people require absolute silence while writing, and so they typically prefer to write in a closed office or in a library. Others need noise to help them focus and are quite content to write in a busy construction site or an open office.[1] Still others prefer soft music of various types.

Some people also prefer an environment that can only be described as organic—papers scattered about, *Post-it Notes* (®3M Company) stuck to every available surface. They seem to thrive on disorder. Other people cannot conceive of beginning to write unless everything is perfectly ordered—papers appropriately filed and reference materials organized. For some, even the pencils and pens must be in their appropriate places. Your preferences likely fall somewhere between these extremes.

Some people also prefer to schedule their time carefully while others find they write best under pressure. Some writers prefer to do most of their research ahead of time while others prefer to just start writing and to fill in the research as needed. And some appreciate being alerted by their word processing packages to potential spelling and grammatical errors while they draft whereas others find these features distracting annoyances.

Writers also vary widely in terms of their prior experiences, which can lead to their liking or disliking writing in general or to their liking or disliking writing specific types of documents or writing for specific types of readers. Depending on their past experiences

---

[1]One of the authors of this text, for example, almost invariably prefers writing with the television or the radio on in the background. He doesn't actually listen to them, but he finds he is much more productive when he has this background noise. In his case, writing in a library leads quickly to writer's block.

and their perception of their abilities in relation to the writing task, some writers may procrastinate or encounter writer's block while others experience no such difficulties.

Given this variety, you are well advised to identify your own preferences and idiosyncrasies. As long as your practices do not interfere with writing effectively and efficiently, changing them is unnecessary. If you need to water your houseplants or organize your files before you start writing, then by all means, water the plants and organize the files.

Of course, in the professional environment, you may not always be able to indulge your idiosyncrasies as much as you might wish. You must develop the flexibility to write in a range of environments and on a range of topics. You can adopt many of the strategies we suggest to deal with a range of situations. But more on that later.

1.2.2   Efficiency and the Writing Process

When working as an engineer or as an engineering student, efficiency in writing matters. One case study indicates that while working on a project, engineers spend at least 10 to 15 percent of their time engaged in writing documents. In other contexts (i.e., where engaged in writing proposals or analytic reports), engineers may spend more than 50 percent of their time writing. Engineers working in virtual teams must write a great deal at all stages of a project. Consequently, the time devoted to writing is a major cost to companies.

For example, the estimates for the design and production of a prototype optical tape recorder by CREO Electronics Corporation (at the time, a small startup company) indicated that approximately 11 percent of the total engineering time was budgeted for documentation. This 11 percent was equal to about 3.3 person-years of time, resulting in a total labor cost exceeding $120,000 (in 1988 dollars) devoted solely to the task of producing project documents. (Note that these figures relate only to the time spent on formal project documentation and do not include the time spent writing letters, e-mails, memos, etc.)

A 10 percent decrease in efficiency could have resulted in a serious cost overrun for such a small company. Conversely, a 10 percent increase in efficiency with writing could have saved the company $12,000. With smaller companies, these sorts of savings add up and can mean the difference between financial success and failure. For the individuals involved, increased efficiency increases their chances for personal success (i.e., promotions and recognition) while they help their company become more profitable.

As a student, your ability to write efficiently and meet assignment deadlines also translates into measurable results. Many instructors award lower grades for late papers, and some simply will not accept late work. Even if you dislike writing, increased efficiency has a tangible benefit. The sooner you are finished, the sooner you may move on to other, more enjoyable tasks.

1.2.2.1   The Writing Environment

The ability to focus on the task of writing is critical to your ability to produce documents in an effective and efficient manner. Although individuals clearly differ in terms of their preferences for particular writing environments, some generalizations are possible. Typically, the ideal writing environment is quiet, free of interruptions, orderly, physically comfortable, and well lit. If noise (from fellow students and office equipment) or interruptions (from the phone or colleagues) interfere with your ability to focus on the task at hand, you simply cannot operate at peak efficiency. Nor are you likely to be especially efficient if you spend excess time searching through mounds of paper for reference notes.

Similarly, your physical comfort is critical to your productivity while writing. If you are squirming in an uncomfortable chair, or if your keyboard is placed too high or low in relation to a comfortable typing position, you must take more frequent breaks than might otherwise be the case. If you are constantly squinting at your computer screen because the screen is too small or light is reflecting off it, you will suffer from eye fatigue and may find it difficult to maintain the focus needed to write effectively. Indeed, the lack of physical comfort while writing can pose serious health risks that must be considered. We will talk more about the issues of health and safety in Chapter 2.

1.2.2.2 Using Computers for Writing

The days when engineers had easy access to secretarial staff to produce final drafts are long gone. In most companies today, you can assume that you will produce entire documents yourself—from the initial outline through the rough draft to the formatted and edited final draft. Moreover, you will also generally be expected to produce any financial estimates or graphics required for a document. For that reason, we recommend that you become familiar with a good word processing package, a spreadsheet program, and a graphics package early in your days as an engineering student.

We assume that you write using a word processor, but are you familiar with its advanced features? You should master the full range of features on your word processor: style sheets and templates, spell-checkers, grammar checkers, equation editors, outlining, annotations, automatic numbering, tables of contents, and cross referencing. Although you can learn to use some of these features through trial and error, with some of the more complex features, you will save time if you read through relevant parts of the manual or use the on-line help file. Similarly, you should ensure that you are completely familiar with the various features of your spreadsheet and graphics packages.

Despite requiring some extra effort initially, your mastery of the various features will pay off over the long term regarding acquiring professional-looking documents and considerable time savings. Further, your in-depth understanding of a few programs will help ensure that you can easily translate your skills to different word processors, spreadsheets, or graphics packages should they be required in the work place.

If you have not yet mastered the skill of touch typing, we recommend that you invest in one of the computer programs available to help you learn this skill. You will focus much more effectively on what you are writing if you are touch typing at 60 words per minute than if you are hunting and pecking for the right keys at 20 words per minute. That is, you will keep up with the flow of your thoughts more easily if you can type quickly and unconsciously. Although over the next few years, technologies for voice recognition and dictation may reduce the need for good keyboarding skills, these programs are not yet commonplace.[2] Fast, accurate typing remains an indispensable ability for all engineers.

A final point that we would like to raise in relation to using word processors involves the surprisingly common practice—even among relatively young engineers—of writing the first draft by hand and then later typing it on the word processor. Although some people follow this practice in order to minimize a tendency to expend excessive time editing

---

[2]One of the authors of this text has switched from typing to using a dictation package for documents that are over a page in length. Because he uses the hunt-and-peck method of typing and only achieves a speed of 20 to 30 words per minute (with poor accuracy), he found that he could greatly increase his speed (to 100 words per minute plus) as well as his accuracy (to over 90%). The other author is a proficient typist who is much more comfortable typing than dictating.

their first draft, other people seem to do so from force of habit. Aside from the rather obvious point that this practice is inefficient, it also fails to optimally use the features of the word processor.

Word processors are not simply typewriters with enhanced formatting capabilities. The outlining feature of the word processor, for example, can be a powerful tool for creating and organizing your documents. Effective use of an on-line thesaurus can assist you with finding the particular words you are seeking to communicate your thoughts and ideas. When working collaboratively on documents, document sharing and use of annotation features can enable you to quickly resolve problems and gain the insights of others.

1.2.3   Developing Flexibility

As we mentioned earlier, you must be flexible when working in a professional environment. You may need to learn to use a different word processor or spreadsheet when you move from school to work, and later in your career, you may encounter new software as you move from one company to another. Although some companies and schools may support a range of different writing applications, others require a standard program in order to reduce costs and to improve document portability and consistency.

Whether you like it or not, eventually you may need to master several different operating environments (*Windows, Macintosh*, and *UNIX*), different word processors (*Word Perfect* and *MS Word*), and different spreadsheets (*Excel* and *Lotus 123*). And who knows what operating systems or applications will be developed in the near future. For this reason, flexibility becomes a key element to your success when writing.

Further, the need to develop flexibility extends far beyond your preferences in terms of operating systems and applications. You must also develop flexibility in terms of the physical environments in which you can work. As mentioned earlier, writing preferences are idiosyncratic. So what are you to do when your preferences clearly conflict with the environments in which you must work or study?

For example, reflect upon the situation faced by people who need complete silence to write productively, but who find themselves working in a busy open office or lab environment (or worse, next to the coffee area). While they cannot close their doors because the open office or lab lacks doors, these individuals nevertheless have a range of options they can employ to resolve the problem.

First, they can wear headphones to screen out the excess noise. Of course, this option creates problems if they must also respond to the telephone. But even then, they could forward their phone calls to voice mail while writing. Or perhaps they could use a telephone with blinking lights (or one that places a message on the computer screen) when calls are incoming.

A second option involves flexible working hours. Some people come to work or school early in the morning or stay late in the evening when most other people are absent and the place is relatively quiet. Or, as working at home becomes more acceptable within various companies, people sometimes work on their PCs at home when they have important writing tasks to accomplish. Third, even the most open of office environments almost always has a boardroom and meeting rooms available to retreat to with a portable computer when a deadline looms. Similarly, a person who objects to working under fluorescent lights may be able to turn them off and use an incandescent desk lamp.

The key point is that you must either adapt to a particular working environment or you must adapt that environment to your needs as a writer. In either case, some flexibility, even creativity, may be required. But before you can make the changes necessary to enhance the effectiveness and efficiency of your writing process, you need to clearly understand the peculiarities as well as the advantages and drawbacks of your particular process. To that end, we have included a series of questions at the end of this chapter and at the end of Chapter 2 that you can use to analyze your writing process in order to more fully understand it.

1.3   PLANNING STRATEGIES

Successful writers begin planning as soon as possible, asking themselves questions about who will read their work and clarifying what they hope their writing will accomplish. Developing a clear sense of audience and purpose helps determine what information to include, how to order it, what to emphasize, what kinds of visual aids or summarizing devices to use, and—where the choice is open—what mode of presentation to adopt (e.g., memo, report, or slide presentation).

1.3.1   Begin as Soon as Possible

Writing, like all other problem-solving activities, often requires periods of hard, conscious thinking followed by periods of rest. Someone playing computer games, swimming laps, or just sitting in the sun may be busy writing. He or she may be consciously thinking about what to write and how to write it, or the person may be working at a less conscious level.

You have probably had the experience of working hard to solve a problem, of giving up in frustration, and then of suddenly finding the solution when you least expected it—while playing a game, taking a shower, exercising, or daydreaming in the sun. It seems our minds continue working on problems even when we consciously give up the effort to do so. If you begin planning a major piece of writing far enough in advance of the deadline, you can afford to put your work on the mind's back burner. There it will simmer while you carry on with other work, indulge in a little needed relaxation, or even catch up on your sleep.

Our first piece of advice, then, is to begin planning as soon as you receive an assignment and, if possible, to spread your work over several sittings. By doing so, you can accomplish more in the same amount of working time and obtain better results than if you wait for the pressure of a deadline to get you started. As Figure 1.3 suggests, a lack of initial planning will often lead to problems later, even for something as simple as printing a sign.

Our second piece of advice is to begin major writing assignments such as lab write-ups, reports, and project documents well before you complete the project. By doing so, you will meet deadlines more easily, and if you dislike writing, you will remove some of the pain from writing. If you start early enough, you can also take advantage of the learning that occurs as you write. Potential problems and areas requiring further investigation often come to light while putting ideas and concepts into words. Taking the time to write sections of a report while still actively involved with your project improves

Figure 1.3 The Dangers of Poor Planning

your chances of recognizing problems while you still have time to correct them. It also encourages insights that can make your work more interesting and productive.

1.3.1.1   Time Management

As soon as you receive a writing assignment, you should note any deadlines you are facing and decide the following:

- When you will do any necessary research
- How long it will take to draft the document
- Whether the document will go through a peer or client review (often a slow process)
- How long it will take to prepare and duplicate the final document
- How much time is required to deliver the document

Allow an additional 15 to 20 percent of the total time for unexpected delays such as difficulties obtaining needed research materials, equipment breakdowns, illnesses, and other urgent tasks.

Table 1.1 provides a breakdown of the time required for one individual to prepare a 50-page report that must go through peer review and has a deadline of 20 working days. Note that this breakdown assumes a relatively simple report reviewed in a timely manner.

Obviously, this breakdown is arbitrary in that the amount of research required will vary from document to document. Similarly, the time required for drafting will vary depending on the efficiency of the writer and whether similar documents exist from which material can be extracted. Likewise, the time devoted to revising depends on the proficiency of the writer. Nevertheless, the relative proportions of time required for researching (5 days), for drafting (3 days), and for reviewing, revising, and formatting (8 days) are fairly representative of an expert writer's process.

For large writing projects, consider creating a Gantt chart as shown in Figure 1.4. This chart provides a simplified example of a schedule for a project allocated about 10 weeks.

The Gantt chart serves two key purposes. First, it provides a graphic representation of the amount of time you plan to devote to each phase of producing a document. Second, it allows you to see how the various elements are sequenced and how they overlap each other. (The reason they should overlap lies in the iterative nature of the writing process.)

Table 1.1   Planning Guidelines for Writing Technical Reports

Organization and research	5 days	(25%)
Drafting	3 days	(15%)
Peer review	1 day	(05%)
Revising and editing	5 days	(25%)
Formatting and duplication	2 days	(10%)
Delivery by courier	1 day	(05%)
Contingency	3 days	(15%)

1.3.1.2   Dealing with Procrastination

For many writers, dealing with procrastination is their greatest challenge. Both novice and experienced writers alike will procrastinate until either guilt or deadlines force them to

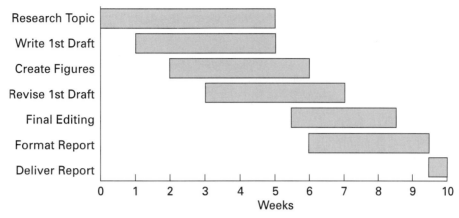

Figure 1.4 Example of a Gantt Chart

put hands to keyboards. As Figure 1.5 suggests, procrastination has unfortunate consequences: incomplete work.

In the case of a student, these consequences may mean late nights, unrevised assignments, and lower grades; in the case of a professional, the consequences may include taking work home, unpersuasive proposals or reports, and lower income or reputation. In either case, the consequences stem from a failure to manage time to allow for sufficient revising and editing. Poor time management is a major problem for engineers whose work will suffer and whose careers will stall if they produce ineffective or unprofessional documents.

Although the causes for procrastination are probably as varied and individual as the writers who suffer from it, this section focuses on the most frequently cited reasons for procrastinating and on strategies for dealing with these particular problems:

- Overcome a dislike of writing by focusing on positive goals.
- Deal with the uncertainty of getting started by asking questions.
- Replace a desire for perfection with a quest for excellence.
- Deemphasize potential criticism by viewing your drafts as provisional.
- Distinguish between procrastination and incubation.

First, writing is hard work that requires a significant investment of time and energy. Most of us tend to put off tasks that are difficult or otherwise unpleasant. But procrastination simply increases the difficulty of the task. Trying to write under pressure to meet that looming deadline is generally both frustrating and stressful. You neither write nor think

Figure 1.5 The Consequences of Procrastination

most productively when stressed out. One strategy that is sometimes used to deal with procrastination involves focusing on the goal and the rewards of a job well done rather than on the work required to reach that goal.

For instance, writing a successful proposal could provide the opportunity to work on a more interesting project; finishing it on schedule could also allow you a few days of needed holidays or time with family or friends. A side benefit of adopting a more positive attitude is often increased satisfaction with the actual act of writing. That is, when writing is goal oriented, it is more challenging, and meeting a challenge is generally satisfying. Consequently, if you focus on the positive, the interesting, and the challenging while writing, you will not only increase your motivation to start writing, you will enjoy it more.

Second, writing often requires that you start without a clear understanding of precisely where you will end up. As frustrating as this uncertainty may be, it cannot be avoided. One of the best strategies for resolving this issue involves making some notes as soon as you are given a writing task. These notes are not intended to fully define the task, but rather are simply intended to focus your thoughts on the task. Ask yourself some key questions: Who is this piece of writing intended for? What do they want/need to know? What is the purpose of this document? How will I achieve that purpose? What tone and perspective should I adopt? When do I intend to write this document? How will I budget my time?

Third, some people have unrealistic expectations of their writing, expecting that they must write the *perfect* report or the *perfect* proposal. If you expect perfection, you likely procrastinate because you fear failing to meet your own (unrealistic) expectations. As painful as it may be to admit, no proposal or report is ever perfect. You should instead consider writing in terms of whether it achieves the intended goal.

Viewing writing in terms of an absolute like *perfection* not only leads to procrastination, but is also a major cause of writer's block. If you suffer from writer's block, you may find that it helps to shift your focus from writing a perfect *product* to the *process* of striving for excellence. Excellence is achievable while perfection is not. More importantly, you can always achieve higher levels of excellence. By shifting your focus in this way, you can find satisfaction in a job well done while being motivated to write more effectively and efficiently.

Fourth, you may be anxious about how the reader will react to your ideas or your writing. So you delay writing your ideas down in a futile effort to avoid criticism. Unfortunately, procrastination is an especially poor strategy for dealing with the fear of failure because it generally results in a poorly revised report that is open to more criticism than a carefully revised one. A better strategy involves viewing writing as *provisional*.

If you are reluctant to have your work read, then try to convince yourself that you can write *anything* in your first draft. You are not obligated to show it to anyone. By starting early and managing your time carefully, you will have a very good chance of producing a final draft that will impress your readers.

Finally, you should distinguish between *procrastination* and *incubation*. Procrastination involves delaying a required task because you do not really want to do it. Incubation, on the other hand, involves delaying a required task because you want to understand the task well enough to do a good job.

Incubation usually occurs during the inventing process when you are mulling over ideas and approaches. Although you may appear to be accomplishing little, a period of incubation is often a necessary part of the planning process. Of course, as a writer, you must always remain alert to the possibility that incubation is changing to procrastination. If this shift from incubation to procrastination occurs, remember to use the four strategies outlined above: focus on your goals, ask yourself questions, aim for excellence rather than perfection, and consider your writing as provisional.

1.3.2 Rhetorical Issues

As you go about planning your document, you can save yourself substantial time by considering a range of rhetorical issues such as audience, purpose, tone, perspective, and form. By spending some time initially examining these issues, you increase the likelihood that your document will achieve your goals and require minimal revision. Those who fail to pay attention to these issues are typically required to revise extensively in an effort to produce effective documents.

1.3.2.1 Audience, Purpose, and Tone

If you are writing a document that is primarily persuasive, analyzing your audience is essential. Ask yourself questions about your readers. What are their values? What are their fears? What is their level of expertise? What is their power relationship to you as the writer? Does your audience include individuals with varying degrees of expertise? If you fail to analyze your audience when writing persuasively, you are unlikely to achieve your goals.

Even if you are writing a primarily informative document, you should analyze your audience prior to drafting. For example, reflect on the challenges of writing about a complex topic for an audience with limited expertise. If you do not consider the audience's level of expertise before drafting, you may fail to sufficiently define the terms or clearly explain the information in the report. This sort of failure inevitably results in extra work when revising. You might also write for an audience that requires a high level of proof for your findings or a high level of support for your recommendations. If you fail to consider this need prior to drafting, you may inadvertently omit important details. Few things are more frustrating than searching for that fact or figure you remember coming across in your reading that would support your case. Inevitably, it's not where you thought it was. Chapter 3 provides an expanded discussion of audience.

Clarifying your purpose for writing a document is important because a clear understanding of your purpose will help you avoid interesting, but irrelevant, digressions. Also be absolutely clear whether your document emphasizes persuasion or information. Expending effort selling someone on an idea, when the person simply needs information, is an ineffective use of your time. Chapter 3 provides more detail about purpose.

The tone of a document is also a critical issue to address prior to drafting. Do you want readers to view you as an authority on the subject? Do you want to come across as willing to negotiate? Do you want readers to know you are angry or would you rather they view you as apologetic? Do you want to be formal or informal? By initially considering the tone you want to convey, you are most likely to choose words with appropriate connotations when you are drafting. If you inadvertently choose the wrong tone, you could end up rewriting the document. Also, considering the tone of a document will help you avoid those unfortunate incidents where, in the heat of the moment, you send a letter or memo that you later regret. Chapter 3 expands on the issues of tone and connotation.

1.3.2.2 Perspective

In the interest of efficiency, you should also consider the perspective that you intend to use in your document. Changing the perspective of a document from third person (e.g., *it, they, no-person*) to first person (e.g., *I* or *we*) is very time-consuming. Table 1.2 presents the four perspectives and some situations where they are useful.

Table 1.2   Four Perspectives and Some of Their Possible Uses

Perspective	Possible uses
1st Person: (*I, We, Organization*)	Letters, memos, reports, proposals
2nd Person (*You*)	Manuals, procedures, letters
3rd Person (*She/He, They, One*)	Reports, legal documents
No Person (*Passive Voice*)	Scientific reports and articles, specifications

Your choice of perspective can also be a strategy for achieving a particular goal. For example, if you are working for an established firm with a solid reputation, using the first person, *we*, makes sense because you can trade on the reputation of the company. On the other hand, if the company is less established, using the third person and carefully employing the passive voice may help convey an air of authority.

You should also note a recent trend in scientific reports and articles toward using the first person rather than the third person and the passive voice. If you have a choice, we suggest using the first person because you can write most clearly and quickly from this perspective. Most readers also find documents written in the first person easier to comprehend than those written using a third-person, passive style. Writing in the first person is particularly useful if English is not your first language or if you struggle to put your thoughts into words. In Chapter 6, we explain the distinction between active and passive voice.

1.3.2.3   Standard Forms

Document formats are useful frameworks for organizing and generating information. You are likely already familiar with a number of standard forms: the five-paragraph theme (*introductory paragraph, three supporting paragraphs, concluding paragraph*), the five-section report or the five-chapter thesis (*introduction, methods, results or argument, analysis, implications*), and the seven-section lab report (*introduction, theoretical background, materials, procedures, results, discussion, conclusion*).

Studying the forms of documents typically produced within a discipline or profession can help you master writing within that discipline or profession. In industry, for example, you will be confronted with documents (such as proposals, user's manuals, and requirement, functional, and design specifications) with which you are initially unfamiliar. If possible, examine samples of previously prepared documents of the same type. Also determine whether templates already exist for common document types.

Attention to format will further help ensure that you are prepared to include all required sections or information. For example, in many technical documents, a glossary is a necessary component. If you fail to compile that glossary when initially writing the document, you face the rather tedious task of rereading the document later to find all the specialized terms that should have been included.

One purpose of this textbook is to teach you how to learn the conventions and organizational patterns appropriate to new types of technical documents before you start work on them. Chapters 6 and 7 provide advice on style and format, and guidance on how to analyze these features in existing documents. Chapter 8 provides more detailed information on some of the standard forms of documents encountered in university and industry.

1.4   INVENTING STRATEGIES

Typically, at the same time that you are involved in planning a writing task, you are also engaged in another process—that of inventing. By *inventing*, we do not just mean coming up with new or novel ideas (although that is sometimes what is needed). The term *inventing*, as we are using it, also means discovering, recalling, and consolidating information that you already know, feel, and believe as well as finding out something new about a subject. Thus, inventing can involve a range of common techniques such as research, discussion, and brainstorming and some techniques you may find more novel such as meditation and heuristics (structured series of questions used to discover information).

You have already accumulated a vast amount of information from your life experiences and your education. If, for example, you were asked to design the user interface for a new program, you might begin by thinking about some of the interfaces with which you are already familiar. You might ask yourself which ones would be easiest for people unfamiliar with computers to learn and which ones are fastest for those people with a great deal of computer expertise. Or if you were asked to write an assembly manual for a particular device, you might begin by considering your past experiences with such manuals. You might ask yourself which ones were easy to follow and which were less than helpful and identify the specific features that caused those differences.

Once you have finished thinking about the topic and reflecting on what it means, you might also start asking yourself questions about what you *don't* know. Perhaps you would then engage in more careful observation or experiments; perhaps you would talk with other people, or perhaps you would undertake some research.

Inventing also sometimes involves collecting and consolidating information that may be spread over numerous pages of a project journal, scattered among a number of sources, or simply *in your head*. Experienced writers use a variety of strategies to help them retrieve information from memory and collect it from other sources. But in writing, as in all problem-solving activities, finding appropriate questions is crucial.

Solutions are always much easier to find once you know what questions to ask. You should, therefore, begin with general questions. But do not stop there. The answers to general questions should lead you to ask more specific questions until you are confident that you have found the best possible approach to dealing with a specific writing situation. Like experienced writers, you should ask yourself a great number of questions when you first start work on a project. Use the answers to these questions to help you judge what information to include, how to organize it, what style to adopt, and even when to break with convention and try something new.

1.4.1   Creativity: A Digression

What does creativity have to do with writing technical documents? Certainly, we all think of creative writing as involving the writing of poetry, stories, and novels. But is technical writing creative? Part of the reason you may not view technical forms of writing as especially creative involves a misperception that technical documents simply report previously discovered information. And yet, in order for any engineering project to be successful, a story must be told. Just as the novelist brings together character, scene, plot, and action in order to create a story, the engineer brings together science, materials, finance, marketing, and human factors in order to create a device or system. Is the device physically possible? Are the necessary materials available? Can the funds be found to produce it? Will anyone want to purchase and use it?

Indeed, even the purposes of the novelist and the engineer are similar: to inform, to persuade, and ultimately, to motivate. Just as novelists provide information about character and scene, so too engineers inform their audiences. Just as engineers persuade others to adopt their proposals, so too novelists persuade readers to adopt their ideas. In both cases, the intent is to motivate the audience to act—usually in order to change our world for the better. But unlike a story created and told through the means of a novel, the engineer's story is told through a range of documents such as proposals, research reports, and specifications. The difference between the writing of a novelist and that of an engineer is not that one is creative and the other is not, but rather that the forms each uses to express creativity differ.

Another difference is that for an engineer, the device or system being built is integrated with the documentation in very special ways. Without the proposal, the project would not get under way. Without the requirements and functional specifications, the problem would not be clearly defined. Without the design specifications, the device or system could not be manufactured. Without the user's documentation, the device might not be used. The engineering design process guides the writing as much as the writing guides the engineering design process.

Although in some cases, a project will be defined in fairly narrow terms, in other cases, you may be provided significant opportunities for creativity—opportunities to design a system or product of your own. While such opportunities provide you with considerable latitude for your imagination, you may also find the need to develop new ideas or new approaches somewhat overwhelming.

In part, you may encounter this problem because *creativity* is often assumed to be an innate ability (i.e., something you are born with) or worse, as arising from some sort of mystical inspiration. Consequently, many people have been persuaded that they are not very creative. That perspective is particularly unfortunate for an engineer.

To enhance your creativity, you must observe the world around you with an open mind. To think creatively, you must also develop a habit of asking *what if*? The following sections discuss specific ways of enhancing your creative powers.

1.4.1.1 Problem Posing[3]

While all engineering schools teach their students how to go about *solving problems*, fewer provide sufficient opportunity for *posing problems*. By *posing problems*, we mean learning to recognize the potential for alternative or new products and systems by observing various aspects of the world around you and then asking questions about what is going on. The following steps provide you with a method to enhance creativity by learning how to pose problems to yourself and others.

1. Observe how existing industrial, commercial, or consumer products and systems work. One trick that some experienced designers use involves collecting old, even apparently bizarre, designs for products. Sometimes the old designs failed not because the concepts were flawed, but simply because the available technology was insufficient to implement them. For example, swing-wing aircraft were designed prior to World War I, even though the technology required to support the design was not developed until 1945, about three decades later.

2. Analyze the purpose of existing products or systems with an eye to finding inefficiencies or problems. Can a device or system be made simpler to use or cheaper to produce? Understanding the psychological, physiological, environmental, technical, and social processes that are involved with using and producing products or systems helps you recognize problems worth resolving.

[3]We would like to thank Doug Girling and Professor Andrew Rawicz (Simon Fraser University) for their contributions to the section on problem posing.

3. Analyze the nature of the problem by considering whether solving the problem is *physically* possible. Also consider whether solving the problem is *fiscally* possible (i.e., is the product potentially useful enough that a market will exist for it?).

4. Determine how best to improve the product or how to resolve or restate the problem.

5. Consider how to develop and market the product.

How might you apply this method? The following experience of a student project team illustrates how the method works. The team recognized the need to design a new type of portable heart monitor by observing that the current monitors were a strain to use for people with heart problems because they were bulky and heavy. By considering the physiological limitations of people with heart problems, they reasoned that it would make sense to miniaturize the monitors using more modern technology.

Once the group had determined that their proposed solution was physically and technically possible, they built a prototype, and then sought funding to further develop their idea. Accompanying and guiding this project were an array of documents: the initial proposal, functional and design specifications, user's documentation, and a formal funding proposal along with many memos, notes from brainstorming sessions, letters, and meeting minutes. Throughout this process, the team continued to employ problem posing as a creative strategy.

1.4.1.2 Observing Nature

The natural world around us can also be an amazing source of ideas. Observing the structures and behaviors of various plants, insects, animals, and landscapes can be a useful way of generating new ideas and alternative approaches to problems. For example, George de Mestral, the inventor of Velcro, came up with the idea in 1948 simply by observing how the burrs from certain types of plants adhere to the fur of animals and to clothing.

Although observing the world around you is no guarantee that you will invent the next Velcro, a walk in the woods will at least be a relaxing and pleasant experience. Because creativity generally requires that you be relaxed and relatively stress-free, inspired ideas are much more likely when you take time to slow down and experience the world around you. For this reason, meditation is sometimes also suggested as a means to enhance creativity.

1.4.1.3 Focused Meditation

Meditation is a strategy that some people find useful for opening themselves to ideas. It involves taking a key idea or problem and focusing on it in a relaxed atmosphere with few distractions. Although providing a complete course on meditation is beyond the scope of this text and requires considerable practice, you might try the following method.

Put on some relaxing music, find a comfortable chair (or even lie on the floor), close your eyes, and breathe slowly and evenly. Relax the muscles in your body starting with your facial muscles and working down through your neck, your torso, and your limbs. Once you are fully relaxed, visualize the ideas that go through your mind as clouds slowly passing by on a sunny day. Then slowly switch your focus to the particular idea or problem you want to consider. Look at it from different perspectives, as if it were a cloud slowly floating by. How does it look from the top and from the other side? If other thoughts and ideas intrude, let them slowly float by, and then return your focus to your topic. After meditating, record any insights or ideas you have had about the topic or problem.

1.4.1.4   Role Playing

Our imaginations are limited by our experiences. To some degree, we can overcome this limitation by seeking out new experiences. If you are designing products for the physically disabled, one way to come up with ideas is to intentionally put yourself in situations where some of your senses or limbs are disabled. You might, for example, function for a day wearing a blindfold or with your hands tied behind your back.

You will, of course, find many everyday tasks are now impossible or very difficult to accomplish. Recognizing the problems you confront and then analyzing them should provide you with a range of revelations. You will also walk away from the experience with a very different perspective on how challenging it is for the disabled to do the things most of us take for granted. (Also note that this kind of role playing is useful for testing products intended for the disabled.)

1.4.1.5   Free Writing

Free writing is aimed at generating ideas through the medium of writing. Because language and thought are indivisible from one another, when we write we often discover or create new insights. This strategy offers a way to generate ideas by letting your mind freely explore a topic while simultaneously recording the ideas that occur to you.

Set yourself a general question about your subject or focus on a specific problem and then find a place in which you will not be interrupted. Write as much as possible about your question or problem in a set period, such as 15 or 30 minutes. Pay no attention to your spelling, grammar, or organization. Avoid revising your writing or censoring what you are writing. Do not stop. If you become stuck, write about whatever comes to mind no matter how unrelated it may seem.

The point is to break through inhibitions and to encourage creativity and imagination. Figure 1.6 provides a short example of free writing. This strategy and the next may also help resolve writer's block.

OK So I need to write about brainstorming. Usually its done in a group but not always sometimes individual. Works best in a group though people feed off each others creativity especially if they have different backgrounds—social professional—etc. So how does it work as a group
   —get people together for a time period $\frac{1}{2}$ to 1 hour
   —anything goes. don't want criticism—insane crazy ideas are ok
   —need to ensure nobody gets censored though so need
   —a group leader to facilitate stuff. should also record
   —Also good to later sort out ideas by being critical and further exploring them
But the biggest point is to make sure that there is no criticism & that people can come up with all sorts of ideas. Should also be fun laughing—sort of like a creative party with stuff written on the blackboard

Figure 1.6 Free-Writing Sample

Figure 1.7 Example of Informal "Scrapbook" Journal

1.4.1.6   Journal Writing

Journal writing is a strategy often used by experienced writers to accumulate questions, observations, insights, and interesting quotations or facts. Note that a journal is not necessarily a diary in which you record the daily events of your lives (although it may be); it can be a place for you to record the ideas and observations you have about life and work. Your journal could, for example, be a place to document original ideas for possible patent applications.

A journal may be *formal* (a small book you carry with you) or it may be *informal* (scraps of paper you jot ideas down on). One of the authors of this text, for example, regularly writes down ideas on little scraps of paper. Figure 1.7 presents an example of something written for this textbook (see Figure 1.3 for an idea of how it became part of the text).

Note, however, that if you use the informal method, you should date and occasionally sort out and file the scraps of paper. A journal becomes more helpful as a creative tool as information accumulates in it. Also note that the informal journal is inappropriate for practicing engineers who require formal journals with numbered pages and dated entries. If you begin keeping a formal journal as a student, you will develop a habit that will serve you well when you become a practicing engineer.

Engineering Journals   Engineering journals serve many purposes. Should you ever want to apply for a patent, an *engineering journal* that chronicles the development of your device or process is not only advantageous, but may provide critical evidence should someone else concurrently develop a similar device or process. Professional engineers also use journals to keep accurate records of their work, their meetings and conversations with clients and colleagues, and even their expenses. They then use their notes to coordinate activities and evaluate progress on team projects, to check the work of other engineers, to verify work performed in the case of legal action, to document services performed for accounting purposes, and so on. Figure 1.8 provides a sample page from a formal engineering journal.[4]

Clearly, engineering journals can serve a number of important creative, professional, legal, and financial purposes, but only if entries are written legibly, if all the necessary information is recorded, if dates can be verified, and if graphic aids are accurately drawn and fully labeled. Of course, these journals are not works of art and may not even be particularly neat. What is important is that notes and numerical values are clearly written and that all necessary information is provided in a logical, clearly dated sequence.

---

[4]We would like to thank Professor John Bird (Simon Fraser University) for permitting us to use a page from his engineering journal as an example.

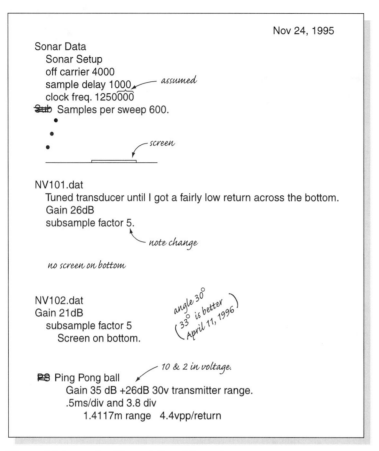

Figure 1.8 Example of Formal, Dated Journal

1.4.2   Structured Problem-Solving Heuristics

The following four strategies are fairly structured methods that combine many elements of the previous inventing strategies into broadly useful systems for generating and organizing ideas. They are especially helpful when you are approaching an unfamiliar problem or when generating novel approaches to an old problem.

1.4.2.1   Create Analogies

Creating unexpected or insightful analogies can be a very powerful way to generate new ideas and thus devise new or alternative products. For example, the analogy of computer screens as desktops and hard disks as filing cabinets enabled the creators of the Macintosh to define a graphical approach that made computing more accessible to the average person. Indeed, that particular analogy has proven so popular that Microsoft Inc. had very little choice but to devise its competing Windows operating environment.

Part of the reason analogy is such a powerful tool is that it allows us to see things from different perspectives. If we view the same event from different vantage points, we tend to see different things. Take one of Einstein's examples, for instance. If someone drops a stone from a moving train, that person sees the stone descend in a straight line while someone watching the train pass by sees the stone falling in a parabolic curve. You can make use of a similar phenomenon by exploring a topic from multiple perspectives. One relatively simple method is to consider a subject from *static, dynamic*, and *relative* viewpoints.[5]

Take three pieces of paper and label each one with a different heading: *Static* (or *Particle*), *Dynamic* (or *Wave*), and *Relative* (or *Field*). On the sheet labeled *Static*, write down anything that pertains to the general description of the subject as an object, such as distinguishing features, details, and definitions. On the *Dynamic* sheet, explore the nature of your subject as a process. That is, consider how it moves, how its parts relate to each other, and what physical and historical changes it has undergone. On the *Relative* sheet, explore relationships between your subject and other things. For instance, classify your subject as part of a larger group of things; compare with other similar things; and create analogies that relate your subject to different kinds of things. For this last view, let your imagination go wild.

If you are unsure what category an idea belongs in, write it down on the handiest sheet. Generating ideas is more important than categorizing them. The goal of this exercise is to stimulate your imagination and to help you recall and discover ideas and insights that might otherwise remain unacknowledged. Do not censor information; write down whatever comes to mind.

1.4.2.2   Generate Contraries

No doubt many of you have seen the following puzzle where you are asked to connect nine dots with four straight lines without taking your pencil off the paper. Figure 1.9 shows the initial layout for this puzzle. Part of the reason this puzzle is so difficult for many people to solve is the assumption that solving the puzzle requires staying within the boundaries outlined by the dots. If those having difficulty solving the puzzle can recognize the underlying assumption they are making—stay *inside* the boundaries—they can solve the puzzle by generating a *contrary* to the assumption—go *outside* the boundaries. Figure 1.10 presents the solution to the puzzle.

Using a *contrary* is simply a technique that involves considering the opposite to an assumption you are making when attempting to solve a problem. For example, a student who recently attended a seminar on résumé preparation was given the advice to make his

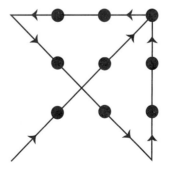

Figure 1.9 Initial Puzzle Layout   **Figure 1.10** Puzzle Solution

---

[5]This method of exploring a topic is adapted from *Four Worlds of Writing*, a textbook written by Lauer, Montague, Lunsford, and Emig (2nd ed. New York: Harper & Row, 1985). They, in turn, adapted their approach from Young, Becker, and Pike's *Rhetoric: Discovery and Change* (New York: Harcourt Brace Jovanovich, 1970). Young, Becker, and Pike labeled these three distinct viewpoints as *particle, wave*, and *field*.

résumé *scanproof*, which involves writing and formatting the résumé to slow down the reader. To create this style of résumé, the student could have used full sentences rather than point form and formatted the résumé to place important information such as the dates of employment on the right-hand side of the page, which would further slow readers down because we read from left to right.

Having previously seen how a senior engineer typically went about reading a stack of résumés, this student noted that the engineer read them very quickly, spending no more than 30 to 60 seconds on each one when initially sorting through them. So this student reasoned that making a résumé *scanproof* would most likely frustrate the engineer reading the résumé because of the need to slow down. And if frustrated, the engineer might be less likely to place the résumé in the group being considered for interviews. The student, therefore, decided that it might be advantageous to ensure the résumé was *scannable* (in other words, to make the résumé even easier to read and comprehend in a brief time frame).

The student went about making the résumé scannable by organizing it so that important information was easily accessible (dates on the left, previous employers and job titles in bold, and job descriptions in point form). Yet the student wasn't completely satisfied that these features made the résumé as scannable as possible, so he decided to talk with the engineer that he had previously seen reading résumés. The student asked the engineer one simple question: *When reading résumés so quickly, what are you looking for?* The engineer replied that she was looking for key words and phrases that indicated the applicant had the skills and qualifications required for the position.[6]

After learning what the reviewer looked for, the student further reasoned that he could assist readers of his résumé even more by bolding key terms that were relevant to the position he was applying for. So if one of the key skills needed for the position happened to be *programming in C++*, he made sure that he bolded these words (along with the words describing a few other relevant skills). Although this approach might not guarantee an interview, by applying the *contrary* to the advice to make his résumés scanproof (and by extending the contrary in a reasoned manner), he increased the chances that the reader would note the key terms and decide he had the required skills.[7]

1.4.2.3   Synthesize Ideas

Synthesizing ideas means bringing seemingly unrelated ideas together in order to see what results. Synthesizing requires being observant of the world around you and keeping an open mind. For example, the experience of the inventor of the Post-It Note (® 3M Co.), Art Fry, illustrates this point (see Figure 1.11).

> *He thought of the idea by accident sitting in church. As a choir member, he was stumped about how to keep the pages marked in his choir book. During a "dull" sermon he mulled over the problem. Back at the office a fellow scientist had found an adhesive that didn't stick well, while trying to invent a super glue.*
>
> *It hit Fry. Could he use the adhesive on a piece of paper as a removable book mark?*

---

[6]Note that another use of the term *scannable* is in contexts where people submit their résumés electronically, and the résumés are sorted by a program that looks for the presence of key words related to the job description. This initial sorting enables interviewers to focus on only those applicants who possess the necessary qualifications for the job. This method is often used by large companies receiving hundreds of applications daily. We provide more information about this type of résumé in Chapter 8.

[7]As it happened, the student was successful in getting an interview for the job in which he was most interested. He also got the job.

Figure 1.11 The First Post-It-Note®

> *He perfected the product and started testing it as a book mark with colleagues. Then, one day he wrote a memo on it to his boss and attached it to a report. The Post-It was born.*
>
> *"We knew we were on to something but it was a difficult sell," recalled Fry. Others in the company thought it was a dumb idea. It was too simple. Office supply dealers said it would never sell.*
>
> *"The people that used them just swore by them," said Fry. The rest is highly profitable history.*[8]

Although the discovery of the adhesive itself is a good example of *serendipity* (an accidental or lucky discovery), inventing the Post-It Note required that Art Fry synthesize several disparate things into a useful product: the adhesive, the paper, and the need for bookmarks in hymnals. By connecting his own need to mark his choir book with the potential of the failed superglue, Fry initiated a remarkably successful invention. All that remained, then, was to sell the idea to his supervisor, others in his company, and office supply dealers (often the most difficult part of the job).

1.4.2.4   Ask Questions

Asking the right kinds of questions is one of the keys to successful writing. A question heuristic is a list of questions that directs your attention to the concerns that should be addressed and to the information that should be included in a particular document. The journalistic formula—*Who? What? When? Where? Why? How?*—is probably the simplest and most widely used question heuristic. Many successful writers create lists of questions appropriate to a particular situation and use these questions as a guide to help them generate and organize ideas. We encourage you to develop the habit of asking questions as a means of generating ideas. For example, a specific audience/purpose heuristic for an informative report could include the following questions:

Who Is My Audience?

- Who will read this report? Technical experts? Administrators? Business people?

- Why will they read it? What motivated them to request the report? What actions will or can they take on the basis of this report?

---

[8]This succinct version of the story is told by Carol Howes, June 9, 1995, "Post-It inventor still sticking with his science" in *The Vancouver Sun*, Pacific Press Ltd. Vancouver, BC, D1. Used with permission.

- What information have they requested? Are their instructions clear or do they need clarification?
- How well informed are they about the subject? How much background information is required? Are they familiar with technical terminology?
- What information do they need? Do I have all the information needed to address their concerns? If not, what do I need to find out and how will I do so?

What Is My Purpose?

- What do I hope to accomplish by writing this report? Can I write a clear, concise statement of purpose?
- How do my goals relate to my reader's expectations? Do they share my objectives? If not, what are the points of disagreement?
- How can I meet both my goals and my reader's expectations? What do I know that they do not and how can I make them aware of it?
- What attitudes or values do they have that must be taken into account?

Remember that any list of questions serves only to direct your thinking. Questions should generate other questions relating specifically to a particular audience and purpose. The trick is to ask the right questions; good questions generate appropriate answers.

1.4.3   Collaborative Inventing Strategies

These collaborative strategies are aimed at using the ideas of others to come up with new or expanded insights into a subject. Members of different social, cultural, and professional communities with varying perspectives can help each other develop new insights by sharing these differing perspectives. The benefit of these different points of view is especially obvious when working in engineering, which is interdisciplinary by necessity.

1.4.3.1   Discussion

Discussion is an especially useful strategy for gaining new insights as well as for testing existing ideas, which may be one of the reasons engineers spend so much time in meetings. By publicly discussing their ideas and approaches to problems, engineers gain the experience and insights of their peers. Discussion allows them to test their ideas and approaches before implementing them, thereby decreasing the risks that might otherwise be involved. Beyond formal opportunities to discuss issues, engineers also engage in many less formal discussions: encounters in the hall or lunchroom, chats on the phone, and so on.

Unfortunately, many university students neglect discussion as a strategy for several reasons. First, we know some students who are excessively concerned about grades and who fear that by sharing their ideas they will help someone else raise their grades and lower their own standing. Unfortunately, students who are excessively focused on competition in this fashion may fail to see the other side of the issue. By cooperating with other students, they may well gain new insights into a topic and may thus actually raise their grades. Cooperative learning can also be an effective strategy to reduce the time required for studying.

Second, some students (and some engineers) are hesitant to subject their ideas to public scrutiny, perhaps out of fear that they will be ridiculed should their ideas seem unworkable or should their questions seem ignorant. Perhaps it is worth keeping in mind the adage that *there are no stupid questions, only stupid answers.* Sometimes asking a question or raising an issue that on the face of it seems simplistic or uninformed reveals that no one else has considered the problem. In fact, this possibility is a good reason for asking questions in class, no matter how trivial or ill informed you fear the question may seem. Frequently, something that you fail to understand is also causing problems for others in the class.

Nevertheless, we recognize that in a competitive environment, you might want to initially discuss your ideas and concerns with a small group of friends or colleagues who will not be excessively critical. Later, of course, you will want to subject your ideas to more rigorous examination by a professor or supervisor.

1.4.3.2 Brainstorming

A good method for generating a variety of different perspectives and ideas about a project or problem is brainstorming. Although sometimes used as an invention strategy by individuals, this technique works especially well in small groups. Consider employing this strategy before using more formal techniques (such as research). All that brainstorming involves is writing down on a blackboard, whiteboard, or piece of paper any ideas that come to mind about a project or problem.

You can increase the effectiveness of your brainstorming sessions by following a few guidelines. First, elect someone as a group leader and appoint someone else to record the ideas generated. The leader is responsible for ensuring that the group stays focused on its task, that everyone has an opportunity to participate, and that no ideas are dismissed.

Second, a good brainstorming session should be as uncritical as possible because criticism inhibits the free flow of ideas. Ensure that everything is recorded no matter how silly it might initially seem. Ideas may range from the practical to the apparently insane. Work quickly, recording ideas in any order. Do not stop to worry about whether you will actually use a particular point and do not censor your thoughts or anyone else's. The less structured your thinking as a group, the more creative it will be. If a brainstorming session is productive, participants should be having fun; brainstorming sessions often generate laughter—and sometimes friends.

Third, people who are brainstorming should ideally come from a wide range of disciplines and have widely divergent social and cultural backgrounds. The more diverse the group, the more likely it will generate unexpected insights, ideas, and connections.

Once the group has exhausted the possibilities of brainstorming, everyone should review the points (perhaps at a later meeting), eliminating those that are unsuitable and further exploring those that are promising. Note that the group must remain open to innovation—which leads us to the final point we would like to make about creativity.

Creativity is more easily fostered in some environments than in others. If you find yourself in an environment where phrases such as the following are common, you may find it difficult to think or work creatively: *It's not possible; No one would buy it; That won't work in our school/company*; or *You gotta be kidding*. These sorts of phrases are most common in companies or institutions with rigid rules and hierarchical structures. Creativity is generally hampered when people are most concerned with job advancement or job security; creativity is usually enhanced in more democratic, less structured environments. You might keep this notion in mind if you have aspirations to invent the next Velcro or Post-It Note.

1.4.4 Research

Research is, of course, one of the most important strategies used within academic and professional communities. By researching the topics about which you are writing, you gain the accumulated insights into the topic of many different people, discovering not only new ideas and approaches, but learning some of the limitations of your initial ideas. Although you may undertake research alone, reading in a cubicle in the library or sitting at a computer exploring a database, research should also be considered collaborative in that you learn from others. Because research involves borrowing other people's ideas, also remember that you must acknowledge their contribution when you discuss them in your work. Chapter 7 provides guidelines for citing and referencing sources.

Not so long ago, if you wanted to research a topic, you would have had to rely largely on hard copies of books and articles. Today, you have many more electronic media to explore; however, books and articles remain a major, reputable source of information and are readily available to most students. To find relevant material for a topic you want to research, begin by visiting the library on your campus. If you are unfamiliar with the on-line catalogue system, ask for an instruction sheet, attend a session on using the library, or ask a librarian for help. You can also explore databases and Internet sources. Note, however, that Internet sources are not regulated by peer review and may not express ideas or facts based on the burden of evidence required in reputable research. Consequently, critically evaluating the quality of the information provided on the Internet is essential. More on that topic in a few pages.

1.4.4.1   Catalogue Searches

The library catalogue system provides several options for finding material. If you simply want to locate a book you already know about, you can use either a *title search* or an *author search* to locate it in the catalogue. Once you have the reference you want on the screen, you will see the title, publishing information, a call number so you can locate it in the stacks, and a subject line that lists subject headings under which the book is cross referenced in the system's subject index. The subject descriptors are useful in finding other books and articles on the same general subject.

You can use the most relevant subject descriptor to begin a *subject search*. The key words in a book that you know is relevant offer a good starting point for finding other sources because subjects are not always described in the key words you expect. For example, if you are interested in researching battery storage systems for electric cars, the *key words* you want may be *Electric Automobiles*; other key words such as *Electric Car* or *Electric Vehicles* may not work. If you do not have a book to help you begin your search, think of possible key words that may be used as subject headings and try them one at a time until you find one that provides a list of books and articles with promising titles.

Read through the listed titles until you find one close to your specific topic. Select that book and read the description provided. If it looks good, record the title, author, date of publication, call number, and location. Check the subject line for headings you have not yet checked. Also check to see if you are provided on-line information on whether a work is in the stacks, out on loan, on reserve, or reported missing. Continue down the screen to locate information on other promising sources.

1.4.4.2   Scanning Books and Articles

Do not worry if the list of promising books and articles is too long. When you locate books and articles and scan them, you will likely discover that only a few sources actually cover the material you are interested in reading. For books, read the table of contents; if you find a heading that fits your topic, turn to that section and read the introductory paragraph to get a better idea of the content. For articles, read the abstract and then scan the headings. After scanning, you should end up with a much smaller list. At this point, you should decide which works are likely to be most useful or provide the best coverage of the subject. Begin reading them first.

1.4.4.3 Taking Notes

Before you even begin reading a book, make sure you have recorded all the information you need for your reference list, including the publishing information. Keep your notes for each work together with the bibliographic information. You might keep this information on good-sized note cards (or create the equivalent in a computer file). The sample note card in Figure 1.12 is part of the research notes for a thesis on gender and academic writing. Note that a key point is summarized and quotations the researcher may use are included.

   If you have read the note card carefully, you may have noticed a problem. The page number for the quotation is not noted. Of course, it may be on the other side of the card, but we draw your attention to this possible omission because of the frustration caused and time wasted when you want to use a quotation but have not recorded the page number(s). Make a habit of including page numbers, not just for quotations but for the source of ideas you summarize in your notes. If you don't, you may realize while drafting that you need to clarify a point and must waste time trying to find the source material. For the same reason, keep your source materials on hand until you finish your project, if possible.

1.4.4.4 Searching Databases

For many engineering topics, especially those relying on literature published in the last few years, you should consider searching databases. Visit the library to find out what is available. Universities with an engineering college should offer a number of options, including field-specific, engineering, and general science databases. You should find out what else is available because you may find material on relevant topics in databases

---

Taylor, S. O. (1978) "Women in a Double Bind: Hazards of the Argumentative Edge," CCC (29), pp. 385–89.

Position Paper: Theoretical & Anecdotal. College Women.

Key Point—Women caught in a double bind: Raised to be conciliatory, but told to write argumentatively.

OK article, but based too much on Robin Lakoff and doesn't provide enough info about the nature of the discourse community. Also tends to over-generalize.

"Women students, when they take composition courses, are at an invisible, though real disadvantage, because both the methods and the goals of such classes are alien to them. This fact is unacknowledged or perhaps is unrecognized by the instructor"

(over)

Figure 1.12 Standard Note Card for Research

covering business, social sciences, government, and so forth. Your library may also have its own databases for maps, special collections, and research centers associated with your university.

As a sample of the sorts of databases you may find, we list four of the many available at our own university:

- *Wilson Applied Science and Technology Index* cites every article of at least one column published in more than 390 periodicals. Coverage includes trade and industrial publications, journals issued by professional and technical societies, and specialized subject periodicals, as well as special publications such as buyer's guides, directories, and conference proceedings.

- *INSPEC* contains abstracts and indexing from thousands of journals, conference proceedings, books, reports, and dissertations, covering all aspects of physics, electrical engineering, electronics, computers, and information technology.

- *ABI/Inform* indexes articles on business conditions in the United States and other countries, covering all management disciplines and specific products, companies, and industries.

- *Humanities and Social Sciences Index* indexes articles from more than 350 periodicals covering about 30 fields, including economics, international relations, law, planning and public administration, policy sciences, public welfare, and urban studies.

Note that some databases, such as INSPEC, provide only abstracts. Reading these short summaries can give you a good sense of how a topic is being treated. You can then obtain full-text hard copies of particularly promising articles. If time and money permits, you can order copies of items not available in your library.

When searching databases, finding the appropriate key words or subject descriptors is particularly important. If you have trouble locating the lists of key words or subjects for a particular database, ask a reference librarian for help. Also determine whether the index searches by subject or by key words in the titles of articles. Each system will provide on-line instructions or the library will provide instruction sheets. You can build a search by using multiple key words. Usually *OR* allows a search for one or both key words while *AND* requires both key words to be present.

You can make a strategy sheet to help guide your search. As indicated in Figure 1.13, begin with your major concept, check how many entries are found, and then limit your search using more key words. Note that while the strategy sheet uses plurals for key words, you can use truncation symbols to allow for all possible endings. For the example above, we could use *car$, batter$*, etc. These truncation symbols vary; some of the common ones are *$, *, ?,* and *!.*

The search strategy for electric cars produced the following results:

1. Electric Cars   28
2. Electric Automobiles   139
3. Automobiles Electric   91
4. 1 or 2 or 3   232
5. 4 and batteries   47
6. 5 and storage   25

Note that the choice of key words and the way they are combined can dramatically affect the number of items recorded. You can use this feature to limit the number of sources you

	Concept 1	And	Concept 2	And	Concept 3	Not	Exclude
	Electric Cars						
Or							
	Electric Autos		Batteries		Storage		
Or							
Or							

Figure 1.13 Database Search Strategy Sheet

want to pursue or to expand the number by including another *OR* followed by an additional keyword.

The best way to learn how to use a database or perform any type of research is by doing it. The best people to help you get started are reference librarians who generally develop courses and information sheets as part of their job. If you have not yet spent much time in the library or have yet to search on a database, check for appropriate introductory sessions offered by the library and check to see what information sheets the library provides. Any time you have a problem, talk to a librarian.

1.4.4.5 Surfing the Net

An increasingly popular form of research involves using the Internet to search for specific topics of interest. Once connected to the World Wide Web, you can use one of the available search engines (*Altavista, Lycos, Excite, Google, Yahoo!*, etc.) to search for topics of interest by using key words. If a search identifies a site of interest, you can connect with that site and read what information it has to offer. Even if the information in a particular site is not particularly helpful, most sites provide links to other sites, some of which may prove useful.

Figure 1.14 shows part of the results of an Internet search for information about the Mars Pathfinder, a mission to Mars undertaken by NASA's Jet Propulsion Laboratory. The search engine used was *Yahoo!* and the key words searched were *NASA* and *Mars*.

Figure 1.15 shows one of the web sites (www.jpl.nasa.gov/mars/) with information about the Mars Pathfinder mission. Also note that this web site provides links to other web pages and the e-mail addresses of individuals who may provide further information.

1.4.4.6 Evaluating Sources on the Web

Although you should analyze the currency, accuracy, completeness, and biases of all information sources, you must be especially critical of materials that you encounter on the Web. Many of the scholarly texts and journal articles that you encounter in university libraries have undergone critical reviews by experts in the field. Thus, when you draw information from *peer-reviewed* (or *refereed*) journals or from books published by well-known academic publishers, you can be relatively certain that the information is accurate and follows the accepted methods of the relevant field.

The same cannot be said for much of the material posted on the Web. No standards yet exist to ensure the accuracy of information on the Web. While some web sites are

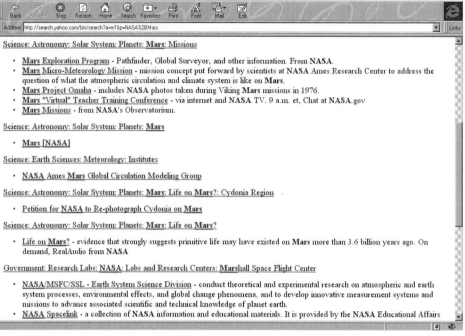

Figure 1.14 Example Results from Yahoo! Search Engine

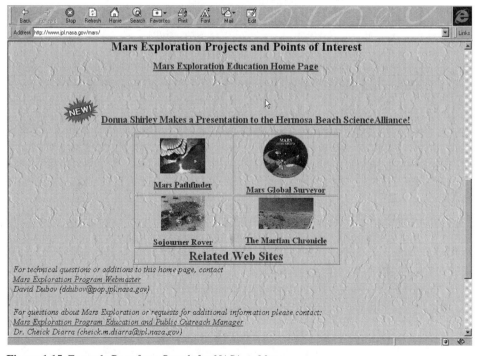

Figure 1.15 Example Page from Search for *NASA + Mars*

reputable, others provide information that is one-sided, dated, or biased. Further, on some web sites, data may have been collected in a haphazard manner, may have been analyzed poorly, or may be interpreted in a misleading fashion.

Before you use any information off the Web for research purposes, you must evaluate the site. We offer five general questions and other, more specific, ones to guide your evaluation.

1. *Is the source appropriate?* Begin by determining whether the Web is where you should be seeking information to fulfill your purpose. Because searching the Web is generally more convenient and enjoyable than a trip to the library, many students fail to assess whether a web search is the most appropriate option. Because the Web is relatively new and does not duplicate library holdings, for a historical perspective on a subject, traditionally published materials may be your best choice. Also, authoritative reference works may provide information you need that is unavailable on the Web or information that is easier to find and more assured to be accurate. Before you turn to the Web to research a topic, ask the following questions: Is your topic suited to a web search? Are aspects of your topic better researched by a trip to the library or by searching a database?

    If a web search is appropriate, then determine whether a particular site meets your needs. Is the content and tone appropriate? Is the topic covered in appropriate depth? Is the site aimed at your level of expertise? Is it a primary source or useful for its links to other sites? How many and what kinds of links are provided?

2. *Is the site current?* Sites come and go on the Internet with awesome regularity. Consequently, the fact that a site is new is not an indication of its value. An older site that is kept current is a better indicator of a useful site. Are you told when a site was last updated? Can you tell when the material on the site was written? Are the links to other sites current? Is the page under construction? How complete is the information? If a site is no longer maintained, has the author explained why it is still available?

3. *Is the site authoritative?* You may have little difficulty determining the authority of official government sites and those sponsored by universities and other educational or research organizations. But what about the myriad pages a search turns up that are not so easily identified? In such cases, learn what you can about the author and the sponsoring organization. What is that author's or sponsor's expertise about the topic? Are the author's qualifications presented or is a link provided to bibliographic information on the author's personal home page? Has the author published refereed journal articles or written books? Who published the author's work? Does the site provide information about the sponsor or link to a site that does? Does this site link to reputable sources? Do you reach this site from a link in a reputable site?

    If the author or sponsor is not identified, turn your attention to the URL (Uniform Resource Locator), or web address. Does the URL indicate a reputable sponsor for the site? To answer this question, you need to pay attention to how URLs are formulated:

    Transfer protocol://servername.domain/directory/subdirectory/filename/filetype.

    The key elements for evaluating a web site are the server name, domain, and directories. As an example, let's use the following URL: *www.jpl.nasa.gov/mars/.* The server name in this case is *www.jpl.nasa*, indicating that NASA sponsors the site. If you weren't familiar with NASA as an organization, you could determine something about it from the domain, in this case, *.gov*, which identifies NASA as a government agency. The six common domain identifiers for the United States follow:

- *.com* indicates a commercial enterprise (or an individual).
- *.edu* indicates an educational institution (or sometimes an individual).
- *.gov* indicates a government agency.
- *.mil* indicates a military agency.
- *.net* indicates an Internet service provider.
- *.org* indicates a not-for-profit organization.

The domain may also include a country code such as *.ca* for Canada or *.uk* for Great Britain. The tilde symbol (~) in a directory sometimes signals a personal web page, so while the server could be a prestigious university, the page could belong to a student and its content might not be sanctioned by the institution. If a site is anonymous or if it lacks a reputable sponsor and the author's credentials are not provided, the site should be treated with a high degree of suspicion.

4. *Is the site objective?* Perhaps the most difficult aspect of evaluating a site is determining biases and the author's or sponsor's motivation for providing the site. Is it geared to help others, to sell a product, or to fight for a cause? What are the goals of the author or organization? Does the site seem to be someone's personal soapbox? Is the presentation balanced or biased? If biased, can you determine the nature of the bias? For example, is it political or economic? Are links provided to web sites with differing viewpoints? Does the site include advertising? If so, is it justified? Do the ads support the information on the page or does the page exist for the ads? Do ads make it possible to offer a service? Is the site funded in some other way? By a National Science Foundation (NSF) grant, for example?

Also note the accuracy of the information provided. Does the site provide fact or opinion? Are sources of statistical information indicated? Are facts verifiable? Are sources of information provided to back up assertions? Has the author left out any information or avoided any points?

5. *Is quality controlled?* The look and feel of a site may reflect its authority. Is the site well written? Do you note obvious typographical, spelling, or grammatical errors that might lead you to question the accuracy of the data or the abilities of the author? Is the site easy to navigate and professional in appearance? Do any visual or audio effects enhance the message?

An author's attempts to control quality are another indication of a site's value as a research source. Are links peer reviewed or does the author check out sources before links are added to the page? Does the author provide a link to submit comments or questions?

No one of the above questions indicates the credibility of a site on its own. You need to consider your answers in context and develop an overall perspective on the site. For more information on web sites, we recommend a visit to the ICYouSee home page for John R. Henderson's upbeat guide to the World Wide Web. Check out the page on Critical Thinking about What You See on the Web. This site was six years old in 2000 and is still being maintained, so we are confident it will still be around when you read this page. The URL is http://www.ithaca.edu/library/Training/ICYouSee.html.

1.5 ORGANIZATIONAL STRATEGIES

Once you have completed the planning and inventing stages, your next task is often to decide what information to use and how to organize it. (For many writers, organization occurs at the same time as planning and inventing.) Planning generally culminates in some form of outline, which prepares you to write the first draft. Depending on the nature of the project, this outline may be a few scribbled notes, a diagram, a topic outline, or a formal sentence outline. For example, a covering letter to a prospective employer may require only a few notes to remind you what points to include and how to order them. A project proposal, on the other hand, will require a substantial, detailed topic or sentence outline that will likely indicate not only what information to include and how to order it, but the degree of emphasis afforded each topic.

1.5.1 Creating an Outline

Although outlining is one of the most common strategies used to organize topics, it is also one of the most commonly abused strategies. All too often, the outline becomes such a controlling force that it prevents the discovery of new information (the same, by the way, is true of document formats—some ideas and approaches do not fit easily into the standard forms). Therefore, maintain flexibility when working with outlines (and with standard document formats). The content of an outline should remain tentative throughout drafting and throughout the early phases of revising.

1.5.1.1 Topical Outlines

A successful topical outline organizes items of information in two fundamental ways: in terms of their relative importance and according to a logical sequence. Determining the relative importance of the points you wish to include is, in part, determined by whether your emphasis is on informing or persuading your readers. To inform your readers, emphasize major technical points; to persuade them, emphasize those points that fill a need for your reader. At this stage, as throughout the writing process, considerations of audience and purpose help guide judgments.

The conventional forms of documents can also be powerful tools for organizing your ideas. For example, the *Introduction-Methods-Results* or *Argument-Analysis-Implications* structure of some kinds of technical reports provides a ready-made topical outline that you can then flesh out by developing subtopics. Once you have decided what information to include and considered the overall structure of the document, you can create a topic outline by writing descriptive headings. You can use indents to indicate the emphasis given to a topic (e.g., whether it is a section or subsection), or, as is most often the case with technical reports, you can use a numbering system to indicate the level of a heading, as illustrated in Figure 1.16.

In shorter documents or in those with only two levels of headings, the numerical system serves little practical purpose. However, numbered sections are often required for cross-referencing and tracking in even short specifications. Also note that with some technically oriented audiences, using a numbering system can serve an important, if subtle, persuasive purpose. It indicates to readers (along with other factors) that you understand the conventions and practices of their engineering field and helps them *identify* with you.

While preparing an outline, you must also consider the descriptive power of your headings. A vaguely worded heading may indicate an equally vague sense of the content

```
1. Fiber Optics in Telecommunications
   1.1. Architectures
        1.1.1. Active Pedestal
        1.1.2. Double Star
        1.1.3. Star Bus
   1.2. Materials
        1.2.1. Fiber Types
        1.2.2. Strand
               1.2.2.1. Core
               1.2.2.2. Cladding
   1.3. Applications
        1.3.1. Residential
        1.3.2. Commercial
   1.4. Advantages and Disadvantages
        1.4.1. Cost
        1.4.2. Flexibility
   1.5. Installations
   1.6. Computer Simulations
```

Figure 1.16 Hierarchically Organized Outline

of that section. To clarify matters for yourself and for your readers, make your headings as descriptive as possible. Headings made up of nouns and adjectives, such as *Digital Plotter* or *Steel Alloys*, generally provide little indication of the actual scope or intent of a section. By adding brief descriptive phrases or by using verbs, you can provide more accurate descriptions while still keeping your titles relatively short. For example, the titles given above might be rewritten as *Limitations of the Digital Plotter* or as *Criteria for Selecting Steel Alloys*.

To find a suitably descriptive heading, first answer the following question: *What am I doing in this section?* If the answer is that you are explaining the advantages or limitations of something, then your heading could begin *Advantages of* or *Limitations of*. If you are evaluating or explaining how to use something, then your heading could begin with the appropriate verbal, *Explaining* or *Performing*. If you have a choice between a noun and a verb, use the verbal form (i.e., *An Evaluation of* should be revised to *Evaluating*). Chapter 7 provides more information about headings.

1.5.1.2 Computer Outlining

When writing relatively long documents, try using the outlining feature of your word processing package. To begin, you can simply type in as many headings and subheadings as you can think of. Later, you can expand these headings into paragraphs and you can also easily add, delete, move, promote, and demote headings and subheadings as you see fit while drafting or revising. Moreover, the outlining feature is particularly helpful when you are revising because you can easily see whether the overall organization of the document makes sense. Further, if you have used outlining appropriately, generating and updating the table of contents becomes a simple task. Figure 1.17 provides an example of a computer-generated outline for one of the early drafts of this text.

1.5.1.3 Alternative Types of Outlines

The traditional topical outline and the more modern technique of creating a computer outline are not the only approaches to outlining. Although the *hierarchical* approach of

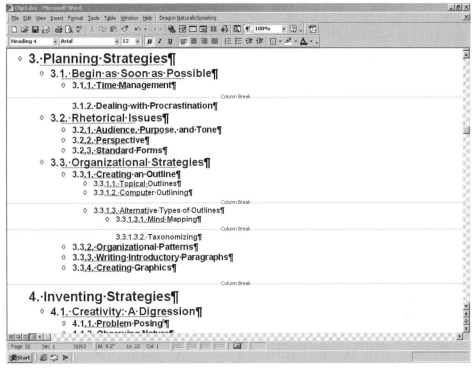

Figure 1.17 Example Document Outline Created on a Word Processor (*Microsoft Word 2000*)

topical and computer outlines works well for many people and with many topics, some find it easier to generate outlines graphically through, for example, *Mind Mapping* or *Taxonomizing*. If you find that standard methods do not work well for you, you may find one of the following alternatives more effective. As you will note, these alternative approaches produce results that can be easily transformed into standard outlines.

Mind Mapping Mind maps are ways of generating and visually organizing ideas to show the relationships between them. Simply take a general topic or problem and write it in the center of the page (or on a blackboard if working with other people); then surround it with more specific ideas as they occur to you. Surround these ideas with even more specific ideas, and so on. If you circle each point and draw lines to the points radiating from it, you end up with something resembling a bug (which is why mind maps are also called *bug diagrams*). Figure 1.18 provides an example of a mind map.

Although this example uses key words, you could instead use symbols or icons to represent the various ideas. You can also try using different colors to represent various levels of importance or the relationships among ideas.

Taxonomizing Taxonomizing is a strategy for taking a range of ideas and grouping them on the basis of shared characteristics. This approach involves taking a piece of paper and dividing it into quadrants. Ideas are then sorted into groups. When you have two or more ideas in each quadrant, write a title or sentence that describes the similarity between the ideas. If you have many ideas in any quadrant, divide them into subcategories. Figure 1.19 presents a taxonomy for the topic used to illustrate mind mapping.

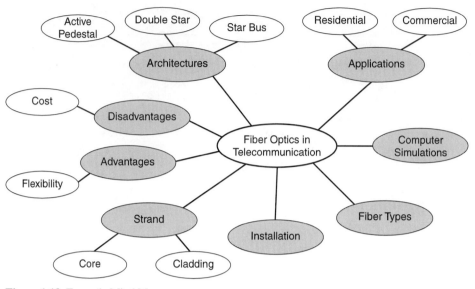

Figure 1.18 Example Mind Map

Fiber Optics in Telecommunications			
Architectures		**Strand**	
Star Bus	**Double Star**	**Core**	**Cladding**
_____	_____	_____	_____
_____	_____	_____	_____
_____	_____	_____	_____
_____	_____	_____	_____
Active Ped.		_____	_____
_____		_____	_____
_____		_____	_____
_____		_____	_____

Applications

Residential | **Commercial**

Advantages

1. _____
2. _____
3. _____
4. _____
5. _____

Disadvantages

1. _____
2. _____
3. _____

Figure 1.19 Example Taxonomy

1.5.2   Organizational Patterns

As well as deciding what points to emphasize, and what major subsections to include in a document, you must consider the best way to organize it. In the vast majority of cases, you will find one of the six organizational patterns outlined in Table 1.3 suited to your needs.

How can you use these organizational patterns? Imagine that you must write a report explaining why a project is running behind schedule. You could organize your entire report, or at least the central sections of it, in one of the following ways:

- By describing the sequence of events that contributed to the delay (chronological order)
- By describing how problems in one area affected others, creating a chain reaction (spatial order)
- By comparing initial expectations, such as availability of parts, against the actual situation, such as supply problems (comparison/contrast)
- By beginning with the most important and working through to the least important reasons for the delay (order of importance)
- By discussing the problems encountered—in order of time, relationship, or importance—and how to avoid them in the future (problem solving)
- By providing a general description of the problems encountered followed by more detailed explanations (general to specific)

You can also use these patterns to organize subsections and even paragraphs. Note that for technical writing, the general-to-specific pattern is the most common way to organize paragraphs. If you have problems with paragraphing, see Chapter 6 for further information about paragraph structure.

Table 1.3   Patterns of Organization

Pattern	Description
Chronological	Sequence of events through time, as for describing a process
Spatial	According to physical relationships, as for describing an object
Comparison/Contrast	Explanation of similarities and differences, as for comparing old and new designs
Order of importance	From most important to least important, as determined by audience and purpose
Problem solving	Explanation of problem, method, and solution followed by recommendations
General to specific	A general point followed by a more detailed explanation and/or specific examples

1.5.3   Writing Introductory Paragraphs

After completing an outline and deciding on an organizational strategy, some writers then write introductory paragraphs for major sections in order to carefully define issues or problems. By writing and rewriting important introductory paragraphs until they are very clear and precise, these writers develop a fuller understanding of their topic and how they intend to approach it. However, some caution must be exercised when using this strategy because perfectionists often have difficulty stopping work on the introductory elements and starting work on the rest of the document. If you find yourself spending an excessive amount of time on an introductory paragraph, move on to another section or subsection that introduces a topic you feel more confident addressing and then return to the first paragraph during revision.

1.5.4   Creating Graphics

If you have not previously done so, you should plan your graphics during the organizational stage of the composing process. Ask yourself what points could be presented most effectively by means of a drawing or diagram, what points might be most effectively summarized with a table or graph, and where these figures and tables should appear in the document.

Creating tables and figures can be a good transition to drafting. Creating the graphics for a document before writing the supporting text is a common strategy used by seasoned engineers, especially when working on functional and design specifications. Further, the mere existence of the graphics may make the task of writing the text seem less daunting

However, as with writing introductions, some caution is in order. Beware of a tendency to spend excessive time on the graphics and thus to leave insufficient time for writing and revising the text that explains the graphics. If you find yourself spending more time on the graphics than seems appropriate, then ask yourself whether you are succumbing to procrastination or perfectionism. Again, the solution to the problem is generally to move on to another section or another task.

1.6   PLANNING CHECKLIST

To conclude this chapter, we provide the following writing checklist for you to use as you plan various documents. We suggest that you initially use this checklist by considering the points that it raises *after* you have written a document. By noting those points that you have not adequately considered (i.e., those that you haven't checked off), you can determine which areas of planning and inventing to work on in order to make appropriate changes to your writing process.

Writing Environment

Were you sufficiently familiar with the various features of your word processing software that you could use it efficiently?   _____

Did you manage your writing environment in such a way that it did not interfere with your efficiency and effectiveness?   _____

Planning Strategies

Did you determine what deadlines were involved with the writing task?   _____

Did you budget your time appropriately?   _____

Did you avoid procrastinating?   _____

Did you analyze your audience and purpose?   _____

Did you determine what tone to adopt in the document?   _____

Did you decide on a perspective (first, second, or third person)?   _____

Did you determine what form to use for the document?   _____

Inventing Strategies

Did you use one or more inventing strategies (if needed)?   _____

Did you pose problems, observe nature, meditate, role play, free write, or use a journal to promote creativity?   _____

Did you use structured problem-solving heuristics such as analogies or contraries, or did you synthesize ideas or ask questions?   _____

Did you engage in collaborative strategies such as discussion or brainstorming?   _____

Did you research your topic, paying careful attention to the appropriateness of your sources?   _____

Organizational Strategies

Did you plan how you would organize the document using document formats or organizational patterns?   _____

Did you prepare an outline of topics that you needed to cover?   _____

1.7   EXERCISES

Ideally, you should apply the concepts, principles, and advice in this chapter to writing that you already do for school or work. We provide the Planning Checklist in Section 1.6 of this chapter for that purpose. The following exercises may help you further understand some of the issues we raise.

1. Thinking about Rules
In a ten-minute period, list all the rules you can think of that you have heard about writing (for example, never start a sentence with *because*, never end a sentence with a preposition, never use the passive voice, etc.). Then identify any places where breaking the rules is a good idea in order to achieve a particular goal or effect. Finally, consider if any of these rules are appropriate and should rarely, if ever, be broken. (This exercise would be particularly effective for a group of three to five people.)

2. Writing Proficiently with Word Processors
Spend a couple of hours working with a medium-length document on your word processor. Pull down every menu and try to explore every feature. See if you can annotate a document, set up a simple style sheet, arrange information in a table, cross reference information, set up headings, insert headers and footers, provide a table of contents, auto-number headings, equations, and figures, set up sections with different kinds of page numbers, play around with the fonts and the format, run the grammar checker (or turn it off), and use the thesaurus. What did you learn about your word processor?

How useful are these new skills and where will you apply them?

3. Analyzing Your Writing Environment
Consider where you do most of your writing. What do you like about that environment? What would you like to change? What could you do to implement those changes? For example, if you write in a noisy environment, could you use earplugs or listen to music through headphones?

4. Time Management and Procrastination
Think back to the last few writing assignments that you have undertaken. How carefully did you manage your time? Did you get started immediately on the assignment, or did you put it off to the very end? Are there particular types of writing that you procrastinate with more? If you procrastinated, can you remember why you did so? Write out a plan for your next writing assignment that describes in some detail when you are going to do it (be specific about the parts of the writing assignment and about times and dates). Keep this plan, and apply it when working on the assignment. Then ask yourself how successful you were in following the plan. What changes do you need to make when planning your next assignment?

5. Engineering Journals
Start keeping a formal engineering journal and record information from work or study that you consider to be important. You might also record things like the writing plan mentioned in exercise 4, the

mind map mentioned in exercise 8, or the creativity strategy mentioned in exercise 6. Review the journal after a few weeks and preferably again after a few months. Do you find anything useful or interesting in it? Do you want to make marginal corrections or additions? Is anything you had forgotten recorded in it? Are you developing a habit of keeping a journal?

6. Strategies for Thinking Creatively

Using the problem-posing strategy outlined on pages 16–17 of this chapter, try to think of an existing device or system that doesn't work as well as you think it could. (If you can't think of anything, then examine how little foil ketchup packages are currently made. They are hard to open and are environmentally wasteful. How could they be improved?). In step 2 of the problem-posing strategy, try applying one of the following substrategies (described on pages 17–24): observing nature, focused meditation, role playing, free writing, creating analogies, generating contraries, synthesizing ideas, or asking questions. What sorts of insights and ideas for improvements are you able to invent?

7. Research Strategies

Research a topic on the World Wide Web (for example, *computer viruses*). As you look at the various web sites, try to answer as many questions as possible from the list on pages 31–32 of this chapter. In particular, try to identify the biases and goals of the various web sites. Which web sites seem most trustworthy to you? Why? Then research the same topic at a university or college library. What sort of differences do you find between the Web and library sources?

8. Outlining Your Writing

When working on your next couple of writing assignments, try applying different outlining strategies. Use the computer outline feature on your word processor, draw a mind map, and set up a taxonomy. Which of these strategies works best for you? Do some strategies seem to work better with certain kinds of writing?

8. Organizational Strategies

Review the table of organizational strategies on pages 33–38 of this chapter, and write three short paragraphs about a topic of interest and use a different strategy for each paragraph. For example, you might write about the *History of the Automobile* using a strategy based on time (*when* was the first automobile built?). Or you might use a spatial organization (*where* was the first automobile built?). Or you might use a problem-solving strategy (*why* was the first automobile built?). Note how the organizational strategy chosen changes the kind of information that you present as well as how you present it.

Chapter 2

Drafting and Revising Strategies

2.1 SUMMARY

This chapter introduces you to the process of drafting and revising technical documents. Our goal is to help you learn a range of strategies for managing your own writing processes in order to increase your efficiency and your comfort while writing. In addition, we describe some of the software resources that can assist in making your writing more effective. The following are some of the topics we cover:

1. **Strategies for Drafting**

   Drafting is that part of the writing process when you put fingers to keyboard. While this part of the writing process is probably the most critical in terms of your efficiency when writing, it is unfortunately also the most poorly understood. We encourage you to observe your own drafting process and to experiment with some of the alternatives we suggest.

   a. **Alternative Approaches to Drafting**

      In general, writers approach the process of drafting in four alternative ways, depending on how this part of the process is combined with the processes of inventing and revising. Although the approach you may most commonly use is typically a matter of individual preference, the length and nature of the document being written are also important factors in determining the most efficient approach for a particular situation. For example, you likely use quite different approaches when drafting an e-mail message than when drafting a formal report.

   b. **Criticism and Creativity**

      All of us have both critical and creative elements to our personalities. Where we engage these parts of personality with respect to the writing process can have a significant impact not only on our efficiency but also on our comfort and confidence while writing. If you are overly critical while drafting a document, you may find that your productivity suffers, and in the worst case, you may encounter writer's block.

c. Writer's Block

At one time or another, nearly all of us must deal with writer's block. This frustrating experience most frequently occurs during the drafting process (although it does sometimes occur during inventing and revising). We outline a wide range of reasons for writer's block and suggest various strategies for dealing with the problem. Generally, if you understand the causes of your block, you will find it relatively straightforward to resolve.

d. Health and Safety Issues

Few people give enough thought to the issues of health and safety in relation to writing. But we have seen far too many 20-year-old students suffering from repetitive strain injuries developed while typing to ignore this problem. We have also seen 40- to 50-year-old writers debilitated by back and neck injuries as well as suffering from eyestrain. We encourage you to carefully consider your own health and safety when writing in order to avoid future problems.

2. Strategies for Revising

In terms of the quality of the document produced, the revising process is probably one of the most important parts of the overall writing process. But it is also one of the most difficult parts of the process to master. Revising effectively requires that you attend to high-level issues such as persuasion, organization, and structure as well as to low-level issues such as style, grammar, and spelling. With careful attention to revising, even a poor first draft can be turned into an excellent document.

a. Various Revising Strategies

Over the past 30 years, we have developed and collected a wide range of strategies for revising our own writing. In order to help you quickly master the revising process, we outline 15 strategies, ranging from oddities such as reading your document backward to very powerful strategies such as speaking aloud while revising.

b. Boredom and Perfectionism

Two problems that many writers encounter while revising are boredom and perfectionism. If you are bored while revising, chances are that your attention will lapse and problems will remain in your writing. On the other hand, if you strive obsessively for perfection when revising, odds are that you will find it difficult to meet deadlines or complete the document. We suggest several strategies for dealing with these problems.

c. Spelling and Grammar Checkers

Spelling and grammar checkers are of considerable value to many writers. Unfortunately, these technologies have not yet been perfected, and worse, many of us remain unfamiliar with their limitations. We describe some potential uses and limitations of spelling and grammar checkers, and outline some of the most common problems encountered when using them.

2.2 DRAFTING STRATEGIES

Once you have done your best to plan your work, you eventually come to that point where you must generate text. Although much research has been directed toward studying the topics of planning and inventing and of revising and editing, far less attention has been directed toward the topic of drafting. What actually takes place in your mind as you set fingers to keyboard or pen to paper? Although the answers to this question remain uncertain, some issues that writers should address when drafting have started to emerge. Nevertheless, you should consider the information in this section as provisional, or temporary. Reflect on what you do as you draft and experiment. See what works for you.

2.2.1   Four Approaches to Drafting

Figure 2.1 represents four distinct ways of structuring the writing process. Note that *I* stands for *Inventing, D* for *Drafting*, and *R* for *Revising*. Although few writers use the same approach all the time, you should determine which one most resembles your usual approach to composing and consider its appropriateness, advantages, and disadvantages. Developing the ability to employ different approaches for different situations will help you become a more efficient, effective writer and help you master various types of documents.

2.2.1.1   Combining Inventing, Drafting, and Revising

Some writers prefer to combine all elements of the writing process in a highly organic fashion. They may simply start writing down points and sentences without a particularly clear idea where they will end up. As some sentences are written, they may be revised or even deleted. New ideas are added, and others are removed. These writers may move from one section of the document to another in a manner that, on the face of it, seems disorganized.

This approach has the advantage that the task is started immediately and can be especially useful when you lack a clear sense of the solution to a particular problem. In that respect, this approach tends to be highly creative and resembles the practice of free writing in that its primary goal is invention (see Chapter 1, page 18).

In practice, however, this approach is rarely used in documents of any length or complexity because it is inefficient. You may spend a significant amount of time writing a section only to discover you have made a false start and must begin again. Nevertheless, this approach has several uses. For example, when writing short, informal documents such as e-mail messages, planning the document in detail, then drafting, and then carefully revising makes very little sense. With short documents, a highly structured writing process that arbitrarily separates the various elements of the writing process is less efficient than this more integrated approach.

Because this approach to writing is often a good way to get started on a writing task, you might consider using it soon after receiving a writing assignment, particularly if you commonly have difficulties getting started. This practice will focus your mind on the topic and help you generate further ideas. Having written even a page or two can encourage you to keep working.

2.2.1.2   Separating Inventing, Drafting, and Revising

At the other extreme, some writers prefer an approach that separates the elements of the writing process. This approach is sometimes used for lengthy documents such as user's

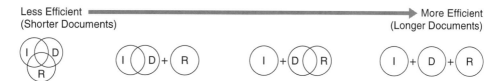

Figure 2.1 Alternative Ways of Structuring the Writing Process

manuals or textbooks, for team-written documents, and by individuals who dictate their drafts. Initially, a document is meticulously planned for its content, form, and organization (sometimes even for its style). Next, the document is drafted as quickly as possible, often in sections. Finally, it is extensively revised, generally in several sweeps.

For example, members of a team working on a lengthy collaborative report meet for several planning sessions to outline the various topics to be included in the document. They decide on an organizational strategy, determine the format and style that is most appropriate, consider what software and computing platform to use, assign the various sections to team members, and set deadlines. The more carefully the team plans the document, the less likely they will encounter problems when integrating the various sections.

Each member of the team then writes the various sections that they have been assigned and typically swaps drafts with another team member to start revising the document. In some cases, individuals other than the authors—professional editors, for example—are responsible for revising. Unless a professional editor is employed, the document is usually given to one team member who integrates the various sections to ensure consistent content, tone, style, and format. In fact, we used this approach while writing this text. Chapter 4 provides more detail about team writing (pages 126–132).

On an individual level, users of dictation programs generally separate inventing, drafting, and revising. In order to produce a document that is reasonably coherent, they will typically outline the document in great detail. Having a detailed outline maximizes their speed and minimizes the need to reread while dictating.

Revising is especially important when dealing with dictated documents because speech tends to be less formal and less coherent than writing. In addition, speech is characterized by more incomplete structures, shorter sentences, and less precise language than writing. In the next few years, speech recognition software will become more commonly used (as the technology is perfected), and many of us will likely move to a writing process that requires us to revise in different ways and for different problems.

Nevertheless, in most circumstances, the approach of separating the various elements of the writing process is not the one that most of us currently use when writing. Nor, for that matter, do most of us frequently use the method that combines the various elements of the writing process in a more integral manner. More commonly, writers of medium-length documents (i.e., 5 to 50 pages in length) use a process that combines either inventing and drafting or drafting and revising.

2.2.1.3   Combining Inventing and Drafting

Some writers typically combine inventing and drafting, leaving revising until later. Prior to the advent of word processing, this approach was probably used more frequently—simply in the interest of efficiency. When writing with a pen and paper (or typing), reorganizing a document requires devoting considerable time to cutting and pasting sections of a document. Rewriting sentences typically involves crossing out work (or using correction fluid) and then rewriting. Writing the document as quickly as possible (perhaps working from brief outlines) and saving most of the revisions for later is generally more efficient. The word processor encourages editing and revising while drafting, which may have made this approach less common.

Nevertheless, some writers prefer to combine inventing and drafting for a variety of other reasons. This approach may well be the most efficient for shorter documents (i.e., 5 to 15 pages) or for sections within longer documents.

The central strategy of this approach is *momentum*. Keep moving. Start writing and do not stop until you have finished your first draft. If you are not certain what to say in a particular section, then simply skip that section. By the time you have finished writing the rest of the draft, you may well have discovered what you wanted to say.

Should you find yourself really stuck while using this approach, you might try free-writing, which involves writing whatever comes to mind, no matter how absurd it may seem (see Chapter 1, page 18). Write about the problem at hand. If you don't know where to start, try writing about how you feel about being stuck and how it would feel to become unstuck. Frequently, after a few minutes of free-writing, you will find yourself back on track and may even discover that you have unexpectedly generated a useful approach to a problem. You can always delete irrelevant material later. If free-writing is not effective, try working for a while on another assignment or another section of the document.

If you cannot think of the word you want, leave a blank space or type *XXX*; you can find it later. Unsure of your spelling or grammar? Ignore them. You can always use a spell-checker or grammar checker when revising. Do not stop to deal with stylistic issues until you have completed the first draft. Rather, focus on the purpose of the report, the needs of your audience, and the ideas you are expressing. Save other concerns for revising and editing.

If you encounter a problem with your logic or organization while inventing/drafting, do not waste time trying to solve it. Write a brief note about the problem, and keep writing. When working with a word processor, you can make these sorts of notes by keeping a notepad handy and jotting your thoughts on the pad, or by typing the idea in the text and boldfacing it so you can easily identify it and deal with it when revising. These two strategies work well for brief documents. Another, perhaps more effective, strategy is to use the *annotation* feature of your word processor to embed extraneous ideas and comments in the text. This last strategy works especially well in longer documents.

Annotations (sometimes called *comments*) are typically formatted in *hidden text*, which simply means that they will not print even though they appear on your screen in a fashion similar to the following: [**SW1**]. In this case, [**SW1**] is keyed to a separate window that contains the note to yourself. Not only is this feature a convenient way to leave yourself notes about various problems, but it is also a good place for free-writing to deal with writer's block.

Further, learning to use the annotation feature may prove beneficial in your future career as an engineer, when you will quite likely be asked to annotate colleagues' documents. We suggest that you format the annotation *style* in your documents in a way that makes it clearly visible on the screen (i.e., uppercase, boldface type and perhaps in red). That way, you or your colleague will not inadvertently miss annotations when revising. Figure 2.2 provides an example of some of the annotations we used when writing this text.

If you combine inventing and drafting in order to write a first draft quickly, then you must pay particular attention to both the planning and revising stages of the writing process. Unless you have devoted sufficient time to planning your approach or researching your subject, you may have difficulty getting started. Further, because drafting quickly allows little time to reflect on what you have already written, you may also wander off into discussions of interesting, but irrelevant, side issues. Consequently, you should plan carefully and comprehensively to help guide your thinking as you write. Writing a tentative title and producing a statement of your purpose can also help keep you on track. But be prepared to remove excessive and irrelevant material from your writing as you revise.

Writers who combine inventing and drafting need to manage their time effectively to ensure sufficient time for revising. If possible, complete your initial draft at least two days before your deadline to allow enough time between drafting and revising to gain some per-

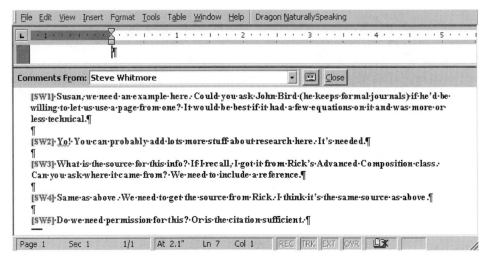

Figure 2.2 Example of Annotations (*Microsoft Word 2000*)

spective about what you have written. Also, when deciding how much time to allow for revision, remember that those who follow this approach must generally revise their work.

Given appropriate time management, the advantages of this approach are threefold. First, the act of writing (of generating text) can itself spawn new ideas, new approaches to issues, and new solutions to problems. Second, a certain amount of satisfaction is gained from completing a draft, no matter how rough it may be. Third, users of this approach may be more likely than others to revise extensively because they have invested less time and effort in their initial draft.

Note that if you use this approach, you should seek an environment where interruptions are minimal. Few things are more frustrating than losing your train of thought because someone wants to chat. If you are interrupted, when you return to your writing, review what you have previously written to ensure you stay on track.

2.2.1.4   Combining Drafting and Revising

Perhaps one of the greatest advantages of word processors is the ease with which they enable us to make changes to our documents while we are working on them. Unfortunately, this advantage is also a major drawback for some writers. For some writers, a word processor makes it too easy to change what they have written, and they become so involved in perfecting certain elements of their writing that they end up wasting time. Nevertheless, this approach may be most effective for writing longer documents (15 to 50 pages) or when writing documents in which the word choices must be very precise (i.e., proposals or legal documents).

Many engineers and engineering students have indicated that they prefer to undertake extensive revising while they draft. These individuals often compose more slowly and more deliberately than those who combine inventing and drafting. Often, they require several sessions to complete their first draft.

Organization is the central strategy of this approach. To use this approach effectively requires a good sense of the organization of a document; otherwise, you risk losing track of what comes next every time you stop to revise. So if you tend to revise while drafting, you must spend sufficient time planning your writing. Planning may include taking notes while reading your texts, carefully researching the areas you intend to discuss, preparing a detailed outline (perhaps using the computer outlining feature), developing lists of questions or issues to guide your drafting, and examining samples of the format you will be following. Failure to adequately plan your writing before using this approach almost guarantees that you will spend countless hours in front of a computer screen agonizing over your inability to write or polishing text that must later be deleted or heavily revised.

If you should become blocked, simply reviewing the past few paragraphs can help because the organization of the past section often points the way to the next idea. In fact, review is another important strategy for writers who use this approach. Where you have been often suggests where you need to go. For that reason, you may find it helpful to spend considerable time on your title, statement of purpose, and introductory paragraph(s), which all serve as signposts. However, be sure to distinguish between time spent on useful refinements and time spent on an unproductive quest for perfection.

If you choose to combine drafting and revising, you should also ensure that you are completely familiar with an advanced word processing package (i.e., a recent version of *Microsoft Word* or *WordPerfect*). You should be entirely comfortable using *outlining, annotations, glossaries, spell checkers* and *computer thesauruses, drag and drop editing, templates* and *style sheets, undo* and *redo*, and the full range of *formatting* features.

The ability to manipulate text easily and quickly on a word processor eliminates the endless rewriting (or retyping) that this approach used to require. A spell checker and a computer thesaurus considerably reduce the need for you to refer to a hard copy dictionary or thesaurus. Further, many writers' reluctance to delete their finely crafted sentences and paragraphs is also alleviated by the possibility of saving what they cut to another file—in case they are needed after all.

Advances in word processing even reduce the time required to correct typos. For example, the *AutoCorrect* feature available on some word processors will automatically correct common typos (e.g., *hte* will correct to *the* and *THus* will correct to *Thus*). Further, slow typists can program the AutoCorrect feature to replace certain letter combinations with commonly used phrases (e.g., *CR* will become *Coefficient of Refraction*). Figure 2.3 shows an example of the AutoCorrect feature.

A major drawback of combining drafting and revising is the tendency to become so focused on perfecting wording, style, or organization that you write far too slowly. If you find yourself constantly rewriting the initial parts of your drafts to create the perfect introduction, you must pay careful attention to the information we provide about perfectionism (see pages 61–62).

We all possess two aspects of personality that come into play during writing: a creative part and a critical part.[1] If you focus too much on criticizing your writing in the inventing or drafting stages, you risk convincing yourself that you lack the ability to communicate your ideas effectively, or worse, that you are a poor writer. Figure 2.4 provides an approximation of the roles that the critical and creative aspects of your personality play in the writing process.

Sometimes all that is needed to increase your confidence in your abilities is to shut down your critic and allow your creativity to come to the forefront while inventing and drafting. Imagining the critical element of your personality or outlining its characteristics may help you recognize how it affects you. Once you recognize these effects, you may more easily direct your inner critic to its appropriate role.

---

[1]We would like to thank Professor Richard Coe (Simon Fraser University) for contributing some of the ideas relating to the appropriate roles of the critic and the creator in the writing process.

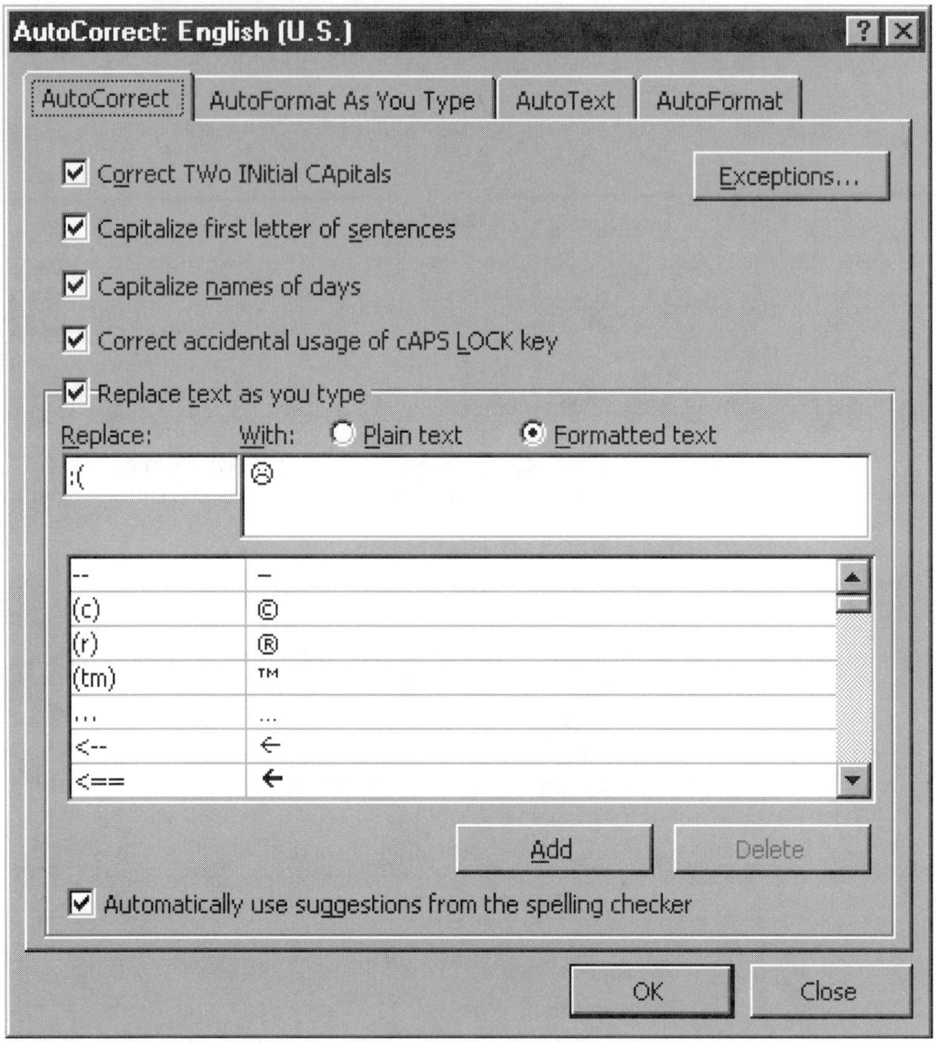

Figure 2.3 Example of AutoCorrect (*Microsoft Word 2000*)

Indeed, the critical part of your personality has some very important roles to play. During revision, for example, the critic helps you focus on potential problems with your writing. Similarly, the critic plays an important role in certain elements of the inventing process (when researching, for example). Our point is not that you should eliminate the critical element of your personality from the drafting process, but rather that you should recognize how it can interfere with creativity.

If, for instance, you find yourself constantly rewriting sentences, reorganizing paragraphs, and generally focused on "getting it right," then you probably need to deal with the critical part of your personality. In fact, you may find it necessary to reevaluate the approach you take to drafting. In some cases, you may want to intentionally separate drafting and revising in order to minimize the effect that the critical elements of your personality have upon your efficiency.

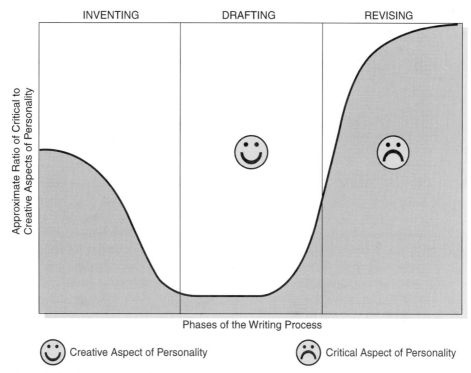

Figure 2.4 Critical and Creative Aspects of Personality When Writing

Further, your internal critic is not the only cause of concern. You may also be affected by your perceptions about the audience for whom you are writing. If you believe that the audience will react very critically to what you have written, you may again find yourself tempted to spend too much time revising while drafting. The simplest solution to this problem is to imagine that you are writing to a different audience, perhaps a friend or a sympathetic instructor. Later, you can revise to adapt to a more critical audience.

Another major drawback of combining drafting and revising is procrastination. A writer who takes this approach, and who has a tendency to procrastinate, confronts an even more serious problem than the procrastinator who combines inventing and drafting. Unlike procrastinators who combine inventing and drafting (who may simply turn in unrevised work), those who combine drafting and revising may well fail to complete a document by the deadline. Therefore, individuals who use this approach must learn to manage their time effectively. If you suffer from this problem, you should probably start writing a document at least a week before it is due (two weeks, in the case of a longer report; several months, in the case of some types of important and lengthy documents).

Nevertheless, despite potential drawbacks, this approach to drafting can be advantageous. First, searching for the right words and the best arrangement for those words can help refine your meaning and can alert you to sections in which you need to refine or clarify your reasoning. Second, some satisfaction is gained from knowing that you have produced a carefully constructed, well-styled, and neatly formatted draft. Third, and perhaps most important, those who use this method generally require less revision after others have reviewed their work than those who combine inventing and drafting.

2.2.2   Managing Writer's Block

Almost everyone experiences writer's block at one time or another in their writing careers. Although quite frustrating, writer's block is a perfectly natural part of writing and can actually be useful insofar as it alerts you to a problem of some sort that must be addressed. Expert writers often become adept at identifying the reasons for a block and then employing specific strategies to resolve the problem. Although not exhaustive, the following sections detail some of the most common reasons for becoming blocked and suggest potential solutions.

2.2.2.1   Blocks Related to Planning

Sometimes writer's block is caused by poor time management. If you find yourself under pressure to meet a deadline (e.g., writing a report the night before it is due), anxiety about the deadline may interfere with concentrating on the task at hand. Although some people write well under pressure, many do not. Fortunately, the solution to this particular block is relatively straightforward: ensure you manage your time in such a way that you leave sufficient time for drafting and revising. If you are engaged in so many other activities that you have little time for completing assignments, then you likely need to reconsider your priorities.

In other cases, a writing task may seem so large as to be overwhelming (e.g., writing a long research report). You may become blocked because you do not know where to start. One possible solution to this problem is to break the work into smaller, more easily managed tasks such as researching a specific problem or writing a particular section.

A third problem involves finding yourself staring at a blank sheet of paper with no idea what to say because you have not devoted enough time to inventing or researching. A solution to this problem is to try one of the inventing strategies outlined in Chapter 1 (pages 15–25).

2.2.2.2   Blocks Relating to Environmental Factors

A very common cause of writer's block occurs when various elements of the writing environment interfere with your concentration. Among the more common sources of interference are distractions: excessive noise, poor word processing skills, poor lighting, and physical discomfort. In this case, you must minimize distractions or move to a different environment.

The time of day when you write should also be considered part of the environment. If you are at your mental peak in the morning, but are trying to write in the evening when you are tired, you may find that you become blocked. Again, the solution to the problem is straightforward: find a way to do your writing when you are most alert (generally the morning, although some prefer to write in the afternoon or evening).

2.2.2.3   Blocks Relating to Topics

If you are bored by a topic, you may well have difficulty starting to write. Quality writing requires a high level of interest, even enthusiasm. Overcoming boredom can be difficult, but you might explore why the topic bores you. Sometimes this exploration will result in useful insights that generate increased interest. Alternatively, focus on those parts of the task that you find least interesting. Doing the most boring tasks before the more interesting ones sometimes helps. Knowing that you have something to look forward to or that the worst is behind you can be a useful motivator.

In other cases, you may become blocked when writing about a topic you dislike. Note that the most common reason for disliking a topic is a lack of familiarity with it. Topics you have not sufficiently explored sometimes seem more complex than they really are. To deal with this problem, try researching the topic in depth. The more you learn about it, the more likely you are to become interested in its subtleties.

2.2.2.4   Blocks Relating to Words

Some people encounter writer's block because they cannot find the appropriate words to use while writing. You may know what you want to say but simply cannot find the right words to express your thoughts. In this case, speaking aloud as you draft is a strategy that often helps. Talking through an issue can be a valuable problem-solving technique. Write what you say. Even if the words are not as precise as you might wish, you can find alternatives that are more appropriate when you revise.

2.2.2.5   Blocks Relating to Form

The form traditionally expected of a particular document constrains the kinds (or amount) of information that can be communicated. For example, a formal laboratory report does not easily accommodate personal anecdotes. Such restrictions can sometimes cause a block. To overcome this problem, you might consider using a form more appropriate to the information you wish to communicate. If that is not possible, then ignore the traditional form, and initially write a document as you prefer. You can then adjust to the form's constraints when revising.

2.2.2.6   Blocks Relating to Audiences

Sometimes you must write for an audience you expect to be hostile to your ideas (perhaps due to a prior negative experience). A useful way to deal with this problem is to write to an audience you believe will be more sympathetic to your ideas. You can then reconsider the audience while revising.

2.2.2.7   Blocks Relating to Expectations

Writers who expect their writing to be perfect in terms of form or style may constantly rewrite sentences or paragraphs to meet this expectation of perfection. Striving for perfection can be an insidious type of writer's block in that the writer seems to be accomplishing something, when in reality little, if any, progress is made. The solution to this block is to accept the reality that no one can produce perfect documents or have a perfect style. You can legitimately strive to be effective, but not perfect. If you encounter this type of block, consider adopting the approach to drafting that separates drafting and revising and then monitor the amount of revising you undertake while drafting. Strive to limit revision to, let's say, 30 to 40 percent of the available time—less for relatively short documents.

You may also suffer from writer's block because you believe that you do not know enough about a subject to write about it. In that case, you may spend an excessive amount of time researching a subject before setting fingers to keyboard. Like perfectionism, this type of block is insidious because it can appear that something is being accomplished when it is not. If you encounter this type of block, separate your inventing and drafting processes and set limits on the amount of research or inventing you undertake. Keep in mind that writing itself is an important process of invention and discovery.

2.2.2.8   Blocks Relating to Crises

Sometimes other problems in life interfere with our ability to concentrate on the task of writing. Obviously, a personal crisis must generally be dealt with before you will regain your ability to focus on writing. But keep in mind that writing can sometimes be used to solve a problem (or at least to understand and accept it). Writing about a problem will help you define how you feel and may even provide you with possible solutions.

Even significant political crises or news stories can cause writer's block. Events such as the bombing of the Federal Building in Oklahoma City or the shootings at Columbine High School in Littleton, Colorado can distract us and draw attention away from more mundane topics. While this type of block usually disappears once the crisis or story ends, you can, nevertheless, minimize the effects by trying to avoid becoming so absorbed in an evolving situation that you leave the TV or radio on constantly.

2.2.3   Attaining a Flow State

Perhaps the most useful advice that can currently be offered about drafting is for you to deliberately observe your drafting process when working on different kinds of documents. Then experiment. See what approach works best for you and with various types of documents. Ideally, when drafting you should aim for what has been described by Mihaly Csikszentmihalyi as the *Flow Experience*—a state of consciousness in which you are so absorbed by the task that you are minimally aware of the passage of time or outside distractions. As Figure 2.5 indicates, achieving this state requires balancing your skills with the challenges of the task.[2]

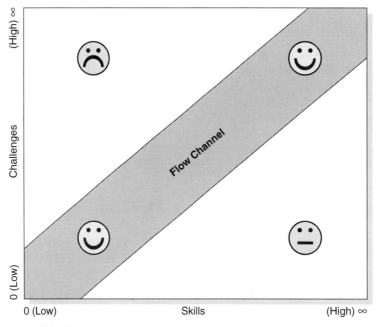

Figure 2.5 The Experience of Flow

---

[2]This figure is adapted from *Flow: The Psychology of Optimal Experience* by Mihaly Csikszentmihalyi (New York: Harper & Row, 1990, 74). Used with permission.

As the position of the unhappy face (☹) indicates, you may experience anxiety if the challenges of a task are significantly greater than your skills in relation to that task. This anxiety may occur, for example, when you have been asked to produce a type of document with which you are unfamiliar. If you find that you are anxious about a writing task, then you must increase your skills in relation to the task, perhaps by finding examples of the unfamiliar document or perhaps by talking with someone who has written that type of document in the past.

As the position of the neutral face (☺) indicates, you may experience boredom if the challenges of a task are significantly lower than your skills. This boredom may occur, for example, when you are editing a document for spelling or grammar (assuming, of course, that you are reasonably skilled with spelling and grammar). If you find yourself bored with a writing task, then you need to increase the challenges involved, perhaps by focusing on higher-level issues such as audience and purpose or form and style rather than just on low-level issues such as spelling and grammar.

Ideally, as the position of the happy face (☺) indicates, you will enjoy writing when you achieve a balance between the challenges of the task and your skills. Although achieving this balance may be difficult, even impossible in some circumstances, when you do achieve it, you will enjoy your work—whatever the task—and be more satisfied with the result.

2.2.4   Health and Writing

When working for extended periods in front of a word processor, you must be aware of some of the potential risks to your health that writing can involve. For example, we know of several engineering students (20 to 24 years old) who have managed to incapacitate themselves by developing *carpal tunnel syndrome* (an inflammation of the tendons in the wrist) while working at their computers. We also know of older individuals who have managed to do significant damage to their necks and spines by bending over their desks and keyboards for extended periods while writing and revising. When working at the computer, you must consider three key health issues: how to avoid repetitive strain injuries, how to minimize eyestrain, and how to reduce your overall stress levels.

As you likely know, repetitive strain injury is damage that is caused to tendons or ligaments through any repeated physical action. It can range from mild but painful inflammations that slow down the writing process to debilitating injuries that require surgery to resolve. To avoid carpal tunnel syndrome, you can take frequent breaks and support your wrists as you type by using a wrist rest that sits in front of the keyboard or by using one of the newer ergonomic keyboards.

To avoid injuries to the spine, you should ensure that, when sitting, most of your body's joints are bent at close to 90 degrees. That is, you should do the following:

- Set your chair so that your arms extend at a 90-degree angle to the keyboard and, if necessary, use a foot rest so that your legs rest comfortably at 90 degrees.

- Ensure that your monitor is at eye level so that you do not need to bend your neck up or down as you look at it.

- Situate notes so that you can see them without swiveling your head excessively or looking down.

Copyholders are readily available that will hold your notes in an appropriate position.

Eyestrain is another problem that writers using word processors often encounter. Ideally, you should use the largest monitor available. If you have a small monitor (14 or 15 inches), use a low resolution such as 640 by 480. Trying to write using too high a resolution on a 14-inch monitor can quickly lead to eyestrain. If you have a 17-inch monitor, you can use a somewhat higher resolution such as 800 by 600. If you have a large monitor (19 to 21 inches), you can use resolutions of 1024 by 768 or higher without straining your eyes. In general, you should be able to easily see the text on your screen from about an arm's length away.

You can further avoid eyestrain by ensuring that your screen is indirectly lit. Few things are more distracting than moving your head about to see around the reflections from light falling on the screen. Antiglare filters that reduce reflections are available for most computer screens. You might also consider regularly removing the dust and fingerprints that accumulate on the screen.

The very act of writing (like most other intellectual tasks) can be quite stressful. You can reduce this stress by taking breaks, personalizing your work environment, and using fast equipment. You should take a 10- to 15-minute break every two hours when working at the keyboard. Writers frequently become so absorbed in the task of writing that they end up sitting in front of the screen for hours at a time. As you write, get up and move around to stretch your legs. Roll your head around and shrug your shoulders to reduce tension in the neck and shoulder muscles. Flex your hands and your arms to relax those muscles.

Frequently, many of us end up working in rather impersonal offices, cubicles, or study areas that are solely functional. This sort of sterile environment can increase your general stress levels. So consider personalizing your work environment by putting up a few pictures or bringing some plants into work or into your study area to create a friendly and relaxed atmosphere.

Using computing equipment that is too slow to keep up with the train of your thoughts when writing can also be frustrating. To reduce the stress of waiting for long documents to repaginate or for large files to load into virtual memory, consider buying or persuading your school or company to provide a suitably fast computer.

2.3 REVISING AND EDITING STRATEGIES

Although the creative part of your personality may be at the forefront when drafting, once you turn to revising and editing your work, your ability to critically review what you have written becomes of key importance. The following sections provide a wide range of strategies that you can use to read what you have written more critically and, thus, to revise and edit more effectively.

2.3.1 Revising Like an Expert

Our experience indicates that there are differences between experienced writers and inexperienced writers in the amount of time they devote to each part of the composing process. For challenging documents, inexperienced writers expend more time inventing and drafting and less time revising than experienced writers. Table 2.1 suggests the nature of differences.

These differences are not too surprising for several reasons. First, it makes sense that as you gain more experience in your field of study, you will require less time planning, researching, and inventing because you will have internalized many of the methods and much of the information from the field. Second, as you become more aware of your own drafting process, both in terms of its strengths and weaknesses, you become more efficient in terms of drafting. Third, as you become more aware of the importance of audience, the

Table 2.1   Differences between the Writing Process of
Experienced and Inexperienced Writers

	Inventing	Drafting	Revising
Inexperienced writer	50%	45%	05%
Experienced writer	25%	35%	40%

limitations of your initial ideas, and the weaknesses in your own style, you begin to recognize the importance of revising your writing and of devoting some of the time saved while planning and drafting to revising.

To become an expert writer, you should expect to spend a significant amount of time revising documents. If you currently devote less than 30 to 40 percent of your time to revising significant documents such as reports, proposals, and term assignments, then you need to seriously consider spending the time necessary to develop strong revising skills. That is, you must develop the ability to read your words as if seeing them for the first time. Doing so is no easy feat. When reading recently written documents, writers all tend to read their intended meaning rather than to see the words they actually wrote. We generally see what we think should be there, filling in background information and making the necessary logical connections in our heads that we omitted on paper.

It takes hard work and considerable practice to train yourself to read critically what you have just finished writing. The task is much easier when you read something you wrote months before. Have you ever found a piece of writing you had forgotten about and taken the time to read it? If so, you may have struggled to understand something that made perfect sense to you when you first wrote it. In this case, you read with the critical eye of a first-time reader. Unfortunately, you can rarely, if ever, afford to leave a document for anywhere near this long. Instead, while still very close to your subject, you must develop the skill of reading as a reader, not as a writer.

2.3.2   Specific Strategies for Revising

While we can offer you no magic formula for developing this skill, we can suggest several strategies to help you master revising. The following 15 strategies have been developed from our experiences with writing as well as from the observations of students and engineers about their writing.

2.3.2.1   Leave Sufficient Time for Revising

Perhaps the most common problem with revising stems from a failure to manage time appropriately. You should budget 30 to 40 percent of your time for revising (more for important documents such as résumés and covering letters and less for short documents such as informal letters and memos). If you find that you have a tendency to procrastinate with writing, then you must make a conscious decision to deal with this problem. Procrastination generally results in minimal revision, which, in turn, generally results in less effective documents.

2.3.2.2   Take Time Out

When planning how to budget your time, also remember that you should put your draft aside for as long as possible before beginning to revise it. The more time between drafting and revising, the greater the likelihood of discovering problems with organization and style. Even one day or a good night's sleep can greatly improve your objectivity, especially when working with short documents. If you have difficulty adopting a reader's perspective, then you should ideally allow a week or two between drafting and revising long or complex documents such as major reports or proposals. On the other hand, a coffee break may provide sufficient time to gain some perspective in the case of a routine letter or memo.

2.3.2.3   Read the Document Critically

During drafting, the critical part of your personality may interfere with the creative process, but during revising, it is necessary and valuable. You should make full use of it. Ask yourself questions about the document. Is this clear? How clear? Is this information accurate? How accurate? How will the audience react to this idea? How can I make this information more easily understood for the reader?

At the same time, try to avoid perfectionism—a major cause of late assignments. Remember to strive for effective, rather than perfect, documents. We touched on the issue of perfectionism earlier and will discuss it in more detail later in this chapter (see pages 61–62).

2.3.2.4   Revise the Document on All Levels

You should *revise* on the levels of substance, persuasion, format, paragraphs, sentences, and words. Unfortunately, inexperienced writers often only *edit* for lower-level issues such as spelling and punctuation (although on occasion, they may also make some stylistic changes by restructuring and rewording sentences). In addition to these types of editing changes, experienced writers also *revise* for higher-level issues by reconsidering their persuasive or informative emphasis, examining the accuracy and completeness of their ideas, reorganizing documents, and restructuring paragraphs.

To become an expert writer requires paying careful attention to revising for organization, substance, and purpose as well as editing for style and correctness. The questions provided in the following sections are intended to draw your attention to a number of issues that you should consider when revising your work.

Content Quality   Think about how true your statements are. Have you appropriately qualified your statements? Or have you made sweeping generalizations? Have you considered alternative views? Or are you perhaps suffering from *tunnel vision*? Think about the assumptions on which your assertions rest. How valid are they? Analyze the degree of detail provided. Is there too much or too little detail? (See Chapter 3, pages 94–105, for more information about informing.)

Persuasive Emphasis   Determine the role persuasion plays in your document. Are you simply providing information? Or do you hope to persuade someone to follow a particular course of action? If so, consider the tone of your writing. Are you trying to persuade by appealing to logic? By appealing to authority? By appealing to emotions? Are you using words with inappropriate connotations or stereotyped language? (See Chapter 3, pages 82–94, for more information about persuasion.)

Format Conventions   Evaluate whether you have followed the expected conventions for a given type of document. Have you formatted your document in ways that help rather than hinder the reader? Does your document look like the work of a professional? (See Chapter 7, pages 245–290, for more information about format.)

Paragraph Organization   Analyze the structure of your paragraphs. Do they move from general to specific? Or do too many paragraphs move from specific to general or lack a general statement that indicates the connection among the other sentences in the paragraph? Have you provided meaningful transitions between paragraphs? Analyze the length of your paragraphs. Are they too long or too short? Are they varied in length? (See Chapter 6, pages 190–194, for more information about paragraphing.)

Sentence Structure   Look at the structure and mechanics of your sentences. Do you use enough conjunctions to clarify the connections among ideas? Does your punctuation help the reader see the connections between ideas or does it seem random? Look at the length and complexity of your sentences. Is the average length between 18 and 24 words? Are they too long? Too short? Are sentence lengths varied? How many embedded clauses have you used? How many prepositional phrases, empty sentence openers, and sentence fragments? (See Chapter 6, pages 194–234, for more information about sentence structure.)

Word Choice   Reconsider the complexity of the language you have used. Is it appropriate for your audience? Have you used jargon unnecessarily? Have you clearly defined any technical terms with which your audience might be unfamiliar? Look at the verbs used in your writing. Are they descriptive or bland? Do you unnecessarily rely upon the passive voice? (See Chapter 6, pages 194–234 for more information about word choice.)

2.3.2.5   Revise in Several Sweeps

Unfortunately, you cannot pay attention to all the elements of your writing at any one moment. Therefore, you will achieve the greatest success if you revise a draft in a number of sweeps, focusing on one or two general concerns at a time. For instance, you might revise first for the accuracy and order of the facts you are presenting. Next, you might focus on persuasive elements. Then, on a third pass, you might revise for sentence structure and word choice. You might also devote another pass to paragraph length and format conventions.

By reviewing for substantive issues first, and for style and grammar later, you can avoid spending time revising a paragraph or sentence, only to realize later that you must totally rewrite the section or paragraph in which it is located. Of course, when you notice problems other than those on which you are concentrating, either deal with them in passing or mark them so that you will remember to deal with them later.

2.3.2.6   Make a Paper Copy of the Document

To view format, we suggest that you print out a hard copy of any documents over two pages in length. Except on computer systems with very large monitors and high-end graphics cards, you cannot see enough of a document at any one time to revise format, paragraph lengths, and figures. Even on advanced computer systems, the complex interactions between graphics programs and word processors are sometimes quite unpredictable. So print a copy and spread out the pages on your desk or on the floor. Stand back and look at the documents. You will more easily spot various formatting problems such as a heading isolated at the bottom of the page, poor-quality graphics, inconsistent font sizes, and missing pages.

We also suggest editing from a hard copy rather than on the computer monitor. Although perhaps not as efficient as working directly with the word processor, most writers see typos and other problems more easily when reading a hard copy of their work. You may also find working with a paper copy will help you focus more intently upon revising and editing—as we did when revising prepublication versions of this text.

Finally, few of us are sufficiently careful about backing-up our files and when our hard disk inevitably crashes or our floppy disk is flawed—*Arrghh!*—a paper copy can function as an important emergency backup.[3] You can then scan the document back into the computer or, if need be, retype it. Although not much fun, retyping a document is generally much less difficult than recreating it from scratch.

2.3.2.7   Read Aloud

The ear is generally better at identifying certain problems than the eye. You can often hear when something is awkward, even if you do not notice it when reading silently. In fact, if you go into an office where many experienced writers are working and listen very carefully, you may hear a lot of mumbling. Try listening for this background noise in a newspaper office; many journalists read their work aloud when revising (and sometimes talk when writing).

As you read your own writing aloud, listen for rough sentence flow, faulty parallelism, incomplete sentences, and excessively long or short sentences (see Chapter 6, pages 202–214, for more information on these topics). Note, for example, how easily you can identify what is wrong with the following sentences if you read them aloud:

- The engineer prepared the specifications, purchased the materials, supervised the construction, and reviewing the completed job. (Faulty parallelism)
- The engineer preparing the budget. (Incomplete sentence)

Further, if you would prefer not to read your work aloud, your computer may be able to read it aloud for you (assuming, of course, that you have a sound card and a program that can read the text back to you). You might be surprised how many problems with your writing become immediately evident when you listen to it being read back to you. Although such programs are far from perfect, with a little adjustment to the program's pronunciation and dictionary you can achieve reasonable-quality speech. No doubt this technology will continue to improve over the next few years.

But whether you read your work yourself or have your computer do it, do not get into the habit of punctuating by ear. Placing a comma where you pause in speech is a poor way to punctuate because it frequently results in errors. (See Chapter 6, pages 234–240, for more information about punctuation.)

2.3.2.8   Make a List of Recurring Problems

If you have recurring problems in your writing (passive constructions, excessive jargon, punctuation difficulties, etc.), making a list of the problems and carefully proofreading your documents for those problems is a particularly valuable revising strategy. Once you have created such a list, you can add new items to your lists as necessary and remove any you have resolved. This system is one of the most effective ways to improve your writing and many engineering students have used it to great advantage. Over a period of six to twelve months, following this practice diligently will enable you to recognize and eliminate the majority of problems that repeat in your writing.

---

[3]One of the authors admits to coming very close to losing the first 100 pages of his Master's thesis because he saved it on a floppy disk that was flawed. Only through the efforts of several computer technicians was he able to recover *most* of the document (all the formatting and about ten pages of text were lost). Since then he has been *obsessively* careful to always make multiple copies of his computer files. One of Murphy's many laws decrees that the likelihood of your hard disk crashing or of your floppy disk having a bad sector is directly proportional to the importance of the information you have just stored on it.

Perhaps the best way to begin creating your list of what to watch for is by noting the comments that your instructors place on your papers. Later, you may want to add some of the points we discuss in the chapters on style and form in this text. Figure 2.6 provides a sample list.

Things To Edit For:

1) ~~Comma Splices~~
2) Unnecessary Passive Voice
3) ~~Empty Openers~~
4) Vague "This" Subjects
5) General Words
6) Talkie Verbs
7) Too many qualifiers
8) "Indeed" & "Really" (Stop Overusing)

Figure 2.6 Example List of Writing Problems

2.3.2.9   Read the Document Backward

Reading a document backward can sometimes help you focus more intently on the task of editing. If you find that you are constantly drawn to the ideas when you are trying to focus on editing grammar or style, try reading individual sentences in reverse order, beginning with the concluding sentence and ending with the introductory one. Because the document makes little sense when read backward, you are forced to attend more closely to the style of the sentences you have written and are less likely to be drawn into the content. You can also apply this technique to paragraphs, reading them in reverse order to check for paragraph structure. (See Chapter 6, pages 190–194, for more information about paragraphing.)

2.3.2.10   Check the Outline of the Document

Checking the outline of a document *after* you have written it is a good way to determine whether the document is logically organized. Because writing is rarely a linear process, our ideas may not end up in the best possible order. As mentioned in Chapter 1 (page 34), you may find it helpful to use the outlining feature of your word processing package when writing longer documents. Examining this outline allows you to check for the consistency of heading levels, and you can also easily move, promote, and demote headings, subheadings, and entire sections as required. We used this method to check our organization when writing this text.

In addition to checking the organization of your document, you may also discover that examining your outline provides an opportunity to edit your headings. Often, when documents are edited, writers forget to check the correctness and descriptive power of their headings. Checking your outline can help you ensure that your headings and subheadings are appropriately descriptive and grammatically correct.

2.3.2.11   Find a Reader

A very useful revising strategy is to find someone who is representative of your audience to read your rough draft. Ideally, you should look for someone who has the same level of expertise and the same values as your audience. But, of course, an ideal substitute is not always available. In that case, you should indicate to your readers who you are writing for, so they can attempt to read from that perspective. If you are a bit sensitive when it comes to accepting criticism, choose a reader you trust. The best candidates are often the engineering students with whom you study.

Ask readers to mark any sections or sentences they had difficulty understanding, to write questions or notes in the margin indicating where they needed additional information, to indicate questions they wished you had answered, and to circle terminology they did not understand. Also encourage them to show you where they wished you had included a figure or table or wondered why you did include one. If you give readers your first draft, ask them to concentrate on content and not to worry too much about grammar or style, commenting only if they find problems that interfere with understanding.

2.3.2.12   Use the Buddy System

Although the buddy system is similar to the strategy of finding a reader, its intent is different. The goal of the buddy system is to practice revising under optimum conditions—that is, for you to offer suggestions for revising as a first-time reader. When you read someone else's work, you gain the perspective necessary for effective revising. You will invariably find it easier to identify the weak spots in someone else's work than in your own. Once you see a problem in someone else's work, you are more likely to notice a similar problem in your own. So swap drafts with another engineering student. Later, when you are working as a professional, follow the same practice with colleagues.

2.3.2.13   Rewrite the Introduction

After you have written a report, particularly a lengthy one, the introduction may no longer represent the rest of the report; sometimes you will have changed your emphasis or reorganized the report while writing. By rereading your introduction and revising it if required, you ensure that your readers are provided with an accurate introduction to your document.

One way that some writers approach the task of revising an introduction is to look at their conclusion to see if it would better introduce the report than the original introduction, which is often the case. They then substitute the conclusion for the introduction, after making some minor revisions to it, and write a new conclusion.

Another issue to consider when rewriting an introduction is tense. Because we anticipate what we intend to write in the rest of a document, we often write our introductions using the future tense (i.e., This report *will* examine the factors that led to the collapse of the walkway). Using the future tense before we have written the report makes perfect sense. But for readers, the document is *present* both physically and in terms of tense. So be sure to check the tense of your introductions (and in some cases, the introductions to various sections) to ensure you have used the present tense (i.e., This report examines the factors that led to the collapse of the walkway.).

2.3.2.14   Completely Review Edited Sentences

After making minor editing changes to the structure of a sentence, we sometimes forget to review the entire sentence before moving on to the next sentence. This oversight frequently results in sentences that have not been fully revised; perhaps a word from the previous structure is left in or a needed word is omitted. The following example shows the sort of problem likely to occur if you fail to carefully review an entire sentence after making a change.

Original	**Revised**
Abracadabra Incorporated now *has ownership of* 60% of the stock.	Abracadabra Incorporated now owns *of* 60% *of* the stock.

As the example indicates, the writer decided to change *has ownership of* to *owns*, but after deleting *has*, and changing *ownership* to *owns*, assumed that the editing of the sentence was complete and thus failed to delete *of*. You can avoid similar problems by developing a habit of rereading an entire sentence or series of sentences after making changes.

2.3.2.15   Break the Rules

Break the rules if you have a good reason to. None are absolute. And breaking the so-called rules can be an effective way to add emphasis or create variety. A proficient writer knows when to follow the rules and when to ignore them. Experiment. It makes the task of revising more interesting![4]

2.3.3   Dealing with Perfectionism and Boredom

In some respects, the problems of perfectionism and boredom represent opposite poles of a continuum; those who suffer from boredom tend not to be perfectionists when revising whereas those who suffer from perfectionism rarely become bored when revising. In a sense, each type of writer needs to cultivate some of the traits of the other. If you suffer from either perfectionism or boredom while revising, you must make a serious effort to deal with the problem because both have serious consequences. Documents submitted by perfectionists are often late; documents submitted by bored writers are often sloppy. In either case, the student's grades suffer. In the professional world, sloppy or late reports can severely impact an engineer's reputation and curtail career advancement.

2.3.3.1   Perfectionism and Revising

Although people are perfectionists for many reasons, in terms of revising, two reasons seem the most common. First, some writers suffer from anxiety about the audience's possible reaction to what they have written. They assume that audiences will invariably view their writing negatively, often in terms of how the ideas are expressed. While revising, these writers often expend excessive energy trying to perfect how they are saying things. They may spend hours rewriting a few sentences or paragraphs in a document simply to meet their expectation of how the audience *might* react.

Second, some writers have unrealistically high expectations of themselves. They assume that unless a document is written to extremely high standards, then it is not worth writing at all. When a document does not meet these self-imposed standards, they view it as reflecting negatively upon their abilities or, in extreme cases, as reflecting negatively upon their self-worth. They expend excessive energy trying to meet their own unrealistic standards.

Perfectionists not only spend so much time revising that they often miss deadlines, they may also procrastinate excessively because the task of writing makes them anxious. Because the experience of anxiety is unpleasant, some perfectionists avoid the task by putting it off. A few fail even to begin to write.

A perfectionist's energy may also be diverted from higher-level tasks (i.e., revising) to lower-level tasks (i.e., editing). Consequently, while perfectionists may allow few grammatical or formatting errors, they may not devote sufficient attention to the critical issues of persuasion and substance. In this case, they may receive sufficiently negative criticism from their audience that they fail to realistically assess their writing abilities.

---

[4]Can you identify the so-called rules we have broken in the last six sentences? How many? Which ones?

Consequently, the drive for perfection continues, with reasonably good writers convinced they are poor writers. This negative reinforcement can lead them to expend even more energy trying to perfect their documents—and so the cycle continues.

If you tend toward perfectionism, you can resolve the problem in several ways:

- Set clear limits on the amount of time and energy you will invest in editing.
- Focus more attention on the high-level task of revising rather than on the low-level task of editing.
- Ask readers to clarify their reactions to documents that you have written, noting the differences between your assumptions and their actual responses.

2.3.3.2 Boredom and Revising

Like perfectionism, boredom while revising has several possible causes and consequences. First, and probably the most difficult issue to resolve, is that editing a document tends to be a repetitive task that a writer may not find especially challenging. Second, if the intrinsic reward for doing the task is minimal, then the writer may not be especially motivated to perform the task to begin with and will perceive it as boring. Third, some writers may claim that they find revising boring because of an unacknowledged difficulty with the task. Claiming that they did not edit their work because they find the task boring may be easier for some writers than admitting to themselves or others that they do not really know how to edit effectively.

Whatever the causes, boredom while revising has consequences. Attention wanders and errors are overlooked. Productivity also suffers because bored revisers are easily diverted by other tasks or distractions. Bored writers are unlikely to be satisfied with the final result.

Fortunately, the problem of boredom while writing has a number of solutions.

- Increase the challenge of the task by focusing on more complex elements of the task (i.e., revising to address readers' expectations rather than editing for correct word choice).
- Increase motivation by considering the task as a small part of something with a more important purpose.
- Do the least interesting tasks first and save the more interesting tasks for later.
- Ask whether the boredom is actually masking a difficulty with the tasks.
- Accept that not everything is interesting and set up a system to help return your attention to the task when it wanders (e.g., use a checklist).

2.3.4 Spelling and Grammar Checkers

The computer revolution has provided writers with a range of computer resources that they can use to improve the quality of their writing. Unfortunately, if these resources are used inappropriately, the benefits you stand to gain are, at best, minimal. The following sections provide a few hints for using spell-checkers, computer thesauruses, and grammar checkers to maximize their benefits.

2.3.4.1 Using Spell Checkers Effectively

English spelling is far from regular, as those of us who have difficulty spelling can attest. Indeed, as Figure 2.7 suggests, wildly inventive spellings of common words are quite pos-

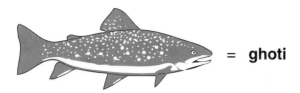 = **ghoti**

where <u>gh</u> as in rou<u>gh</u>
    <u>o</u> as in w<u>o</u>men
    <u>ti</u> as in na<u>ti</u>on

Figure 2.7 Some Absurdities of English Spelling

sible by following how letters sound in certain words.[5] Fortunately for those of us with spelling difficulties, the invention of spell checkers has eliminated much of the tedium and frustration that used to be involved in editing. Unfortunately, the technology is not yet perfect, so you should consider the following when you use spell checkers.

First, the spell checker that is provided in your word processor has a limited vocabulary. As you may already have discovered, it contains few technical terms, proper names, or acronyms. Consequently, you must manually verify that you have spelled these types of words correctly. Initially, you will find this need to double-check spellings a time-consuming process, but if you create your own user's dictionary and add terms that are not in the spell checker's dictionary, you will discover that after a few documents the process speeds up considerably.

Be careful, however, to ensure that any word you add to your user's dictionary is indeed spelled correctly. The misspelled words in Table 2.2 were found in documents intended for audiences external to the company at which they were written. What makes these spelling errors particularly interesting is that the correct spellings were contained within the spelling dictionary for the word processor that was being used within the company. Further, the correct spellings were the first ones suggested by the spell checkers and the documents had been spell checked. So how did the incorrect spellings end up in the documents? Most likely, the authors were a bit too quick to hit the *Ignore* or the *Add* button while spell checking. Unfortunately, when writers add misspelled words to their user's dictionary, the spelling errors recur in future documents. To correct this problem, the authors of these documents must spend a few minutes checking the spelling of words in their user's dictionaries (which are generally simple text files that can be easily edited).

Some of the errors listed in Table 2.2 may have resulted from *last-minute corrections* made after the last use of the spell checker. To avoid such errors, the very last thing you should do before printing out a final copy is to run your document through the spell

Table 2.2 Adding Spelling Errors to Users' Dictionaries

gurantee	= guarantee
effiecient	= efficient
inexspensive	= inexpensive
farady	= Faraday
indepenent	= independent
accross	= across

[5]If we remember correctly, the originator of this rather intriguing spelling for *fish* is George Bernard Shaw.

checker one final time. You should not assume that just because you previously checked
the spelling, the spelling is still correct. Minor last-minute changes to your writing invite
last-minute spelling errors.[6]

You should also avoid adding words to your dictionary that are common typographical errors. For example, adding *hte* to a user's dictionary is not a good idea because *hte* is
a common typographical error for the word *the*. Also note that spell checkers cannot tell
whether a word is spelled correctly in relation to its context. In other words, a spell
checker will not flag the word *to* as an error in a situation where you meant to use *too* or
two (nor will it flag *is* when you meant *in* or *it*, nor *form* when you meant *from*, etc.). Consequently, you must still proofread for these sorts of errors. Note that these errors typically involve small words (i.e., four letters or less). Try reading important documents
aloud or reading relatively short documents backward to catch these errors.

Finally, to emphasize our point about being cautious when you use your spell
checker, we have excerpted the following:

Catching misspilled swords with spilling checker[7]

*As an extra addled service, I am going to put this column in the Spilling Checker, where I tryst
it will sale through with flying colons. In this modern ear, it is simply inexplicable to ask readers to expose themselves to misspelled swords when they have bitter things to do.*

*And with all the other timesaving features on my new work processor, it is in realty very
easy to pit together a colon like this one and get it tight. For instants, if there is a work that is
wrong, I just put the curse on it, press Delete and it's Well sometimes it deletes to the end of
the lion or worst yet the whole rage. Four bigger problems, there is the Cat and Paste option.
If there is some test that is somewhere were you wish it where somewhere else you jest put the
curse at both ends and wash it disappear. Where you want it to reappear simply bring four
quarts of water to a rotting boil and throw in 112 pounds of dazed chicken. Sometimes it
brings in the Cat that was Pasted yesterday.*

*But usually it comes out as you planned, or better. And if it doesn't, there are lots of other
easy to lose options, one of which is bound to do exactly what you want. In no time at all
you'll be turning out prefect artifacts like this one.*

*So join the marsh of progress. Hitch your wagon to a stair. When you become adapt at
world processing there's no end in sigh.*

2.3.4.2   Using Computer Thesauruses

Although computer thesauruses are useful, they suffer from two critical limitations. First,
those packaged with word processors tend to be limited in size and thus do not generally
provide enough alternatives (although larger thesauruses on CD-ROMs have resolved this
problem). If you use a thesaurus frequently, a large, hardcover thesaurus, synonym finder,
or CD-ROM version is strongly recommended as an addition to your library.

Second, computer and paper thesauruses do not provide sufficient context to allow
you to determine the connotations of words. For example, reflect upon the differences

---

[6]A useful feature that was added to *Microsoft Word 7.0* helped to resolve the problem of adding last-minute
spelling errors. The new feature places a squiggly red line beneath spelling errors. We find that this helps us see
those last-minute errors and thus correct them without running the spell-checker again. Some people, however,
find the feature irritating because it forces them to focus on correcting errors while drafting. For people who
prefer to separate the drafting and revising processes, the feature goes against their natural inclination. Of
course, they can simply turn the feature off. (*Microsoft Word 8.0* added a similar feature to help draw attention
to potential grammatical errors, by placing a squiggly green line beneath suspect sentences. Again some writers
find this irritating, but as with the red line under spelling errors, the feature can be turned off.)

[7]©1990 IEEE. Reprinted with permission from *IEEE Trans. on Aerospace and Electronic Systems*, 26:2, 209,
March 1990.

among the words *reproduce, rewrite*, and *plagiarize*. All three are listed under the word *copy* in many thesauruses. Consequently, when you use a thesaurus, chose only words that you know but could not think of without seeing them listed.

2.3.4.3 Using Grammar Checkers Effectively

For many years, grammar checkers were rarely included with most word processing packages and thus had little impact on writers. The release of *Word 2.0 for Windows*, which included a grammar checker, changed that, and consequently more writers are using grammar checkers to assess their writing. As with spell checkers, several cautions are in order. First, grammar checkers cannot completely substitute for careful proofreading. Even the best grammar checkers only catch about 50 percent of errors. Until the accuracy of grammar checkers is significantly improved, you should not rely upon them exclusively.

Second, grammar checkers are necessarily based on the sets of rules provided to the programmers. These rules, quite naturally, reflect the biases of those who decide which rules are most important in *typical* documents. But, as we have pointed out, effective writing is only partly a matter of following rules. In some instances, you may have good reasons, based on considerations of your audience and purpose, to violate some of the grammar checker's rules. For example, grammar checkers will often flag as an error every use of the passive voice or every sentence fragment. In some scientific and technical documents, however, the passive voice is perfectly acceptable and, when used in the dropped lists common to technical specifications, sentence fragments are clearly appropriate.

Of course, if you habitually use the passive voice or write sentence fragments, then you should pay attention to the grammar checker's suggestions. In other words, weigh the advice of the grammar checker against your audience and purpose and your personal writing habits. Do not blindly restructure your sentences in light of its suggestions.

We know of one case where a grammar checker probably caused a business owner major embarrassment and financial losses. Because the grammar checker kept suggesting that his sentences were too long, he started making them shorter. Unfortunately, he did so simply by breaking the long sentences into two. The result of this action was that many of his sentences ended up as fragments.

Sadly, this individual was sending out mass-mailed solicitations to thousands of people to purchase *Registered Educational Savings Plans*. His particular audience from the mailing list happened to be people who were particularly well educated. The question for most of them was probably the same as it was for one of the authors of this text: "If this person can't deal with the details of his writing, how can I trust him to deal with the details of my financial investments?" Most people probably discarded the solicitations, and the business owner not only wasted the cost of the postage and envelopes, but he also failed to generate any of the new business for which he was hoping.[8]

Third, using grammar checkers on lengthy documents or on documents with a complex style can be very time consuming. In those situations, grammar checkers may identify so many "errors" (many of which are not errors) that revising the sentences becomes an overwhelming task. We have found that we can generally proofread and correct errors more accurately and quickly without the grammar checker. In this text, moreover, we

---

[8]We know that a grammar checker was at fault in this case because the author who received the solicitation sent it back to the business person and pointed out the grammatical problems. Subsequent conversations indicated that not only was he misusing his grammar checker, he was also misusing his spell checker. Some of his letters were filled with spelling errors (*there* instead of *their, to* instead of *too*, etc.). Part of the problem was that he was not revising his writing by reading over it carefully.

have intentionally written a fair number of sentences that break the so-called rules (for reasons of idiom, reader interest, and rhetorical purpose). Having a grammar checker repeatedly flag these sentences as errors is somewhat frustrating.

One way to deal with this problem is to turn off many of the grammatical rules in the grammar checker. In this text, for instance, in the interest of efficiency we have chosen to use our grammar checkers only to check for a few specific problems.[9] So we turned off the settings for most of the *grammatical* rules (with the exceptions of "negation" and "punctuation").

In addition, we turned off the settings for many of the *style* rules. We left on only "words in split infinitives (more than one)," "successive nouns (more than three)," "successive prepositional phrases (more than three)," "sentences longer than 60 words," "clichés," "gender-specific words," and "wordiness." These settings provide us with the optimal balance between speed of checking and identifying stylistic problems that concern us.

Of course, the specific settings you choose will depend on your own particular difficulties with writing. If you habitually use the passive voice or inappropriate prepositions, write run-on sentences or sentence fragments, or have trouble with subject-verb agreement or wordiness, then you will want to turn on the settings for those particular rules. As we mentioned earlier, one of the quickest ways to resolve problems with your writing is to maintain a list of those problems so you can revise with them in mind. You should, of course, turn on the setting for anything on that list as well as for anything that a writing instructor points out (assuming your grammar checker is capable of checking for it).

Although using a grammar checker may sometimes seem like a lot of work, especially when you first start using it, keep in mind that you will eventually learn to identify problems on your own. Even if you choose not to use a grammar checker for your entire document, you might consider using it to help you figure out how to rewrite a problem sentence or paragraph. This practice may help you identify particular problems and, in some cases, may save you time.

Finally, we would like to mention a few points about the *readability statistics*[10] provided by most grammar checkers. Although they are a useful way of gauging how well you will likely communicate with your readers, some caution is in order when interpreting readability statistics. These statistics do not consider the reading abilities of *specific* audiences; instead, they are designed based on the reading abilities of an *average* audience. For instance, the statistics could suggest that a technical document would only be comprehensible to someone with a Ph.D. while, in fact, most second-year engineering students would possess the specialized knowledge and technical vocabulary to understand it.

In other words, you need to consider the expertise of the specific audience to whom your document is addressed. Nevertheless, if the statistics tell you that your average sentence length is 38 words and that 80 percent of your sentences are written in the passive voice, then your text has a readability problem. Figure 2.8 provides the readability statistics for an earlier draft of this section on dealing with grammar checkers.

---

[9]The grammar checker we used to prepare this book for publication is the one packaged with *Microsoft Word 9.0*.

[10]Readability statistics are determined by various formulas based on the number of syllables or characters per word and the number of words per sentence. Because the meaning of readability scales differs from scale to scale and the scales that are calculated differ from grammar checker to grammar checker, we recommend you read the documentation that comes with your grammar checker or word processor in order to determine more precisely how to interpret the results.

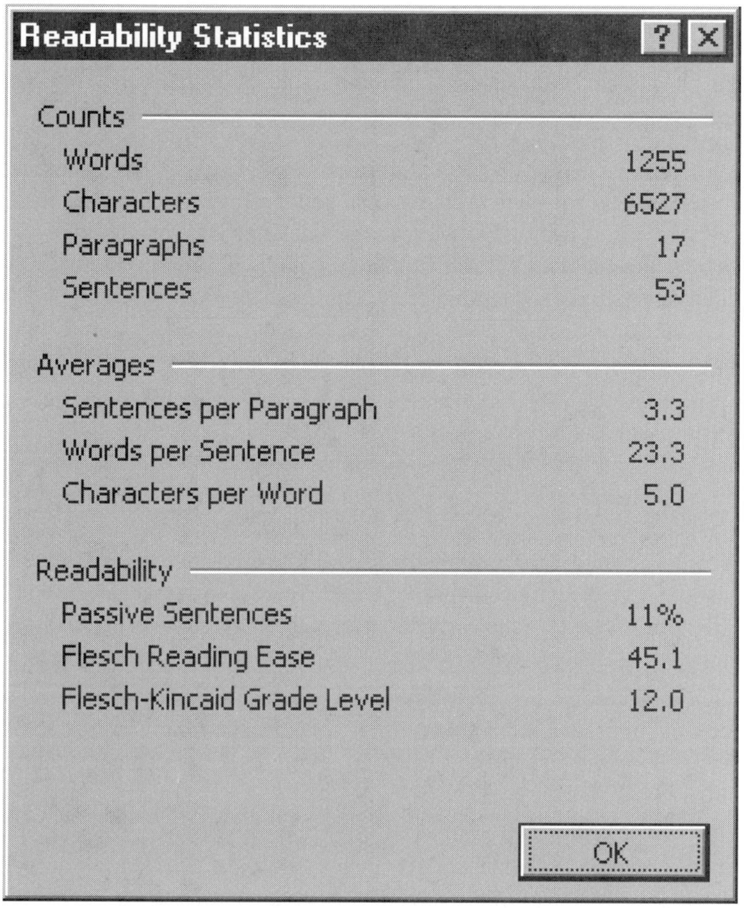

Readability Statistics   [?] [X]

Counts
Words   1255
Characters   6527
Paragraphs   17
Sentences   53

Averages
Sentences per Paragraph   3.3
Words per Sentence   23.3
Characters per Word   5.0

Readability
Passive Sentences   11%
Flesch Reading Ease   45.1
Flesch-Kincaid Grade Level   12.0

OK

Figure 2.8 Example Grammar Statistics (*Microsoft Word 2000*)

2.4   ANALYZING YOUR WRITING PROCESS

The more clearly you understand your own writing process, the easier you can make constructive changes. The following questions address some of the issues you might want to consider as you analyze your writing process. We suggest that you spend an hour or two reflecting on these questions in order to more completely understand how you go about writing. As you answer these questions, consider *how, when, where, what*, and *why* you typically write (you could also consider your last major writing task or writing tasks that seem to cause you particular difficulties). Note that this list of questions is by no means exhaustive; rather, it provides a starting point for considering your writing process.

Because the writing process is individual, these questions have no right or wrong answers. Nevertheless, some of your writing practices may interfere with your efficiency and effectiveness. Maintain the practices that work for you, and find alternative approaches for those that do not. You may also find it helpful to write down some of your observations so that you can reconsider your process at a later date to determine whether you have succeeded in making any changes.

2.4.1 Factors Relating to Your Writing Environment

- What is your typical writing environment like? Quiet or noisy? Organized or disorganized? Home, school, or elsewhere? Do you like your environment? Would you like to change it? How? Why?

- Do you write your first draft on a word processor? Or do you prefer to use paper and pen? Are you a proficient typist or do you use the hunt-and-peck method? How skilled are you at operating your word processor? Do you like using your word processor? Why or why not? Do you use the spell-checker, thesaurus, and grammar checker? Stylesheets and templates? Outlining or annotation features? Why or why not?

- Do you write your first draft at one sitting or do you prefer to write it over a period of several days or weeks? Must you often write to deadlines or can you usually pace your writing to suit yourself?

2.4.2 Factors Relating to Audience and Purpose

- To whom is most of your writing directed? Instructors, colleagues, friends, or others? Do you often write for an audience with less expertise than you have? With more? Do you like or dislike writing to a specific audience? Why? Who is that audience?

- Why do you write? To persuade someone to adopt your ideas? To describe or provide information about a specific situation or issue? To analyze the results of a particular action or the implications of a specific situation? Or a combination of purposes? Do you find writing for some purposes more difficult than writing for other purposes? Which ones? Why?

- What kinds of writing do you undertake? Letters, memos, proposals, reports, essays, other types? To which do you devote the most time during school? During your free time? Do you like or dislike some kinds of writing more than others? Which ones? Why?

2.4.3 Factors Relating to Your Writing Process

- Roughly what percentage of time do you devote to the process of inventing (researching, planning, or otherwise deciding what to say)? What percentage to drafting (i.e., to actually writing material)? What percentage to revising and editing (i.e., to changing what you've previously written)?

- Do you do a lot of revising and editing at the same time as you do your drafting? Or do you write a complete (or nearly complete) first draft prior to revising? Does your approach vary according to the writing task? How? Are there some kinds of writing you don't revise? How much time do you usually leave between producing a rough draft and revising? Do you ask a friend or family member to read over your rough draft and make suggestions? Why or why not?

- Do you become blocked when writing? How often? Can you describe the specific circumstances? When looking for an idea or specific information? When looking for the right words? When looking for a way to organize your writing? Can you describe at what point in the writing process blocks occur? During invention? During drafting? During revision? If you become blocked, what strategies do you use to deal with it? Wait? Try to determine why you are blocked? Keep writing and sort it out later? Do these strategies work?

- What kinds of changes do you most often make when revising? Do you correct spelling and punctuation errors? Do you substitute words? Do you add new information or remove excess information? Do you rewrite sentences or paragraphs? Do you make major changes to your organization? Do you change the format? Why do you make the changes that you do? Do you print a hard copy of your document prior to undertaking the final revisions?

2.4.4   Factors Relating to Your Attitudes about Writing

- In general, do you like or dislike writing? Are there specific types of writing and specific things about writing that you like or dislike? Can you describe them? Is writing boring or interesting? Is it difficult or easy? Do you suffer when writing? Do you have fun? Or are you indifferent? How do you feel when you encounter writer's block? Anxious? Angry? Frustrated? Depressed? Resigned? Do you procrastinate when faced with most writing assignments? Or do you procrastinate only with certain kinds of assignments or audiences? Do you procrastinate a lot or a little? How do you deal with procrastination?

- How important do you think effective writing skills are to your academic success? To your future career? Very? Moderately? Not at all? Why? How much effort are you willing to devote to improving your writing? What sort of writing courses have you taken in the past? What did you think about them?

- Do you think you are a good writer, an average writer, or a poor writer? Has anyone criticized something about your writing lately? If so, what did they criticize? Do you agree with them? Why or why not?

- If you could change any one thing about your writing process, what would that be? Why? If you could master any one writing skill, what would you want it to be? Why?

2.5   CHECKLIST

We provide the following writing checklist for you to use as you go about drafting and revising various documents. We suggest that you use this checklist by considering the points that it raises *after* you have written a document. By noting those points that you have not adequately considered (i.e., those that you have not checked off), you can determine which areas need work and make appropriate changes to your writing process.

2.5.1   Drafting Phase

Did you identify which approach to drafting was most appropriate for the type of document on which you were working?   _____

Did you deal effectively with internal and external critics (if needed)?   _____

Did you deal effectively with writer's block (if needed)?   _____

Did you deal with any potential health issues?   _____

Were you able to achieve a flow state (by dealing with the issues of boredom or anxiety)?   _____

Did you stay on topic?   _____

Did you stay on schedule?   _____

2.5.2   Revising and Editing Phase

Did you leave some time between drafting and revising? _____

Did you read your document critically? _____

Did you create and use a list of recurring problems? _____

Did you make a paper copy so you could revise the format? _____

Did you read the document aloud or backward (if needed)? _____

Did you outline to see if the organization still made sense? _____

Did you have someone else read the document? _____

Did you rewrite or reconsider your introduction? _____

Did you revise the document in several sweeps? _____

Did you completely review any sentences you revised or edited? _____

Did you revise on all levels (substance, persuasion, format, paragraphs, sentences, and words)? _____

Did you check your facts and figures for accuracy? _____

Did you reconsider your audience, purpose, and tone? _____

Does the document look like the work of a professional? _____

Did you spell-check the document? _____

Did you use a grammar checker to check for repeated or common problems? _____

2.6   EXERCISES

Ideally, you should apply the concepts, principles, and advice in this chapter to writing that you already do for school or work. We provide the Drafting and Revising Checklist in the previous section of this chapter for that purpose. If you do not have opportunities to apply that checklist at school or work, then the following exercises may help you further understand some of the issues we raise.

1. Analyzing Your Writing Process

Using the set of questions outlined in the section about analyzing your writing process (pages 68–69), spend one or two hours describing your writing process. Use your word processor for this exercise. While you work, try to pay attention to relatively how much time you spend inventing, drafting, and revising. After you have completed the exercise, discuss this assignment with a friend or colleague who has also completed it. What differences do you note between your writing process and your friend's writ-

ing process? What things do you think you should change in your writing process? What things should you keep the same?

2. Separating Drafting and Revising

On your next writing assignment, try writing at least two or three pages of the document without doing any revising at all. Just type as fast as you can. Then revise this section. How long does drafting and revising take? After that, try writing another two or three pages of the document in which you correct everything as soon as you notice it. Keep revising sentences and paragraphs until you are satisfied. How long does this process take? Which of the approaches do you prefer? Why?

3. Dealing with Writer's Block

Think back to the last time you experienced writer's block. Once you have identified that circumstance, try to determine what was the cause of the block. If you are unsure of the reason, review the information about writer's block (pages 50–52). Were you able to successfully apply any strategies to resolve the block? Could you have done anything else to deal with it?

4. Developing Perspective for Revising

Find a document that you wrote and revised at least a year ago. Read it through carefully. What kinds of errors did you miss in that document that you can now find? What things were clear then, but are not so clear now?

5. Making a List of Writing Problems

Taking three or four recently written and reviewed documents, make a list of the problems that the reviewer noted. Also add any other problems that you note as you read. Before starting work on your next writing assignment, review the list. Then, while you are revising the assignment, refer to the list and deal with the problems. Follow this proce-

dure for the next several months, adding new items as necessary and deleting other items as they cease to be problems.

6. Exploring Your Grammar Checker

Run a 5- to 10-page document through your grammar checker with all the rules turned on. What sorts of things are flagged as problems? How many of those problems can you identify as being correct? What rules would you like to turn off? Also read through the help file or the user's manual for your grammar checker in order to determine how to interpret the readability statistics that are generated. Do the statistics seem intuitively accurate? (If not, perhaps you will get a more accurate reading by eliminating all headings, labels, and lists.) Try making some changes that make sentences longer (or shorter) and that use more complex (or simpler) language. Does that affect the result? How much?

Chapter 3

Rhetorical Strategies

3.1 SUMMARY

In this chapter, we focus on the interactive, interpersonal nature of communication. We hope to demonstrate that communicating requires careful attention not just to the content and purpose of a message, but also to the needs and concerns of your audience, the reasons readers or listeners attend to your message, and the contexts in which your message is received. Throughout this chapter, we emphasize the role your audience plays and the importance of paying close attention to its members' needs and expectations.

1. **Rhetorical Situation**

    Communication is shaped by the interplay of subject matter, writer or speaker, reader or listener, and the surrounding personal, social, cultural, and professional contexts in which it occurs. Appropriate attention to each of these elements helps ensure success as a communicator. When you have a choice between writing and speaking, attending to the elements of your rhetorical situation can help you decide which mode of communication is most appropriate.

2. **Persuasive Writing**

    Persuasion requires attention to what readers know, to their fears, expectations, and values, to how they will respond to you, and to the criteria by which they will judge what you are saying. Constant attention to those people you want to persuade is a key to success. We draw your attention to the notion of persuasive appeals, of appealing to readers' sense of reason and logic, of appealing to their values, and of appealing through your credibility as a speaker or writer. We also draw attention to the importance of building an argument on the basis of shared beliefs and points of agreement and of identifying with your readers and encouraging them to identify with you.

3. **Connotation**

    Many words carry meanings, or connotations, derived from context. For example, at one time, *Dracula* was just a name, but now it carries connotations associated with vampires. In technical writing in general, and for persuasion in particular, the connotations of words are important.

4. **Inclusive Language**

   Referring to all people as *man* or to an unspecified engineer as *he* or to the person in charge of a committee as *chairman* is unacceptable. This chapter offers advice and strategies for using inclusive language that reflect the realities and social norms of our time.

5. **Informative Writing**

   When your emphasis is on informing your readers, your major concern should be conveying information as clearly and concisely as possible. To help you become a more reader-centered writer, we offer eight principles for you to apply to all informative documents. These principles will also prove useful in many persuasive documents and in formal technical presentations.

3.2 RHETORICAL SITUATION

The rhetorical perspective we take throughout this book draws attention to the interactive, interpersonal nature of communication. From this perspective, we approach writing and speaking as dynamic interactions among the writer (or communicator), subject matter, audience, goal to be achieved, and surrounding personal, social, cultural, and professional contexts. This interaction of writer, subject matter, reader, goal, and context forms the rhetorical situation for any act of communication.

Note the representation of the rhetorical situation in Figure 3.1. The circle represents the world around you. This context includes the culture, language, personal experiences, and education that help determine how you address your readers, what ideas and information you communicate, and how you express your thoughts. Within this context, communication is shaped by an interplay of your subject matter and what you know about it, the goal you hope to accomplish, what you know about your readers' background, knowledge, expectations, and concerns, and your relationship to your readers in terms of power and expertise.

In everyday contexts, when the elements of a rhetorical situation are familiar, you meet the demands of that situation by relying on common sense. However, negotiating the demands of less-familiar situations requires conscious consideration of your rhetorical situation and relies on your problem-solving abilities. To ensure you negotiate these demands successfully, keep the following principles in mind:

- Verify your emphasis as primarily persuasive or informative.
- Clarify your goal and write a statement describing how you will achieve it.
- Analyze your audience's values, expectations, experience, and status.

3.2.1 Verify Emphasis

When faced with a new rhetorical situation, you should begin by verifying whether the emphasis is on expressing your feelings or on informing or persuading your audience. Each of these emphases represents a focus on one of the three pivotal elements of the rhetorical situation depicted in Figure 3.1: writer or speaker, subject matter, and audience. If you focus primarily on yourself, on expressing your thoughts or feelings, your emphasis is on self-expression and your communication can be described as *expressive*. If you focus primarily on the subject matter, on presenting facts and

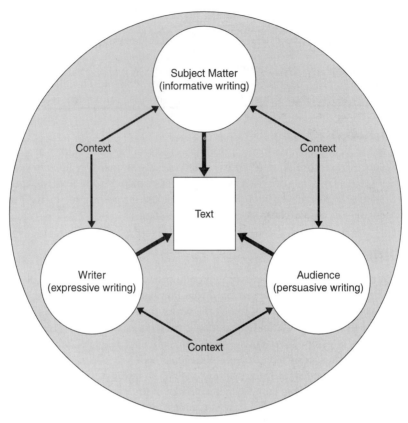

Figure 3.1 The Rhetorical Situation

providing useful information, your communication is *informative*. If you focus primarily on your audience, on assessing and addressing their needs, fears, values, and concerns in order to persuade them to agree with your point of view, then your communication is *persuasive*.[1]

Engineers rarely, if ever, write expressive documents or make expressive presentations that focus primarily on personal thoughts and feelings. However, they sometimes write private drafts of letters, memos, or e-mail messages in which they express their feelings without worrying about readers' reactions. You can use expressive writing to get something out of your system so that you can address your readers more calmly, rationally, and effectively in a later draft.

While in college and during the early years of an engineering career, you can expect most often to write to inform readers. In the early years of your career, opportunities to write primarily to persuade may be infrequent, but keep in mind that the most important writing or presenting you will undertake will focus on persuading an audience to acknowledge a problem, accept your approach, or award you a contract. Developing the ability to speak and write persuasively is one of the keys to a successful engineering career.

---

[1]This notion of categorizing types of writing according to emphasis on elements of the rhetorical situation is informed by James Kinneavy's *A Theory of Discourse* (Englewood Cliffs, NJ: Prentice-Hall, 1971).

3.2.1.1 Attend to Primary and Secondary Emphases

Viewing a piece of writing as primarily expressive, informative, or persuasive is useful in order to focus attention on the most important element of a particular rhetorical situation. That is, when your emphasis is expressive, you focus on personal feelings; when informative, you focus on the facts to be presented; and when persuasive, you focus on readers' needs, fears, and concerns.

If, for example, you are writing a proposal for an engineering project, the ultimate goal is to persuade readers to award you a contract. More specifically, you may want to convince your readers that you understand their needs and possess the necessary expertise, commitment, resources, and professionalism to complete the project to their specifications. Or in another situation, you may want to convince an audience that your project is worthy of support, both filling a need and promising a good return on investment. In any case, a persuasive emphasis focuses attention on your audience. You must decide how much background information your audience requires, how familiar its members are with your subject, and whether they understand your technical language. You must also identify the fears, values, and concerns they will likely have and determine how to address these issues while making your case.

You must also attend to the elements of the rhetorical situation outside of the spotlight. To varying degrees, everything you write seeks to inform *and* to persuade while expressing your personal beliefs, feelings, convictions, motives, and/or perspectives. When you inform, you must also persuade an audience that your information is accurate and worth remembering. When you seek to persuade, you also provide information to support your position and express your commitment and trustworthiness.

In the case of a proposal for an engineering project, persuading your readers also requires informing them. For example, you might need to provide a detailed analysis, recount the history of a development project, survey relevant research, outline your understanding of a problem, or provide a detailed project schedule. To some degree, your proposal will also be expressive, reflecting your faith in a particular technology or approach, your commitment to excellence, your concern with environmental impacts, and your awareness of social implications.

3.2.1.2 Use Emphasis as a Device to Focus Attention

Note that while a proposal contains informative and expressive elements, its persuasive emphasis keeps attention focused on readers. To persuade, experienced communicators order points according to their audience's priorities and determine how much detail to provide based on what the audience needs to know. They provide justifications based on how they think the members of their audience will react and provide explanations based on what the audience wants addressed. Even words are chosen based on the audience's needs, reactions, and preferences.

Note, however, that when writing an informative report rather than a persuasive proposal, experienced communicators focus on subject matter. Rather than attending primarily to readers' reactions, they concentrate on adequately covering the topic and on providing accurate, understandable, useful information. Rather than organizing points to match readers' concerns and interests, they follow a logic inherent to the subject matter.

Our point is that when your emphasis shifts, your attention must also shift so that you attend to different issues and write a very different document. For example, when dealing with the same subject but shifting from a persuasive to an informative emphasis, you must often modify the way you present yourself, alter the information you include, change your organizational principles, alter your focus, and adjust word choice and tone.

3.2.2 Clarify Your Goal

As well as determining an emphasis on informing or persuading, you should also clarify the specific goal you hope to achieve. Imagine, for instance, you are beginning to plan a piece of informative writing in which you need to accurately describe a project's status. After considering the situation and identifying your audience's expectations, your goal might read as follows:

> **Goal**: to show that delays in the first stage of the project resulted from delays in obtaining permits and to present a revised schedule that reassures the client that the next stage of the project will be completed on time.

Note how this goal is specific to a particular rhetorical situation, how it indicates a general approach for accomplishing the goal, and how it takes the audience into account. Because the emphasis is informative, the audience receives less attention than the content.

When the emphasis shifts from informing to persuading, the focus shifts to the audience. For example, suppose you are writing a report you hope will convince your superiors that a change in design is necessary. In this case, your emphasis is persuasive, and your goal must reflect your attention to audience. Perhaps you are certain that your superiors are most worried about penalties the company will pay if bridge repairs are not finished on time. Changing plans at this point will not be appreciated in the head office, but you are convinced that a design change is necessary. In this case, your goal might read as follows:

> **Goal**: to convince my superiors that changing the design is necessary to ensure the problem does not reoccur and to reassure head office that by employing a few extra workers we can complete the job on schedule and maintain the reputation that won our company this contract.

Note that because your emphasis is persuasive, you focus directly on your readers' reactions, concerns, and values.

Develop a habit of writing down your goal. Doing so will help clarify your thinking and create a quick reference you can reread to help maintain your focus while planning, drafting, and revising. Note, however, that as you write or as you prepare a presentation, you may clarify your thinking, discover a problem with your approach, formulate a better solution, or remember a key point you forgot to address. Be prepared to revise or refine your goal as you plan, draft, and revise.

While changes are often useful and even necessary, you can reduce the need for them by analyzing your audience and attending to their needs, expectations, and reactions while in the process of formulating your goal. Although clarifying your goal and analyzing your audience are discussed as if they are two distinct steps, in reality the two are often best addressed at the same time. As you clarify your purpose, pay attention to problems and opportunities relating to your audience. While you analyze your audience, reflect on how the information and insights you are collecting will help you achieve your goal.

3.2.3 Analyze Your Audience

Understanding whom you are addressing is as important as having a clear sense of why you are addressing them. Many students fail to recognize the powerful influence the audience should exert on their writing, if not their speaking, because they generally write for one audience—the teacher or expert—and for one purpose—to demonstrate their general knowledge or understanding of a particular subject. Writing to demonstrate how much you know to experts who know everything you know about a subject—and more—is rarely experienced outside of school. At work, you can expect to write for readers who need the information you are providing or who have the power to grant your request.

Given the test-driven nature of much academic writing, we are not surprised that the importance of focusing on readers is not immediately apparent to all. After all, you need to know very little about a reader when writing to prove that you understand the concepts covered in chapters three through five of a course textbook or if being graded on your explanation of how you verified a particular law in your latest laboratory assignment. Nevertheless, you will encounter very few rhetorical situations outside of school that can be adequately addressed without attention to your audience.

When writing outside of school, the more you know about your audience, the better you can communicate and the greater your chance of success. Assume, for example, that you are writing letters of application, hoping to gain employment. If the only information you have is provided by advertisements that simply describe positions and state the required qualifications, then your sense of audience is vague and limited to what you already know about employers and what they are likely to want in an employee. But readers representing different corporate cultures are unlikely to be looking for the same qualifications and attributes. By learning more about the companies you are applying to, you can address your audience more effectively and increase your chances of gaining employment.

You may learn that one of the companies is relatively small with a flat organizational structure that emphasizes teamwork. You can assume that whoever reads your application will be interested in your experience working as part of a team and will be looking for qualities that make you a good team player. You may also assume that a small company wants creative employees with diverse experience. If you apply for another position offered by a large, conservatively run company with a top-down management style, your assumptions about your readers will change and you must approach them differently.

The more you know about your audience, the better your chances of achieving your goal. Having a clear sense of who will read your documents or listen to your presentations, why they will read or listen, and what you hope to accomplish is essential in determining what information to include, how to organize it, and how best to express and present it. The following sections offer strategies for improving your understanding of audience in any number of rhetorical situations.

3.2.3.1 Ask Questions

To analyze your audience, begin by asking yourself questions: Who is most likely to read what I am writing? Why will they read it? What do they expect to find in my document? How well do their needs and expectations match my goal? What sort of standards do they expect me to achieve? What problems do they expect me to address?

The sorts of questions you might productively ask yourself are endless and depend on the particular rhetorical situation you are addressing. The most important point is to develop a habit of asking questions. You may have heard the saying that *finding the answer is easy if you ask the right question*. Asking questions is the key to analyzing audiences and developing strategies for addressing their concerns.

At the end of this chapter, you will find an audience analysis heuristic—a list of questions to ask about your audience. You can use this list as a general guide, adding questions relevant to the specific rhetorical situation you are addressing.

3.2.3.2 Identify Institutional Values and Concerns

The success of your analysis depends in part on how well you know (or understand) your audience. For engineering communication, keep in mind that both technical and managerial audiences represent institutions or companies. Your readers and listeners will most often respond to what you write, say, or present based on the values, needs, and necessities of the organizations they represent and on the roles they play within these organizations.

If you spend a little time thinking about who comprises your audience, what they want you to accomplish, and their role in protecting institutional values and goals, you should be able to identify a number of points to bear in mind. Each of those points, when turned into a question, becomes a productive means of directing your activities, organizing your thoughts, and solving various problems encountered while planning, drafting, or revising.

Laboratory reports provide an example of how analyzing one particular audience might affect and guide the writing process. (This example suggests that even when writing for school, you can productively analyze your audience.) In this case, your readers are educators responsible for training you for a profession. Given the nature of their position, you can assume that they will judge your work based on established academic and professional standards.

You can assume, for example, that professors read your work principally to evaluate your understanding of the concepts and principles studied in your present and previous courses, to follow established procedures, and to acknowledge the practical applications of what you have learned. For instance, if you are writing a lab report, they will most likely expect you to accomplish the following:

- To demonstrate your understanding of relevant laws and principles
- To describe clearly the procedures you followed
- To present your findings accurately and explain any deviations from expected values or results
- To demonstrate your understanding of the practical value of what you have learned from the lab

Turning such observations into another series of questions can help you begin planning your report even before you start the lab. For instance, the first observation about the audience for lab reports suggests the following questions:

- What are the basic principles covered in this lab?
- Am I sufficiently familiar with them to explain these principles clearly and concisely?

If the answer to the latter question is *yes*, then you are prepared to begin work in the lab. If the answer is *no*, then you should study your textbook more carefully before beginning the experiment.

You should also turn the next three observations about your readers into questions:

- Are my results accurate?
- Have I performed error analysis and explained deviations from expected results?
- Have I indicated how theory applies to reality?

By asking yourself questions, you are more likely to recognize the importance of keeping careful track of your procedures in a notebook or lab journal, accounting for deviations from expected values, recording the accuracy of any device used for measurements, and explaining the practical applications of the procedures used or the knowledge gained from the lab.

Beginning by clarifying your goal and analyzing your audience makes communication not only more effective, but also more challenging and satisfying. Attending to the needs, concerns, values, and reactions of your audience emphasizes the problem-solving nature of communication. Given that engineers enjoy solving problems, applying a problem-solving approach to your writing and speaking will most likely increase your satisfaction in a job well done.

3.2.3.3 Employ Appropriate Conventions

Attending to format conventions is one way to address technical readers' needs. These readers expect you to follow established conventions that function as basic guides to the content and organization of the various types of technical documents they read on the job or for the articles they read in professional journals. Academic readers also expect you to follow conventions. For instance, you must be familiar with the type of lab report a particular course requires. Understanding the format conventions appropriate to a particular rhetorical situation provides you with a good idea of what information to include and how to organize it into major sections.

Following an established set of conventions also identifies you as a member of a particular group. Achieving the appropriate identification can be essential to achieving your goal. Failing to follow conventions can have disastrous results. For example, imagine the sort of grade you would receive if you handed in a lab report with no headings, no figures, and no tables and with results described only in words. Less dramatic violations of conventions can also affect audiences negatively.

Chapter 7 provides general information on technical writing conventions. But you should also study examples of relevant documents or watch relevant presentations when you are required to write a new kind of document or present in a new rhetorical situation. One word of caution, however: when you study sample documents, keep in mind that they may have been written for different purposes, readers, and contexts. If you try to copy a sample too closely, you will not adequately address your own rhetorical situation.

3.2.3.4 Identify with Your Audience

Through schooling, training, and experience, you will learn to communicate in ways that identify you as a member of your field, specialty, and profession. To accomplish this goal, you must identify with your audience and view the world from their perspective. You make this identification every time you acknowledge a concern before it is expressed or answer a question before it is asked.

To illustrate the effect identifying with your audience may have, imagine that you are reading the outline for a project course. The professor offering this course is new and you have not yet met her. Assume you are concerned with how group projects will be graded and wonder what will happen if group members fail to pull their weight. If the outline explained the grading system and addressed your concern with slackers, wouldn't you feel that the professor identified with her students? (We hope you said *yes*.)

Likewise, when you identify with the members of your audience and view a rhetorical situation from their perspective, anticipating their questions and concerns, your words are most likely to receive favorable attention, and you will most likely gain their respect.

3.2.3.5 Determine the Audience's Expertise and Status

One way to identify with your audience is to characterize your relationship to them in terms of expertise and power or status. Determine whether your audience has a high, moderate, or low level of expertise and whether your power and status are greater than, equal to, or lower than theirs. If you pay attention to these issues, you are less likely to adopt a tone that talks down to less-informed individuals. You are also more likely to be respectful when telling superiors that they have made mistakes. Assessing your goal in relation to an audience's expertise and power draws attention to likely concerns and reactions and helps guide your decisions on everything from content and length to word choice and tone.

The need to adjust how and what you write to match your audience's expertise and status is reflected in technical report formats. For instance, technical reports generally begin with an executive summary that caters to people in positions of power. Because senior managers and company presidents are very busy people, the executive summary omits detail, providing just enough information to support the recommendations the writer wants accepted. If executives want more information relating to a particular recommendation, they can read the relevant section of the accompanying report. Because executives tend to be less involved in technical matters and less up-to-date on technological developments, specialized technical language is also minimized in the executive summary.

Conversely, appendices provide very detailed information such as schematics, blueprints, or code that is important to the workers who will implement recommendations, build the device, repair the structure, or debug or revamp the system. Appendices will often be much longer than the report itself. The language will be highly specialized and may be filled with three- and four-letter acronyms.

In general, as you move from technical support workers up the ladder to executives, the length decreases and the language becomes less specialized. While this generalization is useful, keep in mind that it will not apply in all rhetorical situations. You may, for example, write for a senior manager who has remained on the cutting edge of his or her field or for a senior executive who reads reports cover to cover.

You will also write for readers with little or no technical background who can tolerate even less specialized language than you might use in an executive summary. For example, while you could write a users' manual for technical readers, you may also write one for readers with little knowledge of or experience with the subject of your manual. Some writers might be tempted to view these inexperienced readers as having little power and low status, but if you identify with them as customers, you will realize that pleasing them is essential to the continuing success of your enterprise.

If your audience has less expertise, your task involves informing without patronizing: using as few technical terms as possible, explaining the ones you do use, providing everyday analogies, including plenty of pictures or diagrams, and achieving a conversational tone. Most importantly, you want to organize your information and instructions from the user's perspective. Manuals that are organized following the designer's logic frustrate users who approach the task in terms of the sequence of actions they must perform. A user's perspective rarely matches a designer's way of thinking about a device or system. To identify with novice users, you may need to watch them using your device, package, or system, or watch them as they attempt to use a draft version of your manual.

The closer you come to understanding your audience's expertise and status, the better the chances of gaining attention, appreciation, and respect. Of course, when your audience is generic (i.e., users of the manual for a word processing package), you must generalize and aim for a typical reader or listener. But when your audience is comprised of people you work with everyday, you have much more information about your audience—information that may simplify or complicate your task.

3.2.3.6 Other Considerations

While communicating information to a well-known audience is generally easier than addressing unfamiliar or unknown people, dealing with power relationships can be tricky because personality, friendship, even office politics can come into play. When your emphasis is persuasive, assess the needs and fears of your audience in relation not only to power and expertise, but also to money, personal values and ideals, professional responsibility, environment, and politics. Pay particular attention to how your audience will be likely to react to your ideas, suggestions, concepts, and even your choice of words. In other words, addressing your audience successfully requires good interpersonal skills and the ability to view the world from someone else's perspective. We return to some of these considerations in the section on persuasive writing.

3.2.4 Write or Speak?

For many rhetorical situations, whether to communicate in writing or verbally is decided for you. Your teachers tell you they want a report, an essay, or an oral presentation, and your managers tell you they want a written progress report, an oral briefing, or an on-line user's manual. In some circumstances, you have a choice and should decide what is most appropriate based on your rhetorical situation. Deciding whether to write or speak is particularly important in those situations where you have a choice of talking to someone (in person or on the telephone) or of sending them an e-mail message (see Chapter 8, pages 295–302, for more information about e-mail).

Now that so many people have personal computers and access to the Internet at home, at school, and at work, letter and memo writing is giving way to e-mail messages. Colleagues who are literally steps away from one another think nothing of corresponding electronically rather than talking face-to-face. If such electronic correspondence were part of an ongoing discussion that included other participants, then the choice to write an e-mail message rather than go next door to engage in conversation could be appropriate. But what if an e-mail message is from a colleague in the next office and read while both parties are in their offices? If a colleague's message is in the middle of a long list of e-mails, our first impulse is likely to respond in writing. Unfortunately, this automatic response may lead to bad feelings and deteriorated working relationships.

The problem is that many of us think of e-mail messages as a form of conversation and dash off the responses with no more thought than if we were actually speaking to the recipient. But e-mail messages are not conversations; they are permanent written records that are sent and received without the aids to understanding available when people engage in conversation. Consequently, readers can ascribe motives that the sender did not intend or even imagine.

For example, a colleague might send a message asking for a copy of the functional specifications you are drafting. In the sender's mind, the request is a reminder that she needs a copy of the draft when you have completed it. Like many e-mail messages, the reminder is concise and unelaborated: *Please send me a copy of the specifications for the Telco project.*

If you read this message in the context of a hectic week with several difficult or time-consuming tasks hanging over your head, your personal situation might incline you to read her reminder as an unreasonable request for something you are not scheduled to complete for a few more days. If you simply *assume* the sender's motive (i.e., that she is pushing you to finish drafting the specifications before the deadline), your frustration could lead you to respond inappropriately: *I've got TWENTY OTHER THINGS to do that are more pressing. You'll get the specs by the deadline*! Because this message snaps and yells at the recipient, two short sentences turn a neutral request into an emotionally charged incident. In this way, simple misunderstandings can become the source of work-place tensions and poor relationships among colleagues.

Whenever you have a choice between conversation and e-mail, take the time to de-cide which is more appropriate. Speaking with someone rather than sending him or her an e-mail message is strongly advised when feelings are involved. If someone is clearly angry or upset, talk to the person. If you even sense that someone *might* be upset or frus-trated, take the time to speak with him or her. If *you* are the least bit annoyed or frustrated by what you have read, then choose talking over writing—especially if you could possibly have misinterpreted someone's intentions or motives.

This advice may not hold true for group e-mail messages dealing with controversial subjects. In such cases, responding in writing may be your only option. But before you wade into the fray, consider the feelings of the various people receiving the e-mail. Before you send a message, consider carefully the effect it is likely to have on those who will re-ceive it. The information on persuasive writing in the following section will help you make this judgment and, when appropriate, help you craft an effective response.

3.3 PERSUASIVE WRITING

Remember that anticipating a reader's needs and probable responses is especially crucial if your purpose is to persuade. Persuading readers requires close attention not only to what they know but also to their fears, expectations, values, and the criteria by which they will judge your case.

You may, for example, have an idea for a device you want to build. In preparing a proposal, you should begin by considering what criteria will likely be used in assessing the feasibility of your project. You could begin by determining what general goals, re-sources, and constraints govern your design. You should find out what kind of projects have been accepted or rejected in the past and the criteria by which they were judged. Colleagues may offer valuable insights and advice that will help you identify the issues you must address and the criteria you must accommodate.

To write persuasively, you must also sell yourself through the image of yourself cre-ated in your writing. To determine how to best present yourself to a given audience, begin by asking yourself what personal characteristics your reader will most likely value.

For your proposal, you can safely assume that your readers will be influenced by how well your writing portrays you as prepared to deal with design issues. If your proposal is full of modifiers such as *probably* and *maybe*, readers may fear that you lack confidence in yourself and in your proposed project. On the other hand, if you make unsubstantiated claims for the potential success of your project, you will present yourself as overly confi-dent. In fact, you will have the most potential success if you can present yourself as confi-dent, yet aware of the potential problems in your project—as suitably cautious, yet willing to take calculated risks.

Good people skills and common sense are clearly useful assets to draw upon when your emphasis is persuasive. At the same time, developing the ability to write persua-sively can help you deal with people more effectively. If we could offer only one piece of advice about persuasion, it would be to focus on your audience, to scrutinize every aspect of your writing from your audience's perspective—to figuratively walk in their shoes. Adopting the audience's perspective may well be the key to developing your skill as a persuasive writer. However, knowing something about the history and basic principles of persuasion is also useful.

3.3.1   Origins and Principles of Persuasion

Western notions of persuasion have their roots in ancient Greece, and for the current discussion, most directly in the work of Aristotle.[2] Modern notions of persuasion owe much to this early formulation of how to find the available means of persuasion for a particular audience and situation. While our world is very different from that of the Ancient Greeks, much of Aristotle's advice can be adapted to present-day contexts. In particular, we draw your attention to the concept of persuasive appeals, which we have greatly simplified and redefined to suit a modern, engineering context.

3.3.1.1   Persuasive Appeals

Aristotle identified three ways of appealing to an audience: by appealing to their reason, to their emotions, and through the speaker's moral character. Appeal to reason or logic is the most obvious appeal for an engineering audience. While you will rarely appeal directly to the emotions of your readers, you will appeal to their values and address their fears and concerns—an engineering equivalent of the appeal to emotions. Rather than appealing strictly from moral character, you appeal from your credibility, which is achieved by being knowledgeable, competent, and ethical. Note that your credibility as a writer is established *by your readers*, in *their* perception of your character and in *their* acceptance of your expertise and authority.

Keep these three appeals in mind: appeal to logic or reason, appeal to values, and appeal from your credibility. They can direct your thinking so that you are more likely to examine each point of your argument to determine whether readers will view it as logical and well-supported, whether you have appealed to their values and addressed their fears and concerns, and whether they will view you as credible. Viewing your work from the perspective of these three appeals is not simply a guide for determining the effectiveness of what you are writing or have already written.

These appeals are also a way of generating ideas, of *finding the available means of persuasion*. Focusing on the appeal to reason helps determine the points to make in support of your case. For example, if your task is persuading your superiors to allow you to change your work hours so that you can pick up your children after school, you might decide how to appeal to reason by determining why starting work early is a good idea. You might, for example, make the case that starting early will increase your productivity and facilitate contacting clients and vendors in other time zones.

Addressing the appeal to your readers' values, concerns, and fears will likely add more content and clarify your approach. For example, if you want to persuade your superiors to let you start work early, you might determine how sympathetic they will be to your fear for your children's safety. Also identify any fears or concerns they may have about allowing one employee to do something different from other employees, how open they are to change, and how well your request fits with the company's corporate culture. You might decide that the corporate culture is neutral and that while your supervisors will not be particularly happy to set a precedent by allowing you to start early, they are family-oriented and will most likely identify with your concerns and be sympathetic. You might, therefore, assume that their personal values outweigh their corporate concerns and begin by describing the problems at your children's school that prompted your request.

Determining how to appeal from credibility also helps determine content and approach. You might, for example, assess how satisfied your superiors are with your perfor-

---

[2] Our source is Lane Cooper's translation, *The Rhetoric of Aristotle* (Englewood Cliffs, NJ: Prentice Hall, 1960).

mance and decide that they are very satisfied and would not want to lose you. However, continuing along this line of thought, you might decide that reminding them of your worth could appear to be a threat and serve only to annoy them. In this case, you would likely decide to let your record speak for itself and to focus on explaining your personal concerns and outlining the potential benefits of your request.

3.3.1.2.   Identification, Not Division

To persuade an audience, you must build upon a foundation of shared beliefs and points of agreement. Without this common ground, you cannot hope to persuade someone to acknowledge your point of view. Note, for example, the issue of abortion: because the *right-to-life* and *right-to-choice* positions are based on opposing beliefs and values, neither side can persuade the other. Despite years of discussion, debate, and litigation, the controversy continues. A distinct division is also evident between those advocating the right to arms and those advocating gun control. In fact, most long-standing social controversies are just that—long standing and controversial—because the two sides lack sufficient common ground.

Using terms borrowed from Kenneth Burke, we can rephrase the above discussion in terms of *identification* and *division*. That is, common ground helps us identify with others while a lack of common ground divides us from them. To persuade readers to agree with you, they must identify with you. You establish this identification in many ways: by using the specialized language of your readers' field, by using the same kind of logic and burden of proof they do, by following the conventions of their field, by taking their concerns into account, and so on. If you create a sense of division—by relying on feelings and opinions when your readers are concerned with logical arguments or by using terminology your readers do not understand—you will most likely succeed only in alienating, frustrating, and confusing them.

Identification and persuasive appeals can be viewed as two sides of a coin. That is, you want your audience to identify with you by offering them the kind of logical arguments they expect (appeal to reason), by addressing their needs and concerns (appeal to values and concerns), and by demonstrating that you have the personal qualities and abilities they value (appeal from credibility).

3.3.1.3   Other Considerations

Note that persuading others requires flexibility, creativity, and the ability to view a situation from someone else's perspective. Given the increasingly interdisciplinary nature of engineering and the proliferation of international and joint projects, these attributes are worth developing for more than the ability to write persuasively. The humanities or social sciences courses your program requires introduce you to ways of thinking and logic that differ from what you encounter in your engineering courses. These complementary courses ask you to explore different kinds of issues and different kinds of subject matter. They also expose you to unfamiliar terminology and require different conventions and styles of writing.

Learning how to write in ways appropriate to other fields can help you identify with different groups and recognize the differences among fields. This experience should also help you see the degree to which writers identify themselves through details. For instance, if an author uses bulleted lists, cites references using numbers in brackets, avoids personal pronouns, and relies heavily on tables, figures, equations, and passive voice, we are unlikely to identify the writer as, for instance, someone who teaches freshman English.

You may not want to think and write like a historian, psychologist, philosopher, or anthropologist. But keep in mind that being exposed to new ways of thinking, learning about what other people value, and paying attention to the details of writing in a variety of contexts will help you develop skills you will need as an engineer. You also need this sort of awareness and flexibility when working in teams (Chapter 4, pages 117–126, provides more information about teamwork).

3.3.2 Models of Persuasion

We have discussed persuasion at some length without offering models. Those of you who like to work from templates may find the following useful, but keep in mind that communication is a dynamic interaction among writer, reader, subject matter, purpose, and the surrounding context. While models are useful as tools for finding appropriate approaches and organizing your ideas, they should not constrain. Persuasion does not result from following a format; being persuasive requires creativity, problem solving, and a focus on readers' needs, expectations, and reactions.

The following sections describe three specific models of persuasion: adversarial persuasion, collegial persuasion, and constructive criticism. Use them as guides or for inspiration, but avoid following them slavishly. These models should augment the strategies provided earlier and should never substitute for them.

3.3.2.1 Adversarial Persuasion

Adversarial persuasion involves supporting your position while refuting positions that oppose yours. You can take an adversarial stance when persuading a neutral third party to adopt a particular course of action or to accept your interpretation of a situation. Note, however, that if you are attempting to persuade an individual who holds an opposing position, taking this stance will most likely alienate that individual and deepen the divide between you.

As Figure 3.2 suggests, this adversarial model is suited to situations in which two opposing parties attempt to persuade a third party, not each other. This third party is an individual or group with the authority to rule in favor of one of the opposing positions. You can also use this model when you present opposing positions in order to refute one and gain acceptance for the other. For example, you might use this model to persuade colleagues to change a design or to persuade a group to accept your solution in place of the current practice.

The steps in the model for adversarial persuasion follow:

1. Introduce the topic you will be considering.
2. Explain the facts and issues involved and define any specialized terms you will be using.
3. State your case (i.e., present your position or proposal) and explain how it will be organized.
4. Detail the arguments and evidence that support your case.
5. Refute other positions by indicating their flaws or explain why your solution is superior.
6. Summarize your arguments and refutations.

Figure 3.2 Adversarial Persuasion

Note that some of these steps can be combined and dealt with in a single paragraph. For example, journal articles and research papers often combine the first three steps in the introduction. Steps may also be omitted. For example, a summary is not always necessary; another kind of ending may be stronger for a particular situation.

One advantage of this approach is that it is very direct and your position is clear from the start. Because this approach is suited to presenting facts, it is easily adapted to various engineering contexts. However, potential problems with this approach are that it can lead to hard feelings and your colleagues may be uncomfortable with or feel threatened by public disagreement, especially if they are pushed into supporting one side or the other. If handled poorly, this approach can appear more like a popularity contest or ego trip than an honest attempt to solve a problem.

When used to address someone with less or equal power directly, hard feelings and worsening relations are likely to result. How do you think you would react, for example, if a colleague sent you the following e-mail?

> *Robert: The report is due this week and I haven't seen your part of it yet. The boss has demanded that I have the completed report on his desk by Friday morning and everything is finished except your part. This project is the most important thing we are working on. Our company stands to lose a lot of money if the report is late. A week ago, you promised me that your part of the report would be completed in a day. I'm still waiting, and I'm running out of time (and patience). As it stands now, I will have to work late tomorrow evening just to revise and format your part of the report. I must have your report by first thing in the morning. Work all night if you have to. Your job may depend on it.*

While this message may be very mild compared to how the writer feels, Robert will almost certainly find the tone offensive.

For example, the last sentence may have been written to spur Robert on by reminding him of the importance of the contract to the continuing financial well-being of the company. But it will most likely be read as a threat and serve no other purpose than making Robert mad. Although feelings are rarely expressed in technical documents, they are often an issue in interpersonal communications. Note that when interpersonal relationships are a factor, the adversarial model is a very poor choice.

Also note that models are merely guides. You should not follow them too closely and must adapt them to the constraints and opportunities of a particular situation. To help make this point, we use the above example and the ones for the other models of persuasion to point out potential problems. On the one hand, we hope to provide you with a sense of the variety of approaches you can use and a feel for the range of approaches you can employ when your emphasis is persuasive. On the other hand, we want you to remember that models must be adapted to specific rhetorical situations.

3.3.2.2   Collegial Persuasion

Collegial persuasion is our term for an approach developed by Carl Rogers in the early 1970s. This approach is based on Rogers's principle that arguing creates a sense of being threatened, which makes people more likely to hold stubbornly to their original positions. By listening to other people's positions and by communicating that we understand how they could reasonably hold their positions, we reduce their feeling of being threatened and increase the likelihood they will adopt some elements of our position.[3]

As Figure 3.3 suggests, collegial persuasion is most appropriate for reaching compromises and for creating win-win situations. It is suited to situations in which a colleague disagrees with you or when you are attempting to persuade a person with more power than you (such as your supervisor). You can also use collegial persuasion with subordinates, when you have the power to force an issue. Persuading someone to accept your position or allowing them the opportunity to present you with an acceptable compromise goes a long way toward establishing and maintaining good working relationships.

When using this collegial model, the ability to view a situation from another person's perspective is vital. Whenever possible, use this model during face-to-face conversation. Listening carefully to what others say will help you understand their concerns and reach

Figure 3.3 Collegial Persuasion

---

[3]The information about Rogerian Persuasion was adapted from Young, Becker, and Pike's textbook, *Rhetoric: Discovery and Change* (New York: Harcourt Brace Jovanovich, 1970).

an appropriate compromise. If you use this approach to persuade someone by writing to the person, the task can be more challenging. Note that following the advice offered in the first step for this model can help you avoid conflict in the first place:

1. Present the conflict as a problem to be solved rather than as an issue to be disputed.
2. Provide a fair and complete summary of the other person's position, demonstrating you understand by accurately restating his or her position.
3. Describe the contexts in which the other position is valid, demonstrating that you understand how the other person could reasonably hold that position.
4. Provide a fair statement of your own position, avoiding overstating your case.
5. Describe the contexts in which your position is valid.
6. Describe how the other person would benefit by adopting some elements of your position or propose a third position that meets both your needs.

As the last point suggests, this approach has another potential advantage: sometimes the process of seeking a compromise leads to a better solution than the one you were originally committed to implementing.

One disadvantage of this approach, especially for written persuasion, is that your assumptions about other people's positions must be accurate. If you misjudge their concerns, you may unintentionally offend them, creating or intensifying interpersonal problems. Another disadvantage (again when writing) is that judging whether your tone will be read in the way you expect is difficult. Pay attention to the tone of the following e-mail message that addresses the same situation discussed in the section on adversarial persuasion. This time the approach is meant to be collegial.

> *Hi Bob, It sounds like we are going to run into some problems meeting our deadlines for the report. I understand you have a lot of other work to do that is interfering with you completing your part of the report. I know your other commitments are very important and normally it wouldn't matter if you were a bit late. But I really do need your part of the report as soon as possible. Perhaps the deadline we face has slipped your mind. The boss has told me that the entire report must be completed by Friday morning because our company stands to lose a lot of money if it is submitted late and jobs could be lost. If you will get me your part of the report by tomorrow afternoon, I will let him know that you went out of your way to complete it. Thanks Bob.*

Do you find the tone of this e-mail patronizing? Many engineers who have read it think so.

Should you attempt to write using a collegial approach, be sure to have someone at work or someone you trust outside of work read your draft. Someone less involved in the situation is more likely to read it objectively and note problems in your tone. If you wondered why anyone would bother writing when Bob is obviously close at hand, you have a point. Addressing him in person would be more appropriate.

3.3.2.3   Constructive Criticism

Although a collegial approach is useful in contexts where a compromise is possible, circumstances arise where no real potential for compromise exists and persuading colleagues to change requires criticizing their performance or behavior directly. In such cases, especially if you are angry, you might be tempted to adopt an adversarial approach. However, that choice would create more problems because the adversarial approach amplifies the effect of your criticism and, in turn, increases the other person's defensiveness. The more defensive someone is, the less likely you are to encourage positive changes in performance or behavior.

Figure 3.4 Constructive Criticism

An approach adapted from the work of Gracie Lyons offers a potentially effective way of providing constructive criticism.[4] This model combines aspects of both adversarial and collegial models. Constructive criticism can be effective when you must criticize the behavior or performance of a colleague or subordinate, but should be used very cautiously if your complaint is with a superior. As Figure 3.4 implies, the aim of this approach is not to make someone happy, but to encourage a change in behavior or performance.

The formula for this approach involves filling in the blanks in a set of statements:

1. Describe the other's specific behavior or action (**When you do _____**).
2. Describe the specific result of the behavior or action (**_____ results**).
3. Suggest an alternative behavior (**I wish you would do _____ instead**).
4. If appropriate, suggest a way to implement that change (**You can accomplish that by _____**).

Read over the following version of the e-mail to Bob, written as constructive criticism:

Bob: When you give me your part of the report late, I must work a lot of overtime in order to meet the deadlines for the project. I wish you would give me your work at the agreed-upon time instead.

As this example suggests, the four-step formula may require little elaboration, although some situations will require much more elaboration than this e-mail suggests. You can omit the fourth step if you do not have a suggestion for accomplishing whatever is necessary. (The e-mail could have added the suggestion of working all night, but doing so would make the note sound more aggressive without serving a useful purpose.) Situations in which constructive criticism is useful are not pleasant, but when they arise, being able to deal with them is an extremely useful skill.

3.3.3 Connotation

Whatever form of persuasion you employ, pay careful attention to the connotations of the words you chose. A *connotation* is a meaning that a word carries in addition to what it ex-

---

[4]Adapted from *Constructive Criticism* (Oakland, CA: Inkwords Press, 1976). Used with permission.

plicitly describes or names. For example, the word *feminist*, in addition to simply describing an individual who adheres to a particular set of values, also has a negative connotation in some circles and very positive connotations in others. Further, some readers may assume that something that is *feminist* applies only to women. We therefore chose not to discuss constructive criticism as a feminist model, assuming some of our readers might pay less attention to it as a viable model or focus too much attention on their connotations for feminism. If writing for a women's studies course, we might have been more inclined to highlight constructive criticism as a feminist model.

In other words, the connotations of any particular word are heavily dependent on context—on who is writing or reading the word. Words also have connotations simply because language always occurs within specific social contexts. The words you use reflect certain values and attitudes. You must ensure that the image they reflect is accurate and intended. To master the art of persuasion, you must examine your own biases and develop tolerance for different points of view.

Even in technical documents, the connotations of words require serious attention. Imagine that you have been commissioned to write a report for a company seeking the government's permission to increase the level of emissions from their smokestack. Initially, you are inclined to use the term *smokestack* in the report. But company officials do not want you to use that word because it has a negative connotation (i.e., smoke *belching* into the air).

You might be tempted to use the word *stack* because at least that word eliminates the term *smoke*. Unfortunately, choosing *stack* creates another problem; *stack* has many different meanings in relation to libraries, computers, or agriculture, for example. Perhaps you next decide to test the term *flue*. But will you readers be familiar with that word or will they regard it as jargon?

Even though the words *stack* and *flue* are relatively neutral in terms of connotation, you decide against using them. At this point, you might be tempted to invent a technical term to describe the smokestack: *Atmospheric Emission Dispersion System (AEDS)*. However, that particular choice may have an even worse connotation than *smokestack*. Many people would view the invented term as a good example of *double-speak* or euphemism—of using a seemingly innocuous term to hide an unpleasant reality (much as the military uses the term *collateral damage* to hide the fact that they are really talking about *dead civilians*). When people feel that language is being used to deceive them, they are more likely to resist the proposal. So *Atmospheric Emission Dispersion System (AEDS)* is off the list.

Fortunately, you have another option: *chimney*. This term has relatively positive connotations (i.e., cozy fireplaces and Santa Claus). Assuming that the level of emissions falls within legally and environmentally acceptable levels, the word is not being used deceptively. Because the goal of the report is to obtain permission from the government to increase emission levels, you decide to use the term *chimney* in your report. Using this term, you are less likely to generate resistance to your proposal than if you used a term with a more negative connotation. Figure 3.5 visually represents a continuum for the connotations of the various words for smokestacks mentioned above.

You must remain conscious of the potential impact of your language on your readers. If you use a word with the wrong connotation for a particular audience, you risk alienating (even offending) readers. Conversely, if you use a word with an appropriate connotation for a given audience, you are more likely to persuade them. Select words with connotations that will help you achieve your goals, and ensure that the words you choose accurately reflect your values and are in keeping with professional standards and responsibilities.

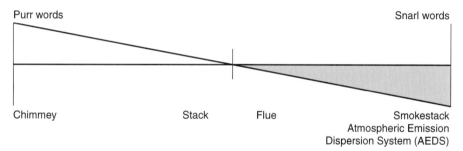

Figure 3.5 Connotation Continuum

3.3.4 Inclusive Language

Language not only helps shape your thoughts about the world, it also reflects your attitudes toward the world. When your attitudes change, so must your language. A case in point is the use of *inclusive* language when writing. In most educational institutions, government agencies, and private corporations, the use of *generic* language is no longer acceptable (i.e., using *he* and *him* to the exclusion of *she* and *her* or using *man* to refer to men and women). Further, most professional journals provide specific guidelines for the use of inclusive language and will not accept articles that violate their guidelines.

Avoid nouns that imply gender (e.g., *businessman, foreman, chairman, policeman, craftsman*); instead use gender-inclusive terms (e.g., *entrepreneur, supervisor, chair, officer, artisan*). Also use an inclusive pronoun (*they*) or pronouns that address both genders (*s/he, he or she*) in your writing.

Original	**Revised**
For example, a user selects "Compile" in order to collect timing signals from the unit. Later, *he* selects "Analyze" to produce a series of graphs for the compiled timing data.	For example, users select "Compile" in order to collect timing signals from the unit. Later, *they* select "Analyze" to produce a series of graphs for the compiled timing data.
Under the new multitasking operating systems, it is possible to run "Timing Analyzer" concurrently with other software that the user may wish to run on *his* computer.	Under the new multitasking operating systems, it is possible to run "Timing Analyzer" concurrently with other software that *users* may wish to run on *their computers*.

We have chosen to shift from singular to plural nouns to avoid the potential awkwardness of repeating *he* or *she* or *s/he* throughout a passage. Note, however, that shifting from a reliance on *he* to the use of *she* is sometimes accepted, as demonstrated below.

Original	**Revised**
The technician then runs the unit through the complete test in order to familiarize *himself* with the unit's normal response. *He* can then . . .	The technician then runs the unit through the complete test in order to familiarize *herself* with the unit's normal response. *She* can then . . .

Why this insistence upon inclusive language? First, and perhaps most importantly, you must use inclusive language as a matter of human rights. Women have been excluded from (or limited in their access to) positions of professional authority and recognition for centuries. This situation is no longer acceptable. As professionals, you are expected to adhere to many ethical standards, and one of those standards prohibits discrimination against anyone because of their race, religion, or gender.

Second, we offer a rhetorical reason for using inclusive language in your writing. Many women (and, it should be added, an increasing number of men) find the use of generic language offensive, or at the very least, irritating. With this change in attitude and with more women entering the professions and being promoted through the ranks, generic language is increasingly likely to create a negative impression.

Third, some readers more easily remember inclusive language. Researchers find differences in the degree to which men and women recall information written in noninclusive language.[5] When information is presented using generic language (*he, him, his, mankind*), women recall the information less well than men. When the information is rewritten using inclusive language (*they, them, their, people*), this difference in recall between the sexes disappears.

Finally, we repeat the pragmatic reasons for using inclusive language. The vast majority of journals will not accept articles written using generic language; editors will return a submission for revision. Government agencies and many large companies have instituted policies requiring the use of inclusive language. Documents that fail to adhere to these policies must be revised.

The following guidelines provide a range of strategies for writing inclusively:

1. *Avoid nouns that imply a gender*. The easiest solution to this problem is to substitute an inclusive noun. For example, *chairman* becomes *chairperson* (or simply *chair*), *salesman* becomes *salesperson*, and *spokesman* becomes *spokesperson, fireman* becomes *firefighter, weatherman* becomes *weather forecaster*, and *middleman* becomes *intermediary*.

2. *Avoid using gendered nouns that have a negative connotation*. Referring to women as *girls* can be insulting. Using the word *girls* to describe women implies that they are immature or frivolous. Similarly, other expressions such as *women's libber* or *spinster* have negative connotations associated with them. Alternatives that are less offensive for the person being addressed would be *feminist* or *single person*.

3. *Avoid gender stereotyping*. Sentences such as "The engineers and their *wives* attended the party" or "The nurses and their *husbands* attended the ceremony" imply that all engineers are men and all nurses are women. These implicit stereotypes can be avoided by substituting *spouses* for both *wives* and *husbands*.

4. *Use neutral or dual pronouns*. Because English lacks a singular neutral pronoun, the use of gendered pronouns in writing is a difficult problem to resolve. Several possible solutions follow:
   a. *Switch to the plural*: "*The user* should then save *his* document" becomes "*Users* should then save *their* documents." (This strategy can be used most often and is the one we recommend most highly.)
   b. *Use both pronouns*: "A good engineer learns from *his* mistakes" becomes "A good engineer learns from *his or her* mistakes." (If used frequently, this approach leads to an awkward style.)
   c. *Combine pronouns*: "A good team leader will also be an effective writing coach. *S/he* may find this to be the most difficult part of the job." (Exercise caution with this approach because it bothers some readers.)

---

[5]For example, see the article by Crawford and English, Generic versus specific inclusion of women in language: Effects on recall, *Journal of Psycholinguistic Research*, 13, 1984, 373–381.

d. *Omit the pronoun or replace it with a definite article*: "The engineer spoke with *his* client" becomes "The engineer spoke with *the* client." (While substituting articles is a good strategy, it is not always possible. Also note that if the engineer speaking with a client has been previously identified as Tom Long, then *his client* is appropriate.)

e. *Alternate pronouns throughout the document*: "Team members sometimes complain about their supervisors, saying, '*He* demands perfection' or '*She* expects too much.'" (Again, this strategy is useful, but not always applicable.)

f. *Use third-person plural pronouns (they, them, their) as the singular*: "You should not discriminate against anyone because of *their* race, religion, or gender." (While this approach is increasingly common, some readers may view it as a grammatical error.[6])

Although the particular approach you choose to ensure inclusive language is, in part, a matter of personal preference, note that switching to the plural—our preferred strategy—is usually a relatively simple revision. In fact, once you learn to write in the plural as a matter of course, you will discover that your writing requires little revision for noninclusive language.

Also note that inclusive language is an issue for informative as well as persuasive writing. While readers' reactions are of primary importance when your aim is to persuade, readers must also be accommodated when your aim is to inform, as the section on informative writing demonstrates.

3.3.5 Ethical Considerations

For more than a millennium, philosophers and others have been concerned by unethical uses of persuasion. In the *Republic* (a work from the third century B.C.), Plato argues that the ability to persuade should not be generally available because people could use it to deceive—to convince others that right is wrong and wrong is right. According to Plato, only a philosopher king with morals beyond reproach should be allowed to wield such power.

In our world, and particularly in engineering contexts, the danger is at least as likely to lie with professionals who fail to persuade others of existing risks and threats as with those who would deceive us. Numerous ethics case studies recount disasters that could have been prevented if engineers had only been capable of persuading those in authority to listen to their warnings.[7] Developing your own ability to persuade is also an effective way to counter those who would abuse their skills as persuaders. If you pay careful attention to the logic of an argument, to the values being appealed to, and to the way a speaker or writer establishes credibility, you will not be easily swayed by misleading arguments.

Developing your powers of persuasion as outlined in this chapter will also help you avoid using your ability unethically. When you understand the influence of the three persuasive appeals, the importance of respecting your audience, and the strength of building your case on shared values and beliefs, you will be unlikely to succumb to a temptation to overstate your point or to deflect attention from issues that deserve attention. Keep in

---

[6]The National Council of Teachers of English (NCTE) and the Conference on College Composition and Communication (CCCC) have both recommended accepting this approach.

[7]C.E. Harris, Jr., M.S. Pritchard, and M.J. Rabins, *Engineering Ethics: Concepts and Cases* (Belmont, CA: Wadsworth, 1995). This textbook provides a large number of ethics cases supported by conceptual frameworks and techniques for analyzing the cases and for resolving the problems they present.

mind that engineers are given a great deal of responsibility in our society, and in return, they have an obligation to protect the public by developing their powers of persuasion to the best of their ability and by being honest and trustworthy at all times.

3.3.6   Principles of Persuasion

We can condense our discussion of persuasion to the following eight principles:

1. Respect your readers and let them be your guides while planning, drafting, and revising.
2. Employ all three persuasive appeals, appealing to reason and logic, appealing to readers' values, and appealing from your own credibility.
3. Promote identification, building arguments on a foundation of shared beliefs and points of agreement.
4. Develop an appreciation for different ways of thinking and different perspectives.
5. Use models of persuasion as guides, where appropriate, being careful to adapt models to your rhetorical situation and to account for your readers' potential reactions.
6. Pay careful attention to the connotations of words.
7. Employ inclusive language.
8. Be honest.

The first of these principles is the most important and informs the others. When your emphasis is on persuasion, readers must be the center of your attention. You cannot write persuasively by following a template; rather, persuasive writing is shaped by focusing on and identifying with your readers and by applying your creativity and problem-solving ability to addressing their concerns. Identifying your readers' needs and fears can help you determine what to put in your introduction and can help you organize the rest of your document, just as imagining readers' reactions to what you've written can help you determine where supporting evidence is required, where you should take the time to recognize the validity of a perspective different from your own, or where your word choice may be problematic. When writing to persuade, let your readers be your guides. As you continue reading, note the range of considerations you must account for when writing to inform and remember that they are also important for persuasion, especially when writing for engineering audiences who are persuaded by facts as well as by theoretical arguments.

3.4   INFORMATIVE WRITING

If your aim is to inform, you must ensure that you provide readers with all essential information and yet avoid overburdening them with too much detail. As technical writers, you will be constantly challenged to keep this delicate balance between too much and too little information. To ensure that you maintain this balance, state your purpose in a sentence. Write it down; keep it in mind as you plan; refer to it whenever you are uncertain about including a point; and read it over before you begin revising.

Keeping your purpose in mind as you plan, write, and revise your work will help you judge what information to include. If, for instance, your task involves writing a quick-and-dirty manual for first-time users of an oscilloscope, your primary goal is to provide the basic information necessary to operate this device. Although you may know a great deal about the inner workings of an oscilloscope and believe that your readers will find what you have to say interesting, such information is inappropriate given your goal.

Determining how your information will be used and by whom will also help you determine what information to include (or exclude) and suggest how to organize it. (You cannot separate considerations of audience and purpose; they always overlap in significant ways.) Even when your focus is on providing information, you must keep your readers in mind as you plan, write, and revise. If, for instance, you keep the novice user in mind as you work on the oscilloscope manual, you will recognize the need to define any terms that may be new to your readers, to explain setup and operation procedures in a step-by-step fashion, and to provide appropriate visual aids.

3.4.1   Readers and Information Processing

Technical and professional writing should convey information as clearly and concisely as possible. The best writers write with such clarity and simplicity that their readers are able to focus attention so completely on the ideas presented that they often fail to notice the writer's style or the format of the document. Your goal, then, is to develop a style and format that allows you to present complex ideas simply and concisely. If you have developed a habit of using big words and of writing long, complex sentences, then you should pay special attention to your style (see Chapter 6). If you have not considered how the format or organization of your document helps or hinders the reader, then you need to reconsider the impact of form (see Chapter 7).

In any case, you should pay special attention to how readers process information. Readers are not simply passive recipients of the content presented by texts; instead, they actively construct meaning from the words, phrases, sentences, and paragraphs that make up those texts. The meanings that you intend to communicate by way of a text are never identical with the meanings that a reader constructs from the text. Each individual comes to a text with a unique set of beliefs, ideas, and experiences derived from that individual's particular social and cultural background.

In general, readers construct the meaning of texts based on a variety of structural and rhetorical expectations about how texts will be organized. When writers present information and structure their sentences and paragraphs in ways that anticipate and fulfill these expectations, they increase the chances of communicating effectively with the reader. On the other hand, when writers violate the reader's expectations, they increase the likelihood that the reader will misinterpret the text.

A useful distinction to draw here is between what has been called *reader-centered prose* and *writer-centered prose*.[8] Reader-centered writers are aware of how readers process the information in a text and are centrally concerned with increasing a reader's understanding. Writer-centered writers, on the other hand, focus on the subject at hand and are centrally concerned with analyzing or describing the material so that *writers* understand what *they* have written. Given the importance of readers, being writer centered is sure to cause problems.

Writing reader-centered prose requires close attention to various stylistic issues such as order, coherence, clarity, and conciseness and to various issues of form such as organization, paragraphing, and format. Attending to these issues, in turn, requires determining how to deal with such issues as jargon, nominalizations, passive constructions, connec-

---

[8]We credit the concept of reader-centered and writer-centered prose to Linda Flower and John Hayes. See, for example, Linda Flower, Writer-based prose: A cognitive basis for problems in writing, *College English*, 41, Sept. 1979, 19–37.

tions, punctuation, headings, generalizations, fonts, and justification (these issues are addressed in Chapters 6 and 7).

To the degree that you account for how such issues affect readers, you increase the reader's potential comprehension of the material. To the degree that you ignore how they affect readers, you risk failing to communicate. Figure 3.6 provides a graphical representation of the differences between writing that is reader-centered and writing that is writer-centered. The more reader centered the writing, the greater the overlap between the writer's intended meaning and the reader's understanding.

A reader-centered document is also easier to read—at least for the intended reader. What counts as a readable style is relative to the reader. For instance, a report containing a great deal of technical language may be read easily by someone who works in the same field as the author. But that same report may be virtually meaningless to someone outside that field. By the same token, a passage that seems terribly dense to someone with limited expertise might seem quite clear to an expert in the field. In other words, the judgments you make concerning style and form depend in large part on the background of those who will read your work.

To further complicate matters, information-processing concerns and rhetorical considerations can sometime conflict. In scientific and technical writing, this conflict is perhaps nowhere more evident than when deciding whether to use the active voice and verb-based *style* (*We completed analyzing the data*) or the more traditional passive voice and noun-based style (*An analysis of the data was completed*). Although the active voice and verb-based style are more easily understood than the passive voice and noun-based style (see Chapter 6, pages 199–202, for a discussion of passive voice), some scientists and engineers resist the more readable style. They accept only a passive, noun-based style as appropriate to the field. Consequently, engineers must be adaptable writers who respect the point of view of their readers and alter their style to adapt to readers' preferences as well as to readers' expertise.

In part, a preference for passive voice and a noun-based style makes sense because this style creates the impression of objectivity, and people trained in the sciences may respond favorably to this apparent objectivity. At the same time, however, some of the resistance is simply resistance to change. Sometimes a supervisor's preferred style will not

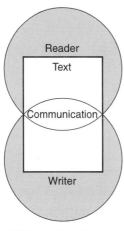

Writer-centered prose

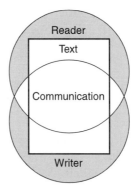

Reader-centered Prose

Figure 3.6 How Style and Form Affect Communication

be the most appropriate choice for a given rhetorical situation. In such cases, the advice on persuasion given earlier should prove useful.

Although many specific details of style and form affect readers' comprehension, employing the following eight principles carries you a long way toward a readable and memorable style:

1. Minimize memory load.
2. Create order.
3. Connect the new with the known.
4. Provide general frameworks.
5. Repeat important concepts.
6. Employ graphics.
7. Keep it simple.
8. Eliminate cognitive interference.

3.4.1.1 Minimize Memory Load

Human short-term memory has very clear limits in terms of the number of items that can be easily remembered. This short-term limit, sometimes referred to as the *magic number*, is 7 ± 2 items. In other words, some individuals will retain as many as nine items in short-term memory while others will retain as few as five items. Note that this memory limitation explains why telephone numbers in North America are limited to seven digits.

Most people can remember a number that they looked up in the phone book long enough to dial the number. (If you frequently fail to remember the number and must look it up again, perhaps your short-term memory is closer to the lower limit.) Further, you will note that North American telephone numbers are broken into two chunks of information (three chunks if the area code is included): (123) 456–7890. By breaking the number into smaller units of information, the number becomes even easier to remember.

Because reading complex technical materials already poses a heavy burden on short-term memory, you are wise to minimize the memory load by carefully crafting the style and form of your writing. Note, for example, the following title:

Mixed Analog/Digital Integrated Circuit Product Delivery Process Benchmarking

By stringing together nine adjectives and nouns, and thus overburdening the reader's short-term memory, the author of this title ensures that the information it contains is difficult for readers to remember.

The title is much easier to remember if prepositions and verbs are included:

Benchmarking the Product Delivery Process for Mixed Analog/Digital Integrated Circuits

Although the revised title is two words longer, most people find it much easier to remember because the words have been grouped into three units: *Benchmarking* (verb), *the Product Delivery Process* (short noun phrase), *for Mixed Analog/Digital Integrated Circuits* (prepositional phrase).

In terms of style, you can minimize memory load by avoiding long strings of nouns or prepositional phrases and by ensuring that lists are parallel in structure. (The grammatical issues raised here are covered in Chapter 6.) Read over, for example, the following conclusion to a persuasive report:

> *The major considerations for review are as follows:*
>
> - *Emphasize team member participation in planning and managing each project.*
> - *Provide formal project management training to all members of the project group.*
> - *Emphasize team building.*
> - *Introduce an effective bid-response system.*
> - *Balanced resource planning.*
> - *Lateral communications between functional departments.*
> - *Marketing to solidify commercial specifications before commitment to build.*
> - *Increase the involvement of senior management in the project process and coordinating the functional departments.*

In part, the difficulty with the above example is that the list is not properly parallel (the first four items and the last item begin with verbs while the other three items begin with adjectives or nouns). Items in a list are inherently easier to remember when they are written in parallel structures (i.e., begin with all verbs or all nouns).

However, the stylistic problem with the above example is probably less important than the problem with the form of the list. By simply listing eight points in no particular order, the writer assures that readers will have difficulty sorting out the more important points from the less important ones. Or worse, the reader may simply fail to remember any of the points, given the limits of short-term memory.

By grouping the eight points into subcategories, the writer minimizes the memory load on the reader and thereby ensures that the reader is adequately informed (and also more likely to be persuaded). Note how the following list not only establishes parallel structures, but also divides the information into subcategories:

> *The major considerations for review are as follows:*
>
> 1. Teamwork
>   - *Emphasize team member participation in planning and managing each project.*
>   - *Encourage team building.*
>   - *Increase lateral communications between functional departments.*
> 2. Coordination
>   - *Develop balanced resource planning.*
>   - *Require marketing to solidify commercial specifications before commitment to build.*
>   - *Increase the involvement of senior management in the project process and coordinating the functional departments.*
> 3. Training
>   - *Provide formal project management training to all members of the project group.*
>   - *Develop an effective bid-response system.*

As this revision demonstrates, writers can minimize the memory load imposed by the form of a list simply by grouping related items and by ensuring that like or equal items are presented in the same or parallel structures.

3.4.1.2 Create Order

One of the most basic characteristics of the human mind is its need to find or create some sort of order in our encounter with reality. Although the approach to organizing information varies somewhat from person to person depending on personal, social, and cultural background, when presented with information, we all seek to order or categorize it. Examine, for example, the following list of items:

- paper clip
- straight pin
- nail
- bobby pin
- staple
- cotter pin

What do these items have in common? One possible organizing principle is that they are all made of metal (composition). They could also be the sorts of things typically found in a junk drawer (location). They are also all used as fasteners (function). If you were to provide readers with such a list without providing any commentary, some would focus on composition and others on location or function.

As a writer, you must guard against readers creating a different order than the one you intended. To minimize the possibility of misunderstanding, present information in ways that clearly indicate to your readers the intended order, connection, and importance of the information you present. You can help ensure that your order is evident to your readers by using an organizational pattern they are already accustomed to following. That is, you can structure a document, a section, or even a single paragraph according to one of these common organizational patterns:

- A sequence of events through time
- A sequence of movements through space
- A comparison of similarities and differences
- A list from most important to least important
- An explanation of a problem and solution followed by recommendations
- A general point followed by a more detailed explanation and/or specific examples

Refer to page 37 in Chapter 1 for a more detailed explanation and examples of these patterns.

One way to ensure that your document clearly communicates your intended organizational pattern to the reader is to provide appropriately descriptive headings and subheadings. You should also ensure that lists reflect the appropriate organizational pattern (i.e. that they are organized in terms of importance, time, etc.). You can also employ the following strategies when creating lists:

- Ensure that what all the items in a list have in common is obvious to the reader.
- Restrict lists to no more than five points wherever possible. (If you have more than five points, attempt breaking up the list by using subheadings.)
- Make lists grammatically parallel when possible (i.e., all items should begin with verbs or all should begin with nouns).
- Clarify how items are related, using bullets (or another symbol) when the items are of equal importance and using numbers to indicate priority, step-by-step instructions, or order of importance.

Within the text of a document, you can also help communicate the intended order of your ideas by using words such as *first, second, before, after,* and *then.* Even within sentences, you should pay attention to the order of instructions, especially when writing user's manuals or procedural instructions. Note, for example, the possible consequences of the following instruction that explains how to bold text on a word processor using short-cut keys:

> To **bold** the selected text, press 'B' while holding down the 'Alt' key.

If new computer users did not read the complete sentence, but instead simply followed the instructions in the order given, they would replace the selected text with a line of B's:

> bbbbbbbbbbbbbbbbbbbbbbbbbbbbbbbbbbbbbbbbbbbbbbbbbbbbbbbbbbbbbbbbb

They would be less likely to make this mistake if the sentence were rewritten as follows:

> To **bold** the selected text, first hold down the 'Alt' key and then press 'B'.

3.4.1.3   Connect the New with the Known

We learn most readily by connecting new concepts with things we already know. Employing certain traditional forms provides the reader with familiar contexts for learning something new. Using conventional forms for memos, proposals, specifications, manuals, and reports therefore helps readers assimilate new information. The more complex the new information you must communicate, the more important conventional forms become.

As well as using standard forms, you also need to bear in mind the accepted, conventional ways of organizing documents. The following, for example, outlines the conventional way of organizing a report:

1. Prefatory Pages
   a. Title Page
   b. Abstract or Executive Summary
   c. Table of Contents
   d. List of Figures
   e. List of Tables
   f. (Glossary)

2. Body of Document
   a. Introduction
   b. Background
   c. Discussion
   d. Recommendations
   e. Conclusion

3. Appended Pages
   a. Glossary
   b. Technical Materials
   c. References

Note that in the above list, the glossary (list of defined terms) appears twice to reflect a changing convention for where to place the glossary in a report. In the past, the glossary was included as an appendix. Currently, the glossary frequently appears in the prefatory pages.

This change reflects growing attention to readers. Not all readers check the table of contents to see whether a glossary is provided, but most will flip through the prefatory section and find the glossary if one is located there. Knowing immediately that a glossary

> In this manual, all terms that appear in buttons are defined in the glossary. To access the glossary, all you need to do is click on the button.
> You can save a lot of time if you set up a `Template` for a document format you use often.

Figure 3.7 Buttons in On-Line Documentation

is provided enables readers to remove or copy it so they can easily refer to it while reading the report.

While you should generally use conventional forms and organize them in conventional ways simply in the interests of assisting the reader, remain aware that conventional forms restrict the kinds of information that can be communicated. Radically new ideas and new media require new forms. For example, using buttons or other hypertext links embedded in the text to access information is a relatively recent invention. Figure 3.7 provides an early example:

Although now commonplace, this way of organizing information was radical when first introduced, at least so far as it considers *how* users actually read manuals: users generally do not read the entire manual, but instead prefer to get started and then seek help as they need it.

In terms of style, you will be most successful in communicating new information to the reader if you ensure that new or important information is placed at the end of most sentences while old or less important information is placed at the beginning. That is, you should write *"Although in the past we have approached the problem in that fashion, we now deal with the problem in this way"* rather than *"We now deal with the problem in this way even though in the past we have approached the problem in that fashion."*

Although placing important or new information at the end of sentences may seem counterintuitive, the ends of sentences actually carry more emphasis than the beginnings. Readers, consequently, comprehend more easily when the important information is at the end. Unfortunately, in their excitement to communicate their ideas, many writers characteristically place the important information up front. (We discuss this issue in greater detail in Chapter 6, pages 194–196.)

3.4.1.4   Provide General Frameworks

Studies in reading comprehension indicate that information is most effectively processed and retained when readers have a general overview prior to reading the more detailed material. Technical documents apply this principle in numerous ways.

For example, the abstract clearly and concisely summarizes the information in a document. By providing a general overview, the abstract increases the reader's comprehension of the full report. Tables of contents and chapter summaries fulfill the same function, which, by the way, is a good reason for reading the table of contents and the chapter summaries of a textbook prior to reading the chapters. The same is true of the introduction to a report: it helps frame the information contained within the rest of the report. In a sense, abstracts and introductions function like road maps: by knowing ahead of time where we are going, we can get there with greater ease.

To illustrate the importance of general frameworks, we offer the example of giving someone directions about how to find your house in a large city. If you were to give them the following specific instructions, they would probably get lost:

> *First, go 8 miles east. Turn right at the light. Then go 1.6 miles south and turn left at the light. Then go 7.8 miles east and turn left. Go one block, turn right. Go two more blocks and turn right at the first set of lights. Then turn left at the next four-way stop sign. Go one block, turn left. I live in the second apartment block from the corner.*

Your instructions would be more efficient and easier to follow if you provided a general framework first and omitted unnecessary details:

Do you know where the Washington General Hospital is? You do? Good. Do you know where the McDonald's restaurant is across from the main entrance? Great! I live in the apartment block behind it.

The general framework provided—Washington General Hospital—makes the specific information that follows—McDonald's—more comprehensible.

Of course, if the person does not know where the Washington General Hospital is, then you must provide an even more general framework. The generality of the framework required is directly related to the reader's depth of knowledge about the topic. An audience of nonspecialists requires frameworks that are more general than an audience comprised of specialists.

3.4.1.5 Repeat Important Concepts

The British Prime Minister, Sir Winston Churchill, uttered the following memorable words during a speech in the House of Commons on June 4, 1940:

We shall not flag or fail. We shall fight in France, we shall fight on the seas and oceans, we shall fight with growing confidence and growing strength in the air, we shall defend our island, whatever the cost may be, we shall fight on the beaches, we shall fight on the landing grounds, we shall fight in the fields and in the streets, we shall fight in the hills; we shall never surrender.

The parallel structure demonstrated in the above example is a particularly powerful way to communicate information because the rhythm created by repeating *we shall fight* helps the reader remember the ideas in the text. Countless people who heard or read Churchill's words during World War II remember them to this day. Indeed, many other speakers make extensive use of parallel structures because of their dramatic impact on memory. Although you are unlikely to make much use of this kind of parallelism in technical documents, you may find it effective if you are ever asked to write promotional literature for a product or to give a motivational speech.

We have provided this excerpt from Churchill's speech to reinforce the value of repetition. Too many writers seem afraid to repeat themselves and thus make their readers work unnecessarily hard. As Churchill demonstrates, repetition enhances memory.

The power of repetition to enhance memory is also demonstrated by effective study methods. Those individuals who review key concepts throughout a course as well as just before exams not only perform well on exams, but also remember what they learned for extended periods. Repeating key points within a document (or an oral presentation) likewise increases the amount of information readers (or listeners) remember. The conclusion of a report serves a similar function: repeating the most important information presented in the report helps readers remember that material after they have finished the document and set it aside.

A well-written report will generally repeat the most important points in at least three places: in the introduction, in the body of the report, and in the conclusion. The importance of repetition is also highlighted in the advice for oral presentations that suggests you should tell them what you are going to say, say it, and then remind them what you said. However, this advice does not mean that you should simply parrot the same information in precisely the same words each time; doing so would bore your audience. Memory retention is increased if you reiterate the same information in slightly different ways.

Moreover, because we comprehend new information best when we connect it with things we already know, repeating information in different ways has the potential to further increase readers' comprehension. Up to a point, the more different ways you communicate the same information, the more likely you are to provide opportunities for readers to connect new information to something already known.

Transitions, or linking statements, at the beginning of paragraphs, sections, and chapters can also remind readers of a previous point, repeating important information while setting a context for what is to come next. Transitions are particularly important following a detailed discussion of a specific point so that the reader is reminded of the appropriate context for the next point.

3.4.1.6   Employ Graphics

Like repetition, the judicious use of tables and figures can help readers more easily comprehend and remember information because they integrate detailed or extensive information into an easily grasped whole. For the significant numbers of technical people who are visually oriented, graphics are particularly helpful. Graphics are especially important for all readers when you are writing about extremely complex or highly technical subjects. As the saying goes, a picture is worth a thousand words. Actually, a picture can be worth far more than that. An accurately labeled figure can mean the difference between total incomprehension on the part of the reader and absolute clarity. The majority of people remember best those things that are shown to them rather than simply explained to them.

Some of the best examples of how graphics can be used effectively are found in well-written user's manuals. You might, for example, look at some of the user's manuals provided by Hewlett-Packard for setting up and maintaining laser printers. Most people find they can complete routine tasks, such as changing toner cartridges, simply by following the diagrams. The text that accompanies the diagrams seems almost unnecessary.

When you use figures in your reports, ensure they are accurately labeled and are good-quality reproductions. A figure that is difficult to understand or hard to read does not help readers; it simply wastes their time. Similarly, you must place your figures as close as possible to the place in the text where you mention them, either on the same page or on the page immediately following the reference. And avoid placing a figure before the textual reference. Few things are more frustrating for a reader than either being required to search for a figure or encountering a figure not explained in the text.

In terms of style, the prudent use of similes (i.e., something *is like* something else) has a similar effect to that of graphics because similes allow readers to visualize new, often abstract, information in terms of already-known, more concrete, information. For example, if you were explaining how to delete a file on a DOS-based system to someone who is only familiar with Macintosh computers, you might use a simile as follows:

Typing "DEL filename.ext" in DOS *is like* dragging that file to the garbage can.

On the other hand, you might simply use Figure 3.8 to communicate the same information.

= DEL (Delete)

Figure 3.8 Using Figures to Explain Concepts

3.4.1.7 Keep It Simple

Although it seems an obvious point, we should nevertheless repeat that readers most easily comprehend what you are writing about if you keep it simple. Of course, you must be careful not to oversimplify or some of your audience may feel you are talking down to them or may dismiss your ideas. In this regard, Einstein had some advice for scientists that also applies to engineers and technical writers: keep your ideas as simple as they can be, but no simpler. As with so many aspects of writing, finding the appropriate level of detail and complexity requires careful attention to audience.

When writing for technical specialists rather than nonspecialists, you can increase the level of detail and the complexity. However, at many points in your career, you can expect to face situations in which you must write reports to mixed audiences comprised of both specialists and nonspecialists. Such mixed audiences can pose considerable challenges.

If you include in the body of your report all the technical information needed to meet the needs of specialists, you will probably confuse the nonspecialists who read your report. On the other hand, if you omit all the technical details in order to meet the needs of nonspecialists, then you will probably fail to provide sufficient information for specialists. One way to resolve this problem, and meet the needs of both audiences, is to place the technical specifications and analyses in appendices.

Nevertheless, in some circumstances the entire report may be so technical that it becomes impractical to use appendices (i.e., the entire report would end up as an appendix). Or it may be necessary to communicate some of the technical information to nonspecialists. In these cases, you would be wise to rely heavily on graphics and concrete examples. Also ensure that you provide sufficient general information so that nonspecialists can place the technical details in the appropriate context.

When writing to nonspecialists, you may also want to use slightly shorter sentences than usual, and you should restrict your use of specialized language and clearly explain any unavoidable technical terms by including a glossary that defines them in nontechnical language. Although the report may still be difficult reading for nonspecialists, the glossary will help readers work their way through the document.

3.4.1.8 Eliminate Cognitive Interference

Another way to assist readers' understanding is by eliminating *cognitive interference*. By *cognitive interference*, we mean the way in which certain elements of poor format or style can interfere with the reader's ability (or willingness) to focus on what is written in a report. We know that readers are best able to comprehend and retain information when they can focus solely on the ideas presented and are not distracted by problems with spelling, grammar, style, or format. Note, for example, what happens as you try to read the text in Figure 3.9. If, like most readers, you struggled to read the small font, then you have experienced cognitive interference. Readers are both less able and less willing to spend time comprehending information when they are distracted by poor format or poor-quality

This paragraph is set in 4 point text and is therefore very difficult to read. You have probably all had the experience of reading a poor quality photocopy where you were required to focus so intently upon deciphering the individual letters and words that it became very difficult to put together the meaning of the text. Because we read most efficiently by recognizing the patterns of the words, when you interfere with this by using a very small font, by using a sans serif font, or by producing poor quality documents in other ways (i.e., dot-matrix printing, poor quality copying), you slow down your readers. Readers who are impeded in this way become frustrated and experience cognitive interference. In extreme cases, they give up.

Figure 3.9 Example of Cognitive Interference

printing or copying, by irrelevant or excessively detailed material, or by grammatical or stylistic problems. Whatever the cause, a high level of cognitive interference will lead to a high level of frustration, which in turn, will lead to a *low* level of comprehension.

Readers should be able to take for granted that you will provide proper margins, headings, pagination, print quality, figure labeling, and so on. Readers should also be able to assume that you will have spell-checked your document and will have carefully edited it for typographical and grammatical problems. To further aid the reader, maintain a consistent format throughout a document. Once readers become accustomed to certain conventions of format (as well as style and spelling), they will find it disconcerting if you switch to others. You minimize cognitive interference by being consistent.

Note that issues of cognitive interference in particular and of information processing in general are also addressed throughout the chapters on style and format (Chapters 6 and 7). Our approach to style and format emphasizes how people read and what you can do to make a reader's task as painless and productive as possible. Finally, note that the rhetorical considerations presented in this chapter are important to all aspects of writing—from planning and organizing to word choice and style.

3.5   AUDIENCE ANALYSIS

The following heuristic (a method or set of questions for discovering needed information) is aimed at helping you understand your audience. Note that if you ask the right questions, you often discover that you know more about a topic than you expected. Some questions may seem irrelevant for a particular situation; for instance, some questions relate to a persuasive emphasis and you may be writing with an informative emphasis. However, given the importance of audience, answer as many as possible. You may discover they are more relevant than you expected. You should also add questions to this list as they come to mind.

Who Makes Up My Audience?

- What is the economic, social, and educational status of the audience?
- What are their political beliefs?
- What value do they place on education, on environment, on religion, and on work?
- How old is the audience? What are the common values or beliefs of this age group?

How Does the Audience Relate to My Subject?

- What issues are most likely to make the audience angry or defensive?
- How can I approach issues that make them angry or defensive without antagonizing them? That is, how can I present unpopular ideas?
- What authorities can I cite to be most convincing?
- Would stories and analogies confuse or help the audience?
- What are the most convincing appeals I can make? Should I try to persuade them by being reasonable and logical? How can I appeal to their values? What concerns and fears must I address?

What Is the Audience's Level of Expertise?

- Should I include a lot of technical information to persuade the audience?
- Do they expect certain content and conventions in what I write?
- What terms must I define and what terms are already understood?

How Does the Audience Relate to Me?

- What are the power relations between the audience and myself?
- Given this relationship, how do I present myself to this audience?
- Do I present myself as an expert?
- How do I demonstrate that I am honest, trustworthy, and sympathetic?

How Do I Relate to the Audience?

- Do I genuinely respect the audience's opinions and points of view?
- Am I simply willing to tell members of the audience what they want to hear, or am I committed to telling them what I believe to be true (in the most persuasive means possible, of course)?

3.6   RHETORICAL ISSUES CHECKLIST

We provide the following checklist to use as you write various documents. We suggest that you use it while you write, but also reconsider the rhetorical issues it raises *after* you have written a document. By noting those points that you have not adequately considered (i.e., those that you cannot check off), you can determine which elements of the rhetorical situation need work and what issues need attention in order to effectively persuade or inform your audience.

3.6.1   Goal and Audience

Have you clarified your goal by writing it out?   _____

When you lose track of your train of thought or begin revising, do you reread your goal?   _____

Have you analyzed your audience using the audience analysis heuristic that precedes this checklist?   _____

Have you determined whether your readers are representatives of an institution or organization and identified the values of that group?   _____

Have you considered your readers' level of expertise and status?   _____

3.6.2   Persuasive Emphasis

Have you employed the three persuasive appeals?   _____

Have you appealed to your reader's reason and to logic?   _____

Have you appealed to your readers' values and addressed their concerns and fears?   _____

Have you appealed to your readers by presenting yourself appropriately?   _____

Did you use these three kinds of appeal to generate ideas and organize your persuasive writing? (If not, did you follow the adversarial, collegial, or constructive criticism model?)   _____

Have you made a conscious effort to identify with your readers and focused on their needs as you write?   _____

Have you assessed how readers will likely react to the connotations of the words you have chosen?   _____

Have you used inclusive language? (Have you considered using plural nouns to avoid potential problems?)   _____

3.6.3   Informative Emphasis

Have you considered the needs of the reader in determining the format and organization of your writing?   _____

Have you provided general frameworks (by using abstracts, executive summaries, and introductions) to provide sufficient context to aid a reader's understanding of the specifics of the document?   _____

Have you made the document as simple as possible for the reader by using graphics, concrete examples, appendices, and a glossary?   _____

Have you appropriately ordered your document (by using a common organizational pattern, by organizing lists, and by logically ordering instructions)?   _____

Have you connected new information with known information by using conventional forms and conventional ways of organizing documents?   _____

Have you considered the needs of the reader in determining the style of your writing?   _____

Have you minimized the memory load for the reader by breaking up noun phrases and lengthy lists into smaller chunks of information?   _____

Have you ensured that new or important information is generally placed at the ends of sentences?   _____

Have you used repetition appropriately (by using parallel structures, by repeating key words and concepts, and by repeating the most important points in the introduction and conclusion of the document)?   _____

Have you used sufficient, well-constructed graphics and similes to explain complex concepts?   _____

Have you eliminated cognitive interference by accounting for how issues of format, style, and grammar will affect the reader's willingness and ability to understand the information you provide?   _____

3.7   EXERCISES

Ideally you should apply the concepts, principles, and advice given in this chapter to writing already required for school or for work. We provide the Rhetorical Issues Checklist for that purpose. If such opportunities are unavailable, then practice exercises are useful in developing your understanding of rhetorical strategies.

1. Explore Rhetorical Emphasis

Write three e-mail messages, each addressing the same problem or situation but with an expressive, persuasive, and informative emphasis. Note the differences in what you write when you address an issue from the perspective of expressing your personal feelings, persuading someone to do something, or providing information. (If you have difficulty identifying useful situations or problems to address for this assignment, brainstorm ideas with a few friends or classmates.)

2. Clarify Your Goal

Write out the goal for a piece of writing assigned for one of your courses; alternatively write goals for each of the three e-mail messages in exercise 1. Keep in mind that you should refer to your goal throughout the composing process and that you may occasionally need to revise the goal to incorporate new ideas or inspirations.

3. Analyze Your Audience

Identify an audience—a person or a group of people—you might want to write to or the actual reader(s) for something you must write. Complete the questions in the Audience Analysis found on pages 105–106 of this chapter, adding questions of your own if any come to mind. Note any questions you cannot answer and consider how you might acquire the missing information. Then write a page of commentary on how you can apply what you have learned about your audience, specifying how your insights apply to your emphasis (persuasive or informative) and purpose.

4. Employ the Three Persuasive Appeals

Choose an audience you know well or one you have analyzed, and identify a point of disagreement or something you would like them to do that they would probably be reluctant to accommodate. Write a paragraph for each of the three persuasive aims, explaining how you would use them to persuade your audience. Also indicate whether a face-to-face conversation or a written document would be most appropriate in this particular situation. (If you have trouble identifying an appropriate audience and point of disagreement, brainstorm ideas with a few friends or classmates.)

5. Address Power Relationships

Identify a situation in which you would want to request that someone behave differently or do you a favor. Write your request in an e-mail message to three slightly different readers: to someone who has less power or status than you do, to an equal, and to a manager or supervisor. (If you have trouble choosing a topic that would apply in all three situations, brainstorm ideas with a few friends or classmates.) Then write a page listing the adjustments you made in your writing to accommodate these different relationships.

6. Use Models of Persuasion

As in exercise 5, identify a situation in which you want to request that someone behave differently or do you a favor. For this exercise, write three different messages to one reader, using the three models provided in pages 85–89 of this chapter. Ask a few friends or classmates to read your messages and to tell you how they think your reader would respond. Then write a page describing the strengths and weaknesses of each model for your particular rhetorical situation.

7. Identify with Readers

Identify readers from a different field or culture that you might need to address as a practicing engineer and locate someone who represents that group who would be willing to talk with you. Complete as much as you can of the Audience Analysis in section 3.5 for that group of readers, and then make a list of questions you want to ask the representative of that group. Interview your contact, and then write a page or two describing what you learned from this exercise and how you might apply what you have learned when addressing people from this particular field or culture.

8. Evaluate the Effectiveness of Persuasive Documents

Identify an e-mail, flyer, solicitation, or piece of junk mail that is trying to sell you something, asking you to donate to a cause, or otherwise seeking to persuade you. Write a page or two in which you critique the author's rhetorical effectiveness in terms of persuasive appeals, identification with the reader, use of rhetorical models, use of connotations, inclusive language, and so on.

9. Evaluate the Effectiveness of Informative Documents

Locate a piece of writing that presents complex information but is relatively easy to comprehend and remember. Analyze the ways in which the writer employs the eight principles listed on page 97. Then locate a piece of writing that is difficult to read and remember. Note the ways in which the writer violates the eight principles, and write an e-mail offering advice to the author on how to revise this piece of writing. Finally, analyze a piece of your own informative writing, noting ways in which you effectively employ and/or violate the eight principles. Revise your document to improve its readability.

Chapter 4

Strategies for Teamwork and Workplace Communication

4.1 SUMMARY

The business of working in teams is increasingly challenging as engineering projects become more complex, multidisciplinary, and international and as teams become cross-functional and even virtual. Today's engineers need highly developed interpersonal skills, including the ability to accommodate diversity, to adjust to linguistic and cultural differences, and to build consensus. Engineers must also be prepared for team writing and be willing to read and write with increasing frequency.

While team-based skills are increasingly important, routine forms of workplace communication cannot be ignored. The basis of these communication skills lies in an ability to make a good first impression on people, to establish and maintain good working relationships, and to chair effective meetings.

1. **Listening Skills**

   Listening is fundamental to your ability to work and write as part of a team. The ability to listen actively reduces misunderstandings, improves comprehension, and helps you demonstrate your interest in what a speaker is saying.

2. **Team Dynamics**

   We provide advice for building a successful team in which all members respect one another, are committed to shared goals, interact frequently, share decision-making power, divide the work equitably, share responsibility, encourage openness, and resolve conflicts appropriately.

3. **Cross-Functional Teams**

   Working in a multidisciplinary, international environment requires accepting diversity and appreciating the central role it plays in making cross-functional teams productive and capable of performing tasks that would otherwise be impossible. Reaching consensus is another challenge for cross-functional teams and another factor influencing how well a team achieves its goals.

4. **Virtual Teams**

   Engineering project teams are increasingly likely to perform their work in cyberspace. Virtual collaboration and virtual team building intensify the need for engineers to develop strong communication skills. In this environment, they must also communicate across a wide range of media and read and write more than in any other project team environment.

5. **Writing as Part of a Team**

   In the section on team writing, we discuss team-writing processes, team planning, and team revision. We also point to the value of peer reviews as a means of improving your own writing while helping colleagues improve theirs.

6. **Routine Workplace Communication**

   Our discussion of routine workplace communications includes face-to-face conversations and telephone communication as well as developing and maintaining effective office relationships, and arranging for and running *ad hoc* meetings.

7. **Chairing Meetings**

   Well-run, well-organized meetings are essential to the day-to-day operations of any company or to the success of any team endeavor. We outline five steps for chairing an effective meeting and provide basic advice on delivery techniques of meeting leaders.

This chapter also includes a team-writing heuristic that lists questions to help your team assign roles and tasks, produce a practical schedule, and overcome potential communication problems. We also include checklists for listening skills, team dynamics, team writing, routine workplace communications, and chairing meetings. The exercises at the end of the chapter encourage you to participate on team projects, to experience writing as part of a team, and to practice the skills necessary to the daily routine of office life.

4.2 WHY COLLABORATE?

The easy answer to the question posed in the heading is that if you want to be an engineer, you must collaborate. Engineers work in teams, and these teams are increasingly complex, comprised of people from many professional, occupational, linguistic, and cultural backgrounds who may be scattered across the country or around the world. Engineering educators are responding to the increasing reliance on teams in the workplace by increasing emphasis on group assignments and team projects, including those that involve students with a variety of specialties and disciplines working together.

If you have not already participated in a team project, you can expect to soon. As a student, you should have frequent opportunities to develop team skills through project courses and other collaborative activities. As an engineer, you may undertake the majority of your work as part of a multidisciplinary team and may even do the majority of your collaborating in cyberspace.

Beside the fact that you must collaborate, we can offer a number of other reasons to engage in team projects:

- Improved interpersonal skills
- Greater confidence
- Increased leadership potential
- More appreciation of the value of diversity and individual differences
- Enhanced creativity
- Stronger working relationships and lasting friendships

As we discuss in the following section on team dynamics, working on a team encourages you to improve your interpersonal skills. This improvement increases both your confidence and your leadership ability. Engaging in collaborative activity also helps you develop an appreciation for different ways of thinking and of approaching problems. Encouraging the full participation of team members with different perspectives greatly increases the potential for creative and innovative solutions to all sorts of problems. For example, when all ideas, however crazy, eccentric, or naive they may first appear, are carefully considered, a team's creative potential is greatly increased and team members open themselves to new ways of thinking that enhance their own creativity.

Collaboration also slows things down, encouraging new kinds of understanding. For example, the give-and-take necessary to achieve consensus allows many opportunities to compare your approaches and processes with those of others, to question your own assumptions, and to become more aware of the potential range of ways to solve problems and reach decisions. By recognizing these differences, you enhance your critical thinking and problem-solving abilities while increasing the potential range and flexibility of your approaches.

Collaborating on a project can also lead to lasting friendships and stronger relationships with colleagues. Working closely with people while striving for consensus is an excellent way to get to know people and their abilities. In many fields of engineering, you can expect to change companies several times during your career, and job opportunities are as likely to depend on knowing someone in a company as on an impressive résumé. In the high-technology fields that we are most familiar with, engineers tend to change jobs relatively frequently. A significant number gain employment or move into better positions in the companies where their friends work. We suggest you keep this advantage in mind when involved in team projects. Being a good team player can have many benefits.

If you have already worked on a project team, you may have enjoyed the experience and have participated in achieving a feat far beyond what you could have accomplished alone. On the other hand, you may have found the experience frustrating and spent more time dealing with personalities than working on achieving your goal. The communication and interpersonal skills of one or more members likely shaped the result.

We have seen teams with great potential fall apart and projects falter because team members failed to listen to and respect one another. We have also seen teams deal very well with difficult individuals, making the experience positive for all members and the project a success. You can help ensure the success of your teams if you develop your ability to listen carefully, to encourage others to participate, to accept criticism, to take responsibility, to resolve disputes, and, when appropriate, to lead. The following sections provide advice and strategies for developing and applying these abilities.

4.3 LISTENING SKILLS

One of the most important skills you must master to work capably as part of a team is the ability to listen effectively. Effective listening skills underpin the ability of any team to resolve disagreements and to meet their goals. When team members fail to listen to each other, goals may be poorly defined and minor disagreements can grow into major disputes. What is sometimes described as a *breakdown in communication* is often simply a failure to listen.

Unfortunately, the traditional educational system devotes far more attention to teaching us to read, write, and speak than it does to teaching us how to listen. Perhaps part of the reason for this neglect is that listening is often considered a passive activity. In fact, listening requires you to take an *active* role in conversations by becoming fully engaged in what is being stated and by attempting to understand what is left unsaid. Make no mistake, listening carefully is hard work. The following section outlines the skills involved in active listening.

4.3.1   Listening Actively

In order to become an effective member of any team, whether engaged in writing or not, you must learn to listen actively. That is, you must interact with the speaker by asking questions and providing verbal as well as nonverbal feedback. You must appear vitally interested in what the speaker is saying, and you must clearly communicate that interest to the speaker.

Learning to listen actively has several advantages. First, it helps ensure that you fully understand what is being said. If English is not a first language for you or the speaker, active listening will reduce misunderstandings and improve comprehension. Second, by actively engaging speakers with questions, you increase the likelihood that they will provide more complete information than might otherwise be the case. Third, your interest in what speakers have to say often encourages them to listen more closely to your responses. You can communicate your interest by using attending responses, open-ended questions, tracking responses, and summarizing responses.

4.3.1.1   Attending Responses

Attending responses are verbal and nonverbal indicators that communicate your interest in what speakers are saying. These types of responses indicate to speakers that you are paying careful attention to them. For example, your tone of voice can quickly indicate whether you are interested. We suggest that you use a warm, informal, and friendly tone. Because we all like to be called by name, you should generally address speakers by their first names. You can also make occasional use of expressions such as *uh-huh, yes,* or *I see* to indicate that you are following what they are saying. But be careful not to overdo the *uh-huh*'s. Attentive silence is more effective than too many grunts.

Unless you know speakers well enough to know their level of familiarity with specialized language, avoid terminology that could be perceived as jargon or *techno-babble*. This point is particularly important when nonspecialists or potential users are explaining their needs for a device or system that you might develop for them. Also keep your statements and questions short and easy to understand and use conversational English. For example, say *talk* instead of *communicate* and *write* instead of *correspond*. Above all, recognize that if you are speaking more than half of the time, you probably are not listening effectively.

In addition to verbal attending responses, you can also use the following nonverbal techniques to indicate that you are paying attention:

- Head nods and a slightly tilted head
- Suitable facial expressions and natural smiles (without excessive smiling)
- An open posture (rather than arms crossed) and open palms (rather than clenched fists or fidgeting)
- Regular eye contact (without staring)
- Suitable gestures for the context of the conversation
- Appropriate distance (usually an arm's-length away from the speaker)

Proper use of these techniques enables speakers to feel comfortable while providing you with information and also helps them recognize when you are not following what they are saying.

4.3.1.2   Open-Ended Questions

An *open-ended question* requires more than a *yes* or a *no* answer or more than a brief statement of fact. These sorts of questions generally start with a stated or implied *what* or *how* to encourage speakers to expand on their subjects. For example, "*What solutions have you thought of?*" is open ended while "*Have you thought of this solution?*" is not. Because the open-ended question allows the speaker to expand on the subject, you gain more information than with a question that may be answered with *yes* or *no*. Because open-ended questions elicit more information, they are especially helpful when you are working with clients or with the potential users of a system or device. Examples of open-ended questions follow:

- Could you tell me what that means?
- What do you imagine . . . ?
- What have you thought of?
- What would it be like . . . ?
- How do you see things changing?
- What would you like to do about . . . ?
- I'm wondering . . . ?
- What's that like?
- What can you think of?

Despite their usefulness, you should be somewhat cautious when asking questions. Asking too many may lead speakers to suspect that rather than seeking information, you are interrogating or challenging them. Also avoid leading questions that require the speaker to agree with you (i.e., *You don't really want to do that, do you?*). Note, for example, the difference between the following sentences:

Why didn't you try to solve the problem that way?

What happened when you tried that solution?

Be particularly careful not to imply to speakers that they should or should not have done something by phrasing your question like the first example. A *why-didn't-you* question is much more likely to elicit a defensive reaction than an alternative such as the one offered in the second example.

4.3.1.3   Tracking Responses

Tracking responses, which help the speaker stay focused on a specific topic, take three forms: *reflecting, clarifying*, and *silence*. *Reflecting* checks your understanding of what the speaker is saying. For this response, you rephrase as a question the feelings, content, or words of the speaker's message. Reflecting statements are often expressed as questions taking the following forms: *You say . . . ? You mean . . . ? You feel . . . ?* For example, if someone on your team came to you to complain that another team member was not doing a fair share of the work, you might respond in one of the three following ways:

1. You say John is lazy?
2. You mean John hasn't completed his part of the design specs?
3. You feel angry because John isn't doing his fair share of the work?

Clarifying gathers further information about something a speaker has said that is unclear or confusing. For example, you might start a sentence with *"Correct me if I'm wrong, but . . ."* or ask a question beginning *"Do you mean. . . ."* These sorts of questions indicate to speakers that you want to understand what they are saying and encourage them to elaborate on a point.

Silence can also function as a tracking response. In most everyday conversations, a one-second pause between the end of one person speaking and the beginning of another person speaking is typical. Slightly increasing the length of pauses conveys your interest in hearing everything the speaker has to say. If you lengthen your pauses more than a little, the speaker will likely find the silence uncomfortable and be encouraged to continue speaking. Note, however, that you must also exercise caution when using pauses because a speaker may interpret excessive silence as a lack of interest or a lack of attention and then bring the conversation to an end.

4.3.1.4 Summarizing Responses

Summarizing responses review the information that has been covered in the conversation and provide an opportunity to move on to another topic or to add to what has already been discussed. Summarizing responses function as an important reality check, allowing speakers to correct any misunderstandings you may have about what they are saying. To employ this technique, briefly summarize the conversation and then ask a question such as *"Is that accurate?"* After asking the question, allow the speaker ample time to respond.

4.3.2 Other Features of Effective Listening

Beyond learning to listen actively, you should also aim for the following:

- Minimize distractions.
- Listen with respect.
- Avoid assumptions.
- Avoid superficial reactions.
- Situate facts in context.
- Remain focused.

4.3.2.1 Minimizing Distractions

Part of being a good listener involves creating an appropriate environment for listening. Sometimes distractions are unavoidable, but you should minimize those over which you have control. If you are having an important meeting, you should select a quiet location and shut the door. Similarly, if you are meeting in your office, you should put your telephone on *call forward* and place a *do-not-disturb* sign on your door. Such actions help create an optimum environment for listening.

You should also avoid fiddling with papers or books when someone is speaking (although you may want to take a few notes). If you attempt to divide your attention between the task of listening and some other task, you will do neither effectively. As much as possible, devote your undivided attention to the speaker. If other tasks are more immediate, then perhaps you should tell the speaker you are busy and set up a time to talk later.

4.3.2.2 Listening with Respect

You should communicate to speakers that you respect them, particularly if they are angry or otherwise upset. Even if you disagree with them or have reservations, you should hear them out and acknowledge the strength of their ideas. You may well discover that, once fully explained, their ideas make more sense than you first thought.

Initially, you should avoid pointing out any contradictions or inconsistencies in a speaker's statements. Later in the conversation, you may express doubts, but note that expressing them in the form of questions is usually most effective. A question such as *"How do you think that will work?"* is preferable to a statement such as *"That won't work!"* which cuts off the conversation and prevents compromises from being explored.

Listening with respect also involves respecting speakers' feelings, even when you do not understand them. Focus on feelings by paying attention to the emotions or passion associated with a speaker's content. The feelings that a speaker attaches to a subject communicate essential information about the content—about its importance and accuracy.

For example, if speakers sound excited about their subjects, then the subjects are usually important to them, and you should respond accordingly. Let them know when you share their enthusiasm. Similarly, if speakers sound frustrated or confused, you can ask questions that help them clarify their ideas. Note that feelings are generally expressed on a nonverbal level that you can identify by listening to their voices and by observing their body language, gestures, and so forth. Also note that unless you are nonjudgmental, speakers may be reluctant to continue a conversation.

4.3.2.3 Avoiding Assumptions

Statements such as *"I know exactly what you mean"* typically have the effect of cutting off the speaker. After all, what is the point of the speaker telling you something if you claim to already know it? Avoid assuming you know exactly what the speaker means, even if you are familiar with the subject. The speaker may have something unanticipated to say, may provide an innovative perspective on a subject, or may communicate details that are new to you. Also beware of the natural tendency to want to show a speaker that you know *something* about the topic under discussion. You may appear rude or even ignorant if you interrupt speakers—especially if you have based your response on mistaken assumptions.

Force yourself to listen to the main ideas, to summarize them in your mind, and to avoid mentally looking for answers as you listen. You should also avoid defining the nature of a problem and proposing solutions too quickly. In the words of H.L. Mencken, "for every problem, there is a solution that is simple, plausible, and wrong." Hasty assumptions are rarely correct.

4.3.2.4 Avoiding Superficial Reactions

You cannot help noting the physical features of someone's style—their clothing, posture, mannerisms, or accents. However, if you are excessively distracted by how someone looks or sounds, you may miss what the person is saying. Be sure you are paying attention to more than a person's appearance.

In addition, you should become aware of your biases and prejudices. We all possess them. When you have an emotional reaction to a speaker or their topic, you may start mentally refuting what you just heard and stop listening. By becoming aware of your biases, you can learn to avoid tuning out when certain issues are raised. By focusing on the issues at hand rather than on your initial reactions to them, you are more likely to discover why a topic matters.

4.3.2.5   Situating Facts in Context

In engineering, you are taught a deep respect for facts, and you are also taught to correlate these facts to determine whether any general conclusions can be drawn from them. Unfortunately, this training may lead you to focus on the facts rather than on the context into which a speaker places them. To place the facts you hear into a proper context, you should listen for several minutes before responding. Gain a general understanding of the subject before you decide what the facts might mean and why they are important.

4.3.2.6   Remaining Focused

Typically, we speak at the rate of about 125 words per minute. Yet we can think at a rate of 400 to 500 words per minute (even faster when we think in images and concepts rather than words). Consequently, you can listen to a speaker and still have time to think about other things, perhaps what you will have for dinner or how you will finish your next assignment on time. Unfortunately, such digressions affect your ability to focus on what is being said.

If you catch your mind wandering, you should immediately return your attention to the speaker. Further, you should also avoid faking attention to a speaker. You can appear attentive by nodding your head and smiling while your thoughts are on other things. In such cases, the speaker may ask you a question that reveals you are not listening. Should that happen, or if your attention has momentarily drifted, you should ask the speaker to repeat the information you missed.

Another way to ensure that you remain focused is by taking notes. The act of writing something down helps you remember it. Summarizing something in writing also requires that you understand what the speaker is saying. But if you are in the habit of *doodling* while taking notes, you should monitor the amount of drawing you do. An excess of *doodles* generally suggests you are not listening very closely to the speaker.

4.3.3   When Not to Listen

Although active listening is a useful skill to employ in most circumstances, you will no doubt discover times when it is more appropriate *not* to listen, but rather to respond in a more direct manner. For example, active listening is inappropriate when speakers are verbally abusive or when they monopolize a conversation in order to prevent you from expressing your ideas or to avoid discussing an issue that they are uncomfortable addressing. In these sorts of circumstances, we recommend being assertive with speakers by pointing out that they are acting inappropriately and stating that you will not continue a conversation unless you are treated with respect. In an extreme case, you may need to walk away after stating that you are only willing to continue the conversation once the speaker has calmed down.

4.4   TEAM DYNAMICS

Effective teams are built through a substantial commitment of time, energy, and goodwill on the part of all members. In addition to team members having a sound foundation in listening skills, they must collectively ensure that they realize the following:

- Respect for one another
- Shared goals that have been clearly articulated
- Frequent interaction
- Shared decision-making power
- Equitably divided tasks
- Shared responsibility for both mistakes and successes
- Free expression of opinions, conflicting perspectives, and constructive criticism

The following subsections offer advice to help ensure your team develops these positive characteristics. Other issues relating to cross-functional and virtual teams are discussed in later sections of this chapter.

4.4.1   Fostering Respect

You have probably heard some variation of the adage *respect is not given; it must be earned*. This sentiment could prove dangerous if applied to team contexts. You cannot form an effective team, especially one that includes people from different professions, without accepting that each member of the team has an equally important contribution to make—whether you can see it initially or not.

Engineering traditions and the rigors of engineering education may lead engineering students to feel that they are better qualified and harder working than students from other disciplines. In the context of a technical project, practicing engineers may also feel that their opinions should carry more weight than those of folks from the communication department or from marketing. Such attitudes undermine collective efforts and can easily derail a project that requires collaboration among several disciplines.

Even when all members of a team are chosen from the same specialty, some members will have more experience or superior skills in one area or another. The apparently stronger members of a team may tend to ignore the suggestions of more junior members or of those with less clearly defined technical skills. Doing so weakens the team in at least two ways: First, teammates who feel their contributions are not respected are less likely to put in the same effort as they would if treated with respect. Second, focusing on individuals' apparent weaknesses deflects attention from their strengths—their ability to find compromises or to relate concepts in novel ways or their artistic abilities, leadership skills, expertise in metal work, and so on. The most successful teams are not necessarily the ones with the greatest talent, but those that function on the basis of respect, encouraging all members to perform to the best of their abilities.

4.4.2   Articulating Goals

When you are forming a team for a school project, you should devote some time to clarifying the general goals of the team *and* the personal goals of each team member. For example, the goals for an engineering project course could include such professional development issues as mastering the design process and producing effective project documentation. They might also include dealing with social or environmental issues such as assisting the disabled or accomplishing a goal without using toxic chemicals. Individual goals for such a course might include developing leadership skills, gaining confidence in public speaking, or learning to program in C++. Given that team projects are an opportunity for students to improve their skills, tasks should be assigned not only on the basis of existing abilities, but also on the basis of enthusiasm and willingness to learn.

The value of articulating both professional and personal goals extends to the workplace. In this context, team members may seek to broaden their management and technical skills while contributing to the team task. For example, some team members may attend workshops that address issues relating to team activities or study a foreign language in order to increase their value to the team and to their company.

As in any other rhetorical situation, the team must focus on the needs of the client, customer, or sponsor when establishing project goals. Once these goals have been articulated, the team should then decide how to achieve them, keeping in mind that its overall strength will be enhanced if members are enabled to develop new skills or to improve proficiencies. For example, if one of the team members wants to develop her confidence with public speaking, then the team might allow her to undertake any presentations that are required.

The team must also prepare to address a series of process goals such as resolving conflicts, improving performance, meeting deadlines, and the like. These process goals often reflect the various stages of team development that Bruce Tuckman (1965) has labeled the stages of *forming, storming, norming, performing*, and *adjourning* (see Table 4.1).[1]

Table 4.1   Stages in Team Development

Stage	Major Processes	Characteristics
1. Forming (Orientation)	• Exchange of information • Increased interdependency • Task exploration • Identification of commonalities	• Tentative interactions • Polite discourse • Ambiguity • Self-discourse
2. Storming (Conflict)	• Disagreement over procedures • Expression of dissatisfaction • Emotional responses • Resistance	• Criticism of ideas • Poor attendance • Hostility • Polarization and coalition forming
3. Norming (Cohesion)	• Growth of cohesiveness and unity • Establishment of roles, standards, and relationships	• Agreement on procedures • Reduction in role ambiguity • Increased "we-feeling"
4. Performing (Performance)	• Goal achievement • High task orientation • Emphasis on performance and production	• Decision making • Problem solving • Mutual cooperation
5. Adjourning (Dissolution)	• Termination of roles • Completion of tasks • Reduction of dependency	• Disintegration and withdrawal • Increasing independence • Regret

[1]Adapted from Barry McNeill and Lynn Bellamy, 1994, *Engineering Core Workbook for Active Learning, Team Training & Assessment*, Arizona State University, Slide 20. They, in turn, developed their material from the work of Bruce W. Tuckman, Developmental Sequence in Small Groups, in *Psychology Bulletin*, 63: 6, 384–399. Copyright © 1965 by the American Psychological Association. Adapted with permission.

In the *forming* stage, team members are often somewhat anxious as they do not yet know the other team members, and their roles may be somewhat ambiguous. The *storming* stage is typified by conflicts over procedural issues and by misunderstandings or misinterpretations of other team members' behavior. In the *norming* stage, a team grows increasingly cohesive, and individual roles as well as team standards become well defined. During the *performing* stage, the team begins to achieve its goals, and the performance of team members is emphasized. Finally, in the *adjourning* phase tasks are completed, and the roles assumed by team members are concluded. Table 4.1 outlines the stages of team development as well as the major processes and characteristics of the five stages.

4.4.3   Interacting Regularly

The most successful teams interact frequently while working on a project. Usually, teams schedule regular face-to-face meetings to establish goals, to plan how to meet those goals, to monitor progress, and to resolve outstanding issues. But such formal meetings are not the only way that team members interact. When a project requires a relatively small team, members often work in close enough proximity to meet informally every day. When working on larger projects with members in a variety of locations, team members maintain regular contact with each other via e-mail, videoconferencing, facsimile transmission, or over the phone. We discuss communicating at a distance in more detail in the section on virtual teams (pages 125–126).

Beyond professional meetings, team members may get together socially, perhaps for lunch, after work, or at a company-sponsored function. Teams with a significant social component generally work together more effectively than teams with strictly professional relationships. However, the frequency with which a team socializes is typically dependent on the size of the team. A large team may meet infrequently, while a very small team may move easily between professional and social activities.

Effective meetings are another very important component of team interaction. Unfortunately, many meetings fail to achieve their goals because they are either poorly organized or poorly run. The following points are some that you should attend to when running meetings.

- Prepare and circulate a detailed agenda in advance of meetings.
- Set a timeline for the meeting.
- Keep the discussion focused and moving, ensuring that no one dominates the meeting and that everyone is heard.
- Evaluate the effectiveness of the meeting.
- Distribute minutes for the meeting, and set the time and agenda for the next meeting.

For more information on running effective meetings, refer to pages 137–143.

4.4.4   Assigning Team Roles

Within teams, people typically fulfill a range of roles, both formal and informal. In general, these roles can be divided into two general categories: *task-oriented roles* and *social roles*. Task-oriented roles include assigned roles such as the *team leader* (who ensures the overall goals of the team are met), the *recorder* (who keeps minutes of meetings, etc.), and the *devil's advocate* (who takes the opposite perspective to the team's usual view). In addition to these sorts of organizational roles, various team members may fulfill specific roles related to their technical expertise (*budget manager, marketing specialist, documentation specialist*, etc.).

Social roles are typically not assigned to specific individuals but are instead assumed by various team members (often as a function of their personality). Although not formally assigned (and generally not officially rewarded), social roles can be essential to the smooth functioning of a team. These roles include the *mediator* (who helps to resolve disputes among team members), the *facilitator* (who encourages other team members to participate), the *comic* (who lightens things up through humor), and the *standard setter* (who expresses the standards and expectations of the team).

The team leader and mediator may occasionally need to deal with problem members such as the *dominator* (who asserts power and manipulates the team), the *aggressor* (who continually disapproves of team decisions), and the *attention seeker* (who constantly draws attention to him- or herself).

Although the positive roles mentioned above help a team function optimally, effective team leadership is often *the* critical factor in determining a team's success. While some small teams function quite effectively without formal leadership, in circumstances where efficiency is a major concern, or where the team is larger than four or five people, a formally designated team leader is a must.

When a leader is not formally assigned, the group must choose a leader carefully, considering who has the traits of an effective manager. Note that effective leadership, particularly in small teams, is generally democratic. Also note that the most strong-willed individual is not necessarily the best choice; indeed, for a team with several strong-willed individuals, the choice of one of them could lead to a power struggle, increased conflict, and in extreme cases, the collapse of the team.

Some of the characteristics that typify effective leaders follow:

Personal Traits of Leaders

- Manage their time effectively
- Recognize the limitations of their authority and expertise
- Delegate tasks effectively and fairly
- Are flexible in their approaches to problem solving
- Make decisions fairly and in a timely manner

Communication Style of Leaders

- Consult frequently with other team members about issues of concern
- Are good listeners (i.e., spend more time listening than talking)
- Encourage the expression of conflicting opinions and alternative viewpoints
- Provide honest and specific feedback about individual and team performance
- Resolve conflicts between team members effectively and fairly

Motivational Ability of Leaders

- Keep the team focused on the tasks at hand, encouraging team members to fulfill their personal and team goals

To a significant degree, effective leadership is also based on experience and expertise. Thus someone with a background in project management may be the overall project manager, someone with expertise in documentation may lead the writing team, and someone with experience in engineering design may lead the design team. This task-orientated approach to assigning leadership roles is a model frequently applied in industry. If student teams encounter difficulty choosing an overall project manager, leadership could be delegated on the basis of task expertise (i.e., assigning leadership roles for a number of functions).

4.4.5 Dividing Tasks Equitably

In order to minimize conflicts and increase efficiency, work must be divided equitably among team members. Overburdening one individual will quickly lead to resentment and conflict. In addition, when assigning tasks to various team members, the team must consider not only the quantity of work to be undertaken by an individual, but also the nature of the work. For example, editing a document is rarely as interesting as drafting it. If someone is consistently assigned the task of editing, they are liable to become bored (and thus ineffective). Where feasible, team members should rotate through both the less challenging and more interesting tasks.

4.4.6 Sharing Responsibility

When things do not work out quite the way we expected them to (i.e., a particular design flops), we often blame external factors. Occasionally, however, the members of a team will choose to blame the problem on a particular team member. Although exploring the origins and consequences of the problem is necessary in order to prevent a recurrence of a problem, you should avoid placing blame. Turning a team member or group within the team into the *scapegoat* is one of the fastest ways to destroy trust and respect and to generate conflict.

Most problems that a team encounters result from a team decision or dynamic and should be accepted as a team responsibility. The goal should not be to lay blame by pointing a finger at anyone; rather the team should determine the nature of the problem, how to correct it, and how to prevent it recurring. Similarly, most successes are the result of teamwork and should not be solely attributed to individuals.

4.4.7 Resolving Conflicts

Conflict is inevitable when working as part of a team and should not necessarily be viewed as negative. One of the keys to whether a team achieves its goals is how well the team deals with conflict. An effective team will demonstrate the following:

- A willingness to acknowledge and accurately define problems
- A focus on problems rather than on personalities
- A respect for the interests of all team members
- A desire to limit conflicts and to find solutions that address everyone's concerns
- A commitment to finding solutions agreeable to everyone

Conflict is positive insofar as it brings a problem to team members' attention so that they start working on a solution. Further, conflict is constructive when it forces a team to explore alternatives that might otherwise be ignored. To reap the positive benefits of conflict, the team must be encouraged to express opinions freely, to present conflicting perspectives, and to provide constructive criticism.

At the same time, excessive conflict impairs a team's ability to achieve its goals. Finding a balance that encourages the expression of conflicting positions while maintain-

---

Six Steps to Dispute Resolution

1. Fully and clearly identify the problem to be resolved.
   a) Describe all sides to the dispute.
   b) Listen carefully to all sides.
   c) Ensure everyone accepts the definition of the problem.
2. Generate alternative solutions to the problems.
3. Evaluate the alternative solutions.
4. Ensure that all members accept decisions.
5. Implement the solution to the problem.
6. Set a target date to evaluate the effectiveness of the solution.

Figure 4.1 Six-Stage Model for Dispute Resolution

ing good working relationships requires effective leadership as well as trust and goodwill on the part of all team members. In order to help you achieve this balance, Figure 4.1 provides a six-stage model for resolving any disputes you encounter while working as part of a team.[2]

4.4.8 Avoiding Groupthink

If you are ever a member of a team that encounters no conflict, you may have cause for concern. The absence of conflict may signal that your team is suffering from *groupthink*,[3] a condition that results when a team becomes too integrated or cohesive. A team suffering from groupthink will place so much value on maintaining loyalty, unity, and agreement that critical thinking and open inquiry are prevented. This problem is particularly common among small teams working intensively toward a well-defined goal. Some of the symptoms of groupthink are outlined in Table 4.2.

One way to prevent groupthink is by encouraging at least one team member to actively play the role of *devil's advocate*. But you can also prevent groupthink by positively valuing and encouraging the expression of each other's individuality.

4.5 CROSS-FUNCTIONAL TEAMS

While groupthink can be a concern in functional teams (i.e., those comprised of members with similar backgrounds and areas of expertise), it is highly unlikely in cross-functional teams (i.e., those comprised of members with disparate backgrounds and diverse areas of expertise). Such cross-functional teams are increasingly common in business and industry and are the norm in some fields of engineering because projects are increasingly complex and beyond the capacity of members of any one specialty to complete. With globalization, these teams are increasingly international as well as interdisciplinary.

Rather than facing a tendency to think too much alike, members of cross-functional teams face problems arising from members not thinking enough alike and not appearing to have enough in common. Turning this potential problem into the team's major asset is the ultimate challenge you may face when working in this multidisciplinary, international team environment. The only way to ensure that diversity is an asset is to embrace it.

---

[2]Adapted from *Leader Effectiveness* by Thomas Gordon (New York: Penguin Putnam, 2001). Used with permission. For further information on leader effectiveness, see www.gordontraining.com.

[3]The information about groupthink has been adapted from Irving Janis's article, Groupthink, in *Psychology Today*, November 1971, 71–89. Used with permission.

Table 4.2   Symptoms of Groupthink

Invulnerability	Group members become convinced that what they are doing has no risks. They may become overoptimistic and take excessive risks.
Rationalization	The members of the group construct rationalizations that discount negative feedback or discrepant information.
Morality	Group members believe in the inherent morality of their actions. The ethical consequences of their decisions may be discounted by the group members.
Stereotypes	The members of the group hold stereotyped views that competing groups are evil, weak, or incompetent.
Pressure	Group members are pressured to conform to the norms of the group. Deviation from the group consensus is not permitted.
Self-censorship	Doubts or concerns about group goals and approaches to problems are minimized, and group members may remain silent about their misgivings.
Unanimity	Group members appear to support the majority position. Minority viewpoints are not given a fair hearing.
Protectiveness	Group members may protect the leader and others from negative information that may adversely affect the confidence of the group in its approach.

4.5.1   Embracing Diversity

No one comes to a cross-functional team with a full understanding of the project, and so team members must work together to develop a collective understanding. Only when all team members have the big picture can the problems and issues raised by complex and international projects be adequately addressed. The success or failure of such projects lies in the degree of cooperation and the level of communication among all members of the team. This cooperation is possible only if team members respect one another and assume that each individual on the team is there because he or she has an important role to play and insights necessary to guide the team.

In this context, the interpersonal skills engineers need most to meet the demands of twenty-first century engineering are an appreciation and respect for diversity, an eagerness to share ideas with people from different backgrounds, and a willingness to seek solutions in conjunction with people who know little, if anything, of the technical side of a project.

Working in cross-functional teams, you can expect many of your basic assumptions to be challenged. People will not respond or react the way you expect, and the things you value most may be low priorities for other members of your team or for your international clients. For example, many years ago, an engineer we know was given the job of setting up a soap plant somewhere in Africa. He began the project eager to do something to improve the standard of living in the area. But he was soon disheartened and frustrated by the local people's negative reaction to the project. The problem was that the soap plant undermined deeply rooted social structures and thus threatened the cultural values of the community. The people's needs would have been better met by a home-based industry. The project failed because the people of the area had not been consulted. People from a developed country had planned the project and had not imagined that Western values of money and jobs would clash with local values.

This particular sort of situation is much less likely today because the planning team would most likely include representatives from the local area. But anyone who works in a cross-cultural environment will inevitably suffer the sort of *culture shock* our friend felt when his contribution to making lives better was rejected. You may also encounter a form of culture shock—the shock of having your most basic assumptions challenged—when working with people from different disciplines or perhaps even from different parts of the country.

Challenges to assumptions remain a major issue for anyone working on a cross-functional team. The greater the diversity, the more frequently assumptions will be challenged. A common reaction to this type of challenge is to oversimplify the situation in an attempt to explain it away. For example, the engineer trying to set up the soap plant might have explained the local people's resistance by labeling them as too backward to do the tasks required of plant workers. If he had done so, he would have dealt with his confusion by creating a stereotype. In responding to a complex social situation, he would have demonstrated his lack of respect for the social and cultural values of the local people. (In actual fact, he learned a great deal from the experience that shaped the rest of his career.)

Examining personal stereotypes is a natural and essential consequence of embracing diversity.

4.5.2   Accommodating Diversity

The best way to counter stereotypes and to develop a respect for diversity is to get to know people from as wide a range of backgrounds as possible. Getting to know people as individuals will not only debunk stereotypes, but will also expose you to new experiences, introduce new ways of thinking, and challenge many assumptions. Taking courses in a wide range of fields is another way to broaden your horizons and increase your appreciation for different ways of thinking and for the talents of people with backgrounds different than your own.

When you work on cross-functional teams, you must accept all sorts of differences. For example, even if the international members of your team are all English speakers, they will not spell the way you do or share all the same grammatical rules. You may say that your team *is* meeting in *two weeks* while a colleague from England would say that the team *are* meeting in *a fortnight*. You may already be familiar with the difference between the American spelling *center* and the British spelling *centre*. But did you know that while Canadians use *centre* like the British, they spell *tire* like Americans? Someone from England would write *tyre centre* while someone from Canada would write *tire centre*. We have also met individuals from Hong Kong who use both *center* and *centre* to differentiate between the middle of something and an institutional center such as a Science and Technology Center. Linguistic courses with a focus on the English language provide a wealth of this kind of information.

The sorts of differences we have just drawn your attention to may seem insignificant. But they can become contentious when a cross-functional team must decide whose English will be the standard for the team. Until a standardized international English is developed to address this problem, you should not be too hasty to assume that variations in grammar and spelling are mistakes. They are increasingly likely to reveal linguistic and cultural differences. If you are in the minority on a team, you should consider adopting—as far as you are able—the conventions of the majority. If the group is mixed, an editor familiar with the client's expectations will likely deal with inconsistencies in any documentation the team produces.

Also, when members of your group are working in their second language, take special care to speak and write clearly and simply, using standard English expressions. Expressions such as *"Yeah, and pigs might fly!"* will be extremely confusing to nonnative speakers of English. If teammates have been educated in another language, then English technical terms may also cause some initial difficulty, as will complex written sentence structures.

Working style is another area of potential misunderstanding and contention for cross-functional and multicultural teams. Not everyone shares the same work ethic, demonstrates their commitment to their work in the same way, or is equally willing to express opinions or to ask for help. The hours people work and the role that socializing plays in doing business may also differ not just culturally but among disciplines.

Given the potential differences from one cross-functional team to the next, each team must develop its own ways of responding to the needs of individual members while meeting the challenges of creating a cohesive team. To play your part, you must embrace diversity and accept that you can accomplish very little alone. A cross-functional team's success lies in its collective perspectives and the equality and interdependence of team members. The contributions of a technical writer or a marketing specialist must be no less valued than the contributions of an engineer.

4.5.3   Reaching Consensus

Another major challenge for those working in cross-functional teams is the need to negotiate solutions that are acceptable to all parties. When working on complex projects, everyone requires some idea of what everyone else is doing so that potential problems can be identified early enough to avoid major setbacks or disruptions. To ensure that problems are identified and dealt with in a timely manner, all members of cross-functional teams must be willing to admit when they have a problem and must be equally willing to accept someone else's suggestion that they are progressing in the wrong direction. No individual or group of specialists has the luxury of working on their own.

Negotiating the give-and-take of this sort of teamwork requires goodwill and respect all around as well as understanding the nature of consensus. Contrary to popular belief, consensus is not the same thing as unanimous agreement. When a team reaches a decision through consensus, the result may not suit everyone. But everyone will have been given an opportunity to make their positions heard and will accept the majority decision.

While everyone will support a consensual decision or compromise, some people will likely have preferred a different outcome. For these people to accept the group's decision requires a strong sense of commitment to the overall project goals. Bad feelings when your solution is not the one favored by the group are much less likely if you can accept the consensus decision as the best way to ensure the project reaches a successful conclusion. A focus on your purpose and on client needs is particularly important to the decision-making process of cross-functional teams. Consensus building, like other aspects of teamwork, requires a willingness to listen to others' perspectives, to respect different points of view, and to compromise as a means of achieving team goals.

4.6   VIRTUAL TEAMS

Engineering work is not only becoming more multidisciplinary and international, it is also increasingly virtual. The members of cross-functional teams are ever-more-frequently spread across the country or around the world. These virtual teams face all the issues confronting teams with members working in close proximity to one another but with the further challenge of collaborating and team building at a distance.

The increasing complexity of the rhetorical situations facing engineers working in cross-functional and virtual teams makes the need to focus attention on identifying and accommodating client needs increasingly acute. The needs of clients and the process of finding the best way to accommodate those needs are the glue that holds a virtual team together. When separated from their teammates by time and space, members must remain focused on clients and overall team goals to avoid the natural tendency to focus only on their part of a project and to share too little information too late in the project.

A willingness to share information and ideas and to admit to potential problems are also essential attributes for anyone working on a virtual team. Information must be shared early and often to ensure that the full complexity of a project is clear to all participants and to ensure that problems are identified as early as possible in the project cycle. From an engineering perspective, a solution may be elegant and cutting edge, but from an economic perspective it may be far too expensive to implement. It may also take much longer to accomplish than the client has time to wait.

Virtual teams rely heavily on team members being team players, trusting one another, and having well-developed communication skills. Written communication skills and the ability to communicate through a wide range of media are particularly important. Members of virtual teams should be prepared to use groupware and videoconferencing and to spend significant amounts of time communicating with teammates and sharing information using the Internet or a company's private intranet system, e-mail, facsimile transmissions, and the telephone. Note that because the majority of communication is likely to be computer mediated, in most virtual work environments, written communication skills are relied on much more heavily than oral skills.

Working at a distance does not change the challenges a team faces so much as intensify them. Understanding different linguistic conventions, recognizing stereotypical reactions, respecting different ways of working are all the more difficult. Getting to know your colleagues as people is obviously more difficult as well, which is why many companies arrange regular events that bring members of virtual teams together for meetings and social events. When working on a virtual team, you should consider these opportunities to interact and socialize as essential for team building.

Another challenge of virtual teams is that they generally receive little direct supervision. Instead, when a virtual team forms, the members are typically allowed to work out their own approaches to dealing with myriad issues, including how to schedule conference calls with members in different time zones, how to share information electronically when members use a range of different platforms, how to negotiate consensus without face-to-face meetings, how to deal with conflict, and how to ensure that everyone is kept informed of progress and changes.

Solving these problems often requires a great deal of self-discipline on the part of all team members. Everyone must be willing to share information regularly, which generally means developing a habit of writing frequent updates and sharing concerns with other members of the team. It also means developing a habit of consistently reading what others have written to keep abreast of changes, to ensure everyone has and maintains the big picture, and to provide timely feedback when one member sees a problem in someone else's assumptions. A virtual team also relies heavily on the ability of members to write persuasively in order to convince people from a wide range of backgrounds to work together for the sake of the team and to meet the needs of their clients.

All of the interpersonal and communication skills you require to be a successful engineer will be put to the test when you participate in a virtual team.

4.7   TEAM WRITING

Next to the rigors of working on a virtual, cross-functional, international team, the notion of writing with a partner or two may seem not much of a challenge. However, while many people enjoy writing as part of a team, others find team writing exasperating. Those who would like to avoid it complain that writing documents collaboratively is time consuming and frustrating. Different people have different ideas concerning how to organize information, what is important, how quickly a draft should be completed, what words to use, what counts as good style, and even how to punctuate.

Those who appreciate the process of team writing generally recognize that it offers substantial benefits. Like all team activities, team writing provides an opportunity to improve interpersonal and leadership skills, increase creativity, and arrive at better solutions than you could on your own. It also allows you to produce better documents. When a writing team functions well, the increased brain power, diversity, and creativity can lead to results much superior to what could be produced by any one member working alone.

Of course, not all rhetorical situations are appropriate for team writing. An individual generally produces short, routine documents more quickly and effectively than a team. Documents well suited to a collaborative process are generally long; however, team documents may be of moderate length when no one person has the necessary expertise to address all issues or when the object of the document is to reach consensus. In fact, writing teams are sometimes formed specifically to build a cohesive unit by creating a situation in which a group must achieve agreement in order to accomplish the task at hand. More often, teams are formed to accomplish a task, and both teambuilding and team writing are secondary—if essential—concerns.

While a growing number of companies employ technical writers to take responsibility for producing documentation, some engineering firms expect engineers to produce much, if not all, of the documentation for team projects. Even when technical writers are employed, engineers may be expected to produce rough drafts. Consequently, you should not expect that someone else will do the writing for team projects. All engineers should be familiar with team-writing processes and gain experience writing as part of a team.

4.7.1   Team-Writing Processes

Approaches to team writing vary widely in response to a broad range of rhetorical situations. At one extreme, writers produce documents by working together to plan, draft, revise, and edit. We call this approach *team-composing*. You are unlikely to engage in this approach on the job very often, because it is generally very time consuming. However, it can prove effective when a piece is short and two (or very few) people with similar ideas are committed to a shared goal—especially when one is a good writer and the other has significant contributions to make in terms of ideas.

Team composing is also useful when differing viewpoints on how to address a problem or issue must be resolved in order to complete a project or a document. By writing collaboratively, two or three individuals with differing viewpoints can work through their objections and reach a compromise that all can accept. More frequently, team composing is a useful means of clarifying or refining an issue, approach, or position, such as preparing a statement of purpose for the team. But keep in mind that effective collaboration requires good listening skills and attention to team dynamics so that all members are encouraged to participate—by pointing to possible problems, raising objections, offering suggestions, negotiating wording, and otherwise contributing to a team effort.

At the other extreme from team composing, one writer drafts and others revise, edit, and perhaps even format the document. You may encounter this *single-file* approach on the job, but we discourage you from using it when you can avoid doing so. As the name suggests, this approach provides little, if any, opportunity for team interaction and so negates all the benefits of collaboration. Another drawback is that the person who drafts the document often has no control over the sorts of changes made by whoever revises it. This lack of control can adversely affect the drafter's motivation to produce his or her best work.

When engaged in team writing at school, we suggest that you choose a *combined approach*, which will help you develop interpersonal skills while introducing you to the kind of collaborative writing frequently encountered on the job. The most common variation of this approach involves collaborating with team members to plan and revise, but assigning individuals to draft portions of the document on their own. Note that on occasion, some particularly important sections may be collaboratively drafted, but individual drafting is the norm.

In industry, a technical writer often leads the team through the writing process. But other team members maintain an interest and say in the content and approach of the project documentation. The following subsections address issues relating to this *combined* approach.

4.7.2   Team Planning

The planning process for a team document is generally initiated with a brainstorming session to determine how to deal with various aspects of the writing task. To keep as many ideas and points visible to your team as possible, consider working in a room with a large black or white board or using a flip chart and taping finished sheets to the wall. Also appoint someone to write down ideas on the board or chart, and choose someone else to keep notes on the meeting so the team has a record of points to remember, a record of scheduling decisions, a list of questions that need answering, and so on.

Practicing engineers sometimes begin these brainstorming sessions by identifying members familiar with the type of document being written and calling on them to lead the session. If the document form is new to team members, they may begin by reviewing documents of the same type. In many cases, they will work from templates provided by the company. In this case, team members must keep in mind that these templates should be adapted to a specific rhetorical situation.

4.7.2.1   Establish Objectives and Analyze Readers

Two essential steps that are too often omitted from the team planning process should also be considered while brainstorming:

1. Clarifying the objectives the document must fulfill to be successful
2. Analyzing the readers for whom the document is intended and deciding on an effective approach

In the previous chapter, we discuss the importance of identifying your objectives and analyzing your audience and suggest strategies for doing so. Your writing team must also have a clear idea of what you collectively hope to achieve and of what your readers want. Some time should be devoted to clarifying objectives, and you should write them down in a sentence or in point form so that all team members can refer to them while drafting and during future meetings.

Your writing team should also devote part of its first brainstorming session to determining what the team collectively knows about its readers, what information your readers want, what concerns readers may have, what will sell, and so on. In fact, analyzing your audience may help clarify your goal, and so your objectives should be revisited after the audience analysis is completed. (At this point, your team may want to decide if the meeting should continue or another should be scheduled.)

4.7.2.2   Plan Content and Generate an Outline

Once objectives are clarified, the audience analyzed, and everyone is familiar with the type of documentation required, the team should move on to the following steps:

1. Deciding upon content, organization, graphics, format, and style
2. Generating a topic outline for the document

Discussions of graphics and format are provided in Chapter 7 and style is covered in Chapter 6. Note that, on the job, a company template or stylesheet could dictate many issues relating to style, format, and graphics. In college, you may be given precise instructions or an example of the kind of document required. If such guidance is not provided, you must decide what you want your document to look like and should find an appropriate model to guide formatting decisions.

Begin brainstorming about what topics to include and how to organize them, how to break them into sections, what sort of graphics would be useful, what general formatting conventions you plan to adopt, and any elements of style you can think of. Note that you can put off a discussion of style to a later point in the planning process if having one would waste time and take your session off-track at this stage. If necessary, the team can assign one person to act as editor and stylist.

Once you have a clear idea of the content and general organization of your document, assess the relative importance of your points and how they relate to one another. Lay out general sections for everyone to see: add specific topics under the appropriate headings, indicating which ones are main topics and which more-specific ones fall under them. If you are fortunate enough to have access to a computer during your session, consider having someone enter the resulting outline in the computer so that you can easily move items and quickly produce a copy of the completed topic outline for all members of the team.

4.7.2.3   Finalize Approach and Assign Roles

Once the outline is completed, you can turn your attention to determining what roles each member of the team will play (see pages 119–121), what portion of the work each will do, and a schedule for completing the various tasks. The following steps should be included in this portion of the planning process:

1. Decide which approach you will use for drafting and revising the document.
2. If possible, agree on a common operating system, word processor, and graphics program.
3. Assign individual tasks equitably.
4. Set the deadlines required to complete the document in time to meet overall project deadlines.

While a common procedure is for individuals to draft sections, the team may decide to work together on a particular section. Occasionally, two individuals may also want to write together because they find it easier and faster than writing alone. While this preference is not common, writing partnerships can be particularly productive.

As well as determining who will write which sections, decide how the various sections will be merged and how you will produce or import graphics. If the equipment and software available to the team is varied, then consider how you will produce the final draft before you begin work. Assess how much work you will generate for particular team members. Determine whether any research is required and who will be responsible for what aspects of it. Most importantly, balance the workload among team members.

Your choice of an approach for revising will also play a role in determining each member's responsibilities. In general, everyone should participate in revising. That is, everyone should read the completed draft and prepare comments and questions for a team meeting. After everyone's concerns and suggestions are presented and discussed, members may decide that they will each revise their own sections. Note, however, that one person should be appointed to undertake a final edit so that the style of the document will be consistent throughout. After the team meets to discuss revisions, that person (rather than the individual members) takes responsibility for revising the entire draft. Finally, all members of the team should read the revised draft because all share responsibility for the final product.

At some point in the planning process, the team should also discuss management issues. While some teams function quite well without an official team leader, others require clear lines of authority. Take the time to consider whether your team would benefit from a project manager, a dispute arbiter, a writing editor, or some other form of leadership.

4.7.3.4   Review and Clarify Agreements

Before you move on to individual tasks, take time for the following steps:

1. Review the outline to ensure that all content requirements have been met and the topics are logically ordered.
2. Ensure that everyone is clear on and agrees to the approach, task division, and schedule.

Taking time to recap and clarify agreements during the planning process helps ensure that no points are missed and lessens the chance that disagreements will be overlooked. Much time and frustration may be avoided later on by taking the few extra minutes required for this kind of review.

Note that further planning sessions may be required once certain tasks are completed. In fact, regularly scheduled meetings are often useful to discuss progress and to ensure that everyone is working to meet their deadlines. Also note that the suggestions for planning offered above are not the only viable approaches; teams often develop their own strategies or variations to suit their needs. A number of the strategies for planning presented in Chapter 1 also apply to group situations. (See pages 9–14).

4.7.3   Team Revision

Once an initial draft is completed, your team should meet to begin the revision process. As discussed in Chapter 2, revising is handled in a top-down manner, concentrating first on content and organization at the document, section, subsection, and paragraph levels, and then on sentence structure, style, grammar, and punctuation. Revise in several sweeps or stages, focusing first on content and organization and then on style, correctness, and formatting details.

If your team has planned well, revisions to content and organization should be fairly straightforward. The detailed revision for style, word choice, punctuation, and grammar tends to be more time consuming and contentious. One way to avoid getting bogged down in this part of the process is to appoint an editor who listens to everyone's comments but is trusted to choose the best solution and is given responsibility for achieving consistency.

In some cases, you may also want to appoint members to take on the role of the intended audience, and suggest revisions from the audience's perspective. This special attention to audience can be key to producing a successful document.

4.7.3.1   Peer Reviews

Before we end our discussion of team writing, we turn your attention to another form of team revision commonly employed in the workplace: the *peer review*. First, let us distinguish this type of collaborative review from another form of document review. In the workplace, document reviews are often undertaken by supervisors who have the authority to demand changes. In some cases, supervisors will discuss their concerns and suggestions with the writer and negotiate changes. But frequently, supervisors impose changes, and writers must adjust content and style to accommodate the supervisor's preferences. Such reviews are not collaborative. Reviews when peers read each other's work are by nature collaborative. Advice for providing such collaborative peer reviews follows. When you become a supervisor, we hope that you will maintain a collaborative review process.

As a student, you can engage in a peer review by asking a friend or classmate to read a draft of something you are writing. On the job, peer reviews are often standard practice. *As a writer*, you should approach a review as an opportunity for useful feedback. Either meet with reviewers or write out a list of things you want them to consider. For instance, if you are not satisfied with a section or wonder whether the tone is appropriate, tell reviewers your concerns so that they will pay particular attention to those aspects of the document. You should also explain any features that are unusual or potentially controversial so that reviewers understand your choices. Also state the purpose of your document and provide relevant information about your intended readers.

As a reviewer, you should keep in mind that too much negative criticism will not help the writer. The following guidelines suggest ways of ensuring that you provide *constructive criticism*.

- Evaluate from the reader's perspective.
- Focus on the writer's purpose.
- Comment on a second reading.
- Avoid telling the writer what to do.
- Respond to positive as well as negative features.
- Do not force your style on another writer.

The most important thing to remember is that you are expected to evaluate a document as if you were the intended reader. Familiarize yourself with the writer's audience and purpose. If the writer has not provided you with a clear sense of context, ask for clarification. Also direct your comments toward the writer's specified purpose. For example, comment on whether the writer's purpose will be clear to and is appropriate for the intended audience.

However, do not begin making comments too quickly. If the document is a manageable length, read it through once simply to clarify the gist of the argument and its relevance for the target audience and to gain a general sense of the content. During a second reading, either make marginal notes on the document or organize your comments on a separate sheet of paper. If you make marginal notes, avoid using a red pen; documents that look like they are bleeding are often more discouraging than helpful.

Unless you have been asked to do so, do not respond to grammar and style beyond noting a need to address these concerns. However, point to any wording or sentences you find unclear or ambiguous. When responding to issues of content, audience, or purpose, avoid telling the writer what to do. That is, rather than writing, *"You must address the concerns or fears of your readers,"* consider a statement along the lines of *"If I were your reader, I would be concerned by the cost of the pump."*

Of course, you can make suggestions, but be sure to back them up with reasons to explain why your approach might work better. For example, you might phrase a suggestion as follows: *"If you explained the extreme conditions in that area and what would happen if the pump broke down, the cost would be less of an issue."* When responding to the content of the document, phrase your criticisms as questions rather than as statements. For example, *"This is wrong!"* is much less effective than *"Are you sure that's the correct tolerance?"*

Make a special effort to respond to the positive elements of the document as well as to the negative ones. However, be honest with your comments. Writers need to know both what works and what does not. If you see a problem, do not avoid telling the writer simply to avoid hurt feelings. But at the same time, take a moment to consider whether you are acknowledging a real problem or simply pointing out the way you would have done something. As you will likely learn when you encounter a supervisor whose notion of document review is to impose changes to your document, writers rarely appreciate having someone else's style forced on them.

Through engaging in peer reviews, you should discover the variety of ways in which writers express themselves, develop a certain amount of tolerance, and learn when compromise is appropriate.

4.7.3.2   Additional Benefits

As a final point, we urge you to seek opportunities to review other people's documents and to engage in team writing. As anyone who spends significant time reading and responding to other writers knows, reviewing and commenting on other people's work is a very effective way to improve your own writing. We also encourage you to seek opportunities to write collaboratively, not just because team writing is common in the workplace, but also because, in some contexts, it is a particularly productive way of writing. Our personal experience writing as a team has been very positive. Working together, we capitalize on strengths, minimize weaknesses, and keep each other motivated. If you have not yet done so, give team writing a try. You may find it much more productive and enjoyable than you expect.

4.8   ROUTINE WORKPLACE COMMUNICATION

While the communication skills required for team projects are increasingly important, other forms of interpersonal communication are essential to the daily routines of the workplace. Engaging in conversation, networking, making phone contacts, and organizing *ad hoc* meetings are some of the routine forms of oral communication that deserve attention. These are the forms of communication you use most often in the workplace to interact with the broadest range of people. How well you handle these everyday situations will play a major role in how you are perceived by colleagues, superiors, and clients. They will judge what kind of person you are and how competent you are based in large part on routine and casual encounters.

Routine forms of communication also deserve attention because when you strengthen your communication skills in these areas, you add to your skills in other areas. For example, becoming more comfortable making casual conversation can improve your formal presentations. If you engage in easy conversation with your audience just prior to a formal presentation, they will listen more appreciatively and behave more cooperatively because they have already begun to form relationships with you. Once they perceive you as an individual, they will treat you with the dignity and attention you deserve.

4.8.1   Making Conversation[4]

Throughout your engineering career, you will have many opportunities to meet and talk informally with other engineers and business people. These informal interactions are essential to developing contacts both within and among companies. Such contacts lay the foundation for recommendations and opportunities that could shape your career.

And yet, many people avoid conversation because they feel awkward or shy. Some believe they have nothing of value to say and cannot imagine that others might enjoy speaking with them. Such thoughts are misguided, however, because the real art of conversation and of building relationships is not having something to say, but rather listening and encouraging others to speak. As illustrated by the following example, a good conversationalist helps others talk by asking questions, listening attentively, and providing appropriate responses.

> *Joel arrives to deliver his presentation to a Korean client group. Two of the Koreans arrive just as he finishes setting up his visuals. Joel immediately moves to greet them; he introduces himself and offers his business card. He accepts their cards and carefully examines each before placing it in his cardholder. He then asks about their visit to his city. One of the Koreans mentions a golf course he noticed as he traveled from the airport. Joel asks about his interest in golf and quickly finds out that both clients are keen to test their skills locally. He mentions his own interest in golf, asks about golfing in Korea, and learns it is a very expensive sport there. By the time the other meeting participants arrive, Joel has offered to have someone check on a tee time at a nearby course.*

As this example illustrates, good conversationalists focus on those they are chatting with and not on themselves. Note that Joel asked questions, listened, and linked responses to his next question. He spoke very little about himself while creating a favorable impression with his listeners. When he did speak about himself, he did so to indicate his personal interest in the topic of conversation. (See pages 111–116 of this chapter for more information about listening skills.)

Effective conversationalists also use appropriate nonverbal signals to enhance conversation. What our example does not indicate is that as Joel talked, he stayed a respectful distance from his clients. When he shook their hands, he stayed in a perfect range of two or three feet from them. Joel also used eye contact to indicate his interest in what his contacts were saying, but when he noticed the discomfort of one of the Korean gentlemen, he reduced the duration of his eye contact.

Joel is naturally a very physical person who uses a lot of gesture when he speaks, and yet he calmed his gestures during this conversation to match the style of his contacts. He also spoke more slowly than he normally would because he knew his contacts were speaking in a second language. This attention to and concern for the comfort of his contacts makes Joel an exceptional conversationalist.

4.8.2   Communicating on the Telephone

Making contact and conversing on the telephone remain critical forms of oral communication because so much business is still conducted by phone and the telephone is often your first point of contact with potential clients. In addition to real-time conversations,

---

[4]We would like to thank Margaret Hope for her significant contributions to the rest of the sections in this chapter.

phone contact includes the greetings you leave on your own voice mail system and the messages you leave for other people.

4.8.2.1   Communicating by Telephone

When you pick up the phone, approach each call as an opportunity to make a good first impression. If callers find you too loud, too quiet, or difficult to understand, they may not want to speak with you again. To ensure you leave a favorable impression, monitor the speed of your speech, enunciate clearly, and use pauses to separate one piece of information from another.

Given the international nature of much engineering business, one goal of each call is to ensure your listener understands you. If a caller asks you to repeat information or responds in ways that indicate misunderstanding, then you must speak relatively slowly and repeat yourself. To aid your caller and ensure you understand each other, rephrase what is said to you and repeat any points that cause the caller difficulty.

Because you cannot see your partner in a telephone conversation, you must pay careful attention to auditory cues and stop periodically to review content and invite interaction. For example, a deep intake or an auditory exhalation of air might be your only clue that your phone partner is confused or uncomfortable. Your first reaction might be to ask "*Are you following this?*" or "*Do you agree with me?*" But instead consider asking your caller to review their understanding of the call so far. A simple request such as "*Just so I'm sure we agree, can you tell me what you understand about the purpose of our meeting?*" followed by silence will usually encourage your contact to speak. Then listen for areas of misunderstanding or points of concern.

You also want to ensure your phone conversations leave a good impression. While you cannot see your callers and they cannot see you, your voice will often give them clues about your energy level and attitude. To avoid sounding half-asleep or disinterested, sit or stand in a relaxed but erect posture and use gesture as you speak. While good posture and movement improve your breathing and the quality of your voice, facial expressions give your voice texture. Something as simple as smiling while you greet your caller projects warmth. If you observe good conversationalists talking on the phone, you will likely note that they smile, frown, and make many other facial expressions and gestures as they speak. Figure 4.2 illustrates one approach to speaking on the phone.

Figure 4.2 Being Dynamic while Speaking on the Phone

4.8.2.2   Recording a Voice Mail Message

Your voice mail message should also make a positive first impression. This message should be clear and simple, inviting callers to leave information and indicating when they can expect a response. You should have one voice mail message covering a normal work pattern and variations to indicate that you are working away from your office or are on vacation. If you are rarely at your phone, a daily message can help callers decide what message to leave. A message such as "*I'm in meetings most of today but will be available from four to six p.m.*" helps your callers decide which phone number they should leave or what callback arrangements to suggest.

If you are going on vacation and will not be answering your voice mail, provide callers alternative contacts. Give phone numbers slowly and observe phone-number rhythm: "*I'm at area code (pause) six-zero-four (pause) three-two-zero (pause) seven-six-one-three.*" Avoid alternate expressions callers may not understand. For example, a phone number could be presented as "*I'm at area code (pause) six-o-four (pause) three-two-o (pause) seventy-six, thirteen.*" But will callers understand that *o* means the same as *zero* and will they be familiar with the convention that turns *seven-six* and *one-three* into *seventy-six* and *thirteen*? With so much business conducted internationally, you must be constantly attentive to language and cultural differences.

Also imagine your caller trying to write down your number. If you suspect they will need to replay your message, record it a second time, slowing down and repeating each number. Few things are more annoying than needing to replay a voice mail message several times in order to decipher a phone number.

4.8.2.3   Leaving a Voice Mail Message

Voice mail provides an opportunity to increase efficiency, but only when used properly. Busy people do not appreciate receiving long messages or ones that must be replayed several times in an attempt to catch significant details. When you place a call and connect with someone's voice mail, leave a brief clear message that includes

- Your first and last name (enunciating clearly with a pause between names)
- Your company name and the date and time of your call
- The reason for your call
- Advice on the best times to return the call
- Your phone number (twice) and alternate phone numbers, if appropriate

If your message is unavoidably long or if you are contacting this person for the first time, be sure to leave your name at the end of the recording as well as the beginning. When making an initial contact, you can also improve the value of your message by indicating a mutual contact or other information that will identify you and put your message in context. Note how the following example follows this advice:

> This is Jean (pause) Purvis (pause) of Toynbee and Associates. I'm calling at three pm on December 5th. George Alexander suggested I contact you regarding the next AES meeting. Could you please let me know where the next meeting will be held? You can leave a message on my voice mail at six-six-four (pause) twenty-two (pause) thirty. That's six-six-four (pause) two-two (pause) three-zero (pause) and again, I'm Jean Purvis.

4.8.3   Developing and Maintaining Office Relationships

Despite the increasing use of e-mail for routine communication within companies, face-to-face communication still plays an essential role in establishing and maintaining working relationships. Something as simple as "*Good morning, Nadia. How was your ski weekend?*" indicates that the speaker is interested enough in her colleagues to remember the details of a conversation that took place on the way to the parking lot last Friday evening. Establishing this personal contact may pay dividends when Nadia and the speaker interact professionally later in the day.

People will respond to your requests more readily if they feel you respect and are interested in them. If you have trouble remembering the details of casual conversations, jot them in your journal and then review them regularly until remembering such details becomes a natural part of your thought patterns. But note that you can overdo a good thing.

A casual comment or question in the hall is rarely a problem, but if you interrupt people while they are concentrating on a problem or writing to meet a deadline, your reception will be less than friendly. Because your goal is to build relationships *and* get work done, be sure to consider the needs of your colleagues and respect their privacy. Also be careful to avoid informal discussions in places that will disturb other workers.

To ensure your communications are as productive as possible, observe your co-workers' reactions and if they seem uncomfortable, limit yourself to a simple greeting. With the exception of a handshake when you first meet, also think twice before touching any individual in your workplace. As a general rule avoid touching co-workers because you never know when a friendly pat on the shoulder will be misinterpreted or make someone uncomfortable. (Some forms of touching—such as patting someone on the head—are socially unacceptable to members of certain cultural groups.)

If someone tells you your behavior or communication is a problem, take the criticism seriously but not personally. While remaining friendly, ask for specifics and, where possible, stop doing whatever the person found irritating. Conversely, if you are uncomfortable with other people's behavior or communication, either tell them directly or involve someone to assist you in communicating your displeasure and in resolving the problem. In most cases, you can simply state the problem behavior, explain why it is a problem, and suggest what you would like by way of cooperation to rectify the situation.

When dealing with a problem on your own, try using this formula:

WHEN YOU (describe the specific behavior).

I FEEL (label your feeling).

PLEASE (describe what you'd like to have happen).

For example, you might say,

> *When you wave your index finger in front of my face like this (demonstrate the action), I feel angry because you seem to be lecturing me as if I've done something wrong. Please try not to wave your finger at me as you talk.*

When you ask others to adjust their communication, be direct and assertive, but not unpleasant. Your co-workers may well have no idea they are irritating you and most likely do not want to do so. However, if the problem is serious and ongoing, involve someone else, preferably someone with a human resources background, who can offer you expert advice and support.

A word of caution on asking others to adjust to your needs: make sure your request is not frivolous. Your career will suffer if you are overly reactive to the behaviors of others. One mark of a good communicator is the ability to adjust to the distinctive needs and styles of others. Deal with important issues, but make allowances for minor irritations.

What counts as a major or minor irritation varies from individual to individual and from one context to another. The example of an index finger waved in the face may seem trivial or serious depending on the context and the individuals involved. If this habitual gesture affects someone's relationship with the person making it, then they face a problem worth resolving.

Finally, note that because some people view informal office communication as a time waster, you should keep communications brief to avoid appearing to waste company time. If you find yourself involved in a lengthy story or discussion that is unrelated to work, offer to meet your co-worker on a break to hear more or use body language or a direct statement to conclude the conversation. While a statement such as *"Great to talk to you, Nadia. Guess we'd better get back to work"* may look formal when written, it can sound both polite and natural when delivered with a smile and a wave of your hand as you head toward your work area.

4.8.4 Arranging *Ad Hoc* Meetings

When work is the subject of your communication, be sure to respect the value of your co-worker's time. For example, note the difference between the following two approaches:

1. *We need to meet in seminar room one right now.*

2. *Do you have time to join me in seminar room one to discuss the dredging contract for the airport job? Deeks is having problems with a subcontractor that you've dealt with before. I'm meeting George Gorman this afternoon and could really use your help in figuring out how best to handle the situation.*

Asking for someone's time rather than demanding their attention helps establish and maintain good working relationships. Providing sufficient context for an *ad hoc* meeting also helps co-workers focus on what will be discussed and what they need to bring to the discussion.

Just as you should check on co-workers' availability to attend an *ad hoc* meeting, you should give them a chance to agree on the agenda for the meeting or to suggest changes. Keeping such unscheduled meetings brief and focused avoids wasting time and helps ensure that your co-workers remain willing to accommodate you next time you need their immediate cooperation.

4.9 CHAIRING MEETINGS

The ability to plan, organize, and run an efficient, effective meeting is a vital communication skill for all professionals. A well-organized, well-run meeting can save time, increase productivity, and boost morale. Because so much engineering work is undertaken in teams, you can expect to attend many meetings. Further, you should be prepared to organize and to chair your share of these meetings in ways that achieve predetermined objectives, encourage participation, and avoid wasting participants' time.

Few things are more frustrating than taking time out from a busy schedule to attend a meeting that serves no discernible purpose, for which no one is prepared, and at which nothing is accomplished. Even an important, well-organized meeting can become frustrating when one or two participants are allowed to monopolize the meeting, to keep others from speaking, or otherwise to distract attention from the issue at hand. Consequently, to chair an effective meeting, you should learn how to encourage and control discussion and how to deal with disruptive participants. The following sections provide advice to ensure the meetings you chair achieve their goals while making good use of participants' time.

4.9.1   Five Steps to Chairing an Effective Meeting

The following five steps will help you achieve your goals and demonstrate your leadership abilities:

1. Clarify the purpose of the meeting
2. Select appropriate participants, location, and visual aids
3. Prepare and distribute an agenda
4. Demonstrate leadership
5. Distribute minutes and follow up on outstanding issues

These steps will also help you avoid the common pitfalls that make so many meetings tedious and unproductive.

4.9.1.1   Clarify the Purpose of the Meeting

Effective meetings have a well-defined purpose and clearly identified goals. One way to ensure that you are clear as to the purpose and goals is to complete the following sentences:

1. *If this meeting is successful, those who attended will* . . . (finish the sentence by explaining what they should learn, what information they should share, what problems they will address, what they will do as a result of the meeting, etc.). *If this meeting is successful, those who attended will recognize the importance of backing up their data files from their local hard disks to the network on a daily basis.*

2. *If this meeting is successful, I will* . . . (finish the sentence by describing what you will have accomplished). *If this meeting is successful, I will not need to spend as much time and money salvaging data from local hard disks that have crashed.*

The more specific and action oriented your statement of purpose, the more likely a meeting will achieve its goals.

4.9.1.2   Select Participants, Location, and Aids

One key to a successful meeting is having the appropriate people attend. You want to be inclusive while not inviting people who have no real need or interest in attending but who would feel obliged to come. Restricting the list to only those people who are interested in attending and those who must be present encourages participation and reduces frustration.

You should also locate the appropriate facility for your meeting. A small round table is good for discussion and consensus building among a small number of participants, while a boardroom table or smaller tables arranged in a hollow square encourages a larger group to participate. Theatre-style seating is adequate when the goal is to provide information to participants. But this arrangement hinders participation and should be avoided when the goal involves sharing information and exploring possible solutions to a problem.

Also identify what sort of visual support will be most useful. While computerized visual aids are the flashiest and may seem the most professional choice, lower-tech options are often as appropriate. A white board, black board, or overhead projector can be used for recording ideas or laying out options. If a major brainstorming or planning session is required and a large number of points must be displayed for review, then a flip chart may be the best option so that pages can be removed and displayed around the room. Finally, if more than one person is presenting, be sure to check on everyone's audio-visual needs.

Agenda

Selene Engineering (Mechanical Systems Division)

May 22, 2000
10:30–11:30

1. Approval of Agenda
2. Approval of Minutes (April 24, 2000)
3. John Dunstone—Report on X2-8000 Pump Problems at KPL
4. Mike Markelson–Holiday schedule for July and August
5. Mike Markelson—Replacement for Mary Abrams (Retiring)
6. Frank Winston—Possible Locations for Company Picnic
7. Other Business

Figure 4.3 Sample Agenda

4.9.1.3   Prepare and Distribute an Agenda

Provide participants with as much advanced warning as possible of an upcoming meeting and distribute agendas that indicate the meeting time and place prior to the meeting. When preparing the agenda, consider carefully whether all the items you want on the agenda can be covered in the allotted time. Also determine the order in which items will be discussed, arranging them logically. Note on the agenda those people responsible for introducing or explaining a particular item and also note whether coffee or other refreshments will be served so participants can plan accordingly. Figure 4.3 provides an example of an agenda for a meeting.

4.9.1.4   Demonstrate Leadership

You set the tone for a meeting by the way you arrange the seating, the audio-visual aids you provide, the refreshments you arrange for (at least water for presenters), the way you greet participants, and how promptly you begin the meeting. To ensure you set and maintain the appropriate tone throughout a meeting, plan how you will address the following issues:

- Establishing the appropriate level of formality
- Playing the host
- Arranging for minutes
- Promoting participation

Establishing the Level of Formality   As chair of a meeting, you control, at least to some degree, the level of formality and should adjust it to suit the size and purpose of the meeting. For example, when one or two people gather in an office to discuss a team project, the meeting style will be much less formal than when 35 people meet to learn about the implications of a new government ruling.

Generally, the larger the group, the more formal the meeting. But the level of formality should also reflect the objective and climate of the meeting. Note that a formal style may be appropriate with even a small group when the subject warrants, such as a meeting to discuss a breach in company security. An informal meeting may also be appropriate for a large group, such as a meeting to announce record profits and to thank employees for a job well done.

T-I-S Approach to Introductions

1. Introduce the *topic* the speaker will address.
2. Clarify why the topic is of *interest* or value to the group present at the meeting.
3. Explain why the *speaker* is qualified to address this topic.

Figure 4.4 T-I-S Approach to Introducing Speakers

Hosting a Meeting   When the people attending a meeting are not already acquainted with one another, your first role is that of a genial host who introduces people or chats with them as they arrive. Your second role is to manage time so that the meeting begins and ends on time. Some agendas list the time allotted for each item, but more often time is managed by the chair who keeps discussions moving along, sums up when it is time to move on, and then introduces the next agenda item.

At the beginning of a meeting, you should state the objectives (even if they were circulated prior to the meeting) and give the participants an idea of the flow of the meeting. Also introduce any special guests or guest speakers, keeping introductions brief and projecting your interest and enthusiasm in the subject of a guest's presentation. When introducing a speaker, consider using the T-I-S (*T*opic, *I*nterest, *S*peaker) approach shown in Figure 4.4. Unless you know a speaker well, prepare notes for your introduction to ensure that you accurately present the title of the talk, the speaker's biography, and other details you may not otherwise remember accurately.

Promoting Participation   Throughout a meeting, promote participation where appropriate by encouraging discussion, by calling on people who have appropriate experience or expertise, and by showing appreciation for participants' comments. You can also control participation by reminding people of the purpose of the discussion and by bringing discussions to an end. Note the following example of how to conclude an agenda item:

> *Jo raises an important point we should all keep in mind, but I'm still not clear what we should do with Rav's proposal. Are there any other potential problems we should discuss? (pause) Nothing more? (pause) Then I assume everyone is satisfied with the proposed solutions, and we can tell Rav to work out the details with you, Kate.*

Sometimes you must remind participants of the time so you can move on to the next topic: *We've used most of our scheduled time for this item. Let's hear from Gina and Don and then move on to the next item.* Or you may want to postpone further discussion: *We obviously need more information before we decide. Let's move on to item 2 and return to this issue next meeting, after we've answered these questions.* At the end of the meeting (or at the end of a lengthy item), be sure to summarize decisions and obtain agreement on who is responsible for any actions or follow-up required.

Arranging for a Minute Taker   Someone other than the leader should record the proceedings. If this role is not routine, then a minute taker should be identified well in advance of the meeting. Choose someone you can rely on to produce the minutes in a timely manner.

Short action minutes are generally most appropriate. This style of minutes lists items discussed, the decisions made or actions to be taken, the name or names of those responsible for follow-up, and deadlines for completing tasks. Action minutes eliminate confusion and serve as the basis for appropriate follow-up. Figure 4.5 provides an example of action minutes.

Selene Engineering

Minutes of Mechanical Systems Division Monthly Meeting

May 22, 2000

Present: Mike Markelson (Chair), Mary Abrams, John Dunstone, Patricia Goring, Eric Mallory, Joan Petersen (Recording Secretary), Peter Simpson, Alex Waring

Guests: Frank Winston (VP Human Resources), Sally Marr (Internship Student)

Absent: Hannah Franklin (with regrets)

1. **Approval of Agenda**
   Approved.

2. **Approval of Minutes (April 24, 2000)**
   Minutes from the last meeting were approved with minor corrections.

3. **John Dunstone—Report on X2-8000 Pump Problems at KPL**
   John Dunstone reported that the bearings on the X2-8000 pump supplied to KPL are up to spec and the problem lies with the balance of the main shaft connecting to the pump. After some discussion about possible ways to fix the problem, it was decided that the most reliable and cost effective approach would be to redesign the main shaft. John Dunstone and Peter Simpson will put together a proposal over the next week that can be submitted to KPL.

4. **Mike Markelson—Holiday Schedule for July and August**
   Mike Markelson introduced Sally Marr who is currently working on her Master's at Rockcliffe University. She is helping fill in for some of the holiday absences over the Summer. She is sharing office space with Alex Waring.

5. **Mike Markelson—Replacement for Mary Abrams (Retiring)**
   Mike Markelson noted that Mary Abrams will be retiring on August 31, and a search is currently underway for a replacement. John Dunstone suggested that her work as systems administrator has been of such high caliber that it will probably be impossible to replace her with a single person.

6. **Frank Winston—Possible Locations for Company Picnic**
   Frank Winston was introduced by Mike Markelson. Due to poor attendance at last year's picnic, Frank suggested that the location for the picnic be changed this year. Several people pointed out that the problem was not with the location, but rather with the time because many people are on holiday in August. It was suggested that the date for the picnic be moved to early September from its current late August date. Frank agreed this sounded like a good idea (if the facilities were still available at that time) and said that he would revisit the issue after consulting with people in the other divisions.

7. **Other Business**
   There being no further business, the meeting was adjourned at 11:25.

Figure 4.5 Sample Action Minutes

4.9.1.5 Follow-Up

If you chair a meeting, you must review the minutes and then circulate them as soon as possible after the meeting. As well as recording decisions and actions, the minutes should include the date the meeting was held, a list of those present and those sending regrets, the time and place of the next meeting (if appropriate), and who to contact for clarification. This record should be sent to each participant, to absentees, and to anyone else involved in the ongoing affairs of the group or of the project under discussion.

You should also express your thanks to anyone who made a particular contribution to the meeting. For example, if someone made a special effort to be present, helped deal with a potentially disruptive situation, or volunteered to follow up on an issue, acknowledge the assistance and express your appreciation.

4.9.2   Delivery Techniques for Meeting Leaders

Some meeting leaders take control from the moment they enter the room; others never seem fully in control. Those in the first group have developed effective delivery techniques that help them maintain or re-exert control of the meeting as required. Effective leaders also rely on their delivery skills to encourage and manage discussion and to control difficult participants.

While presentation skills are the focus of the following chapter, general advice on delivery techniques for taking charge and for encouraging and controlling discussion in meetings is provided here.

4.9.2.1   Taking Control

If you need to gain control at the beginning of a meeting, stand at the front of the group and make eye contact with two or three supportive individuals. Avoid trying to catch everyone's attention because glancing about quickly makes you look less than assured. Begin with a relatively loud, firm voice that makes listening easy and conveys your commitment to lead the session. If possible, also address participants directly in your opening remarks and involve them in discussion.

4.9.2.2   Encouraging Discussion

Throughout the meeting, you may need to stimulate discussion. At appropriate places, explain what you expect from the group, and then ask a specific question, keeping your attention on participants but remaining silent and waiting for someone to begin talking. Invite discussion not only with words but also with your gestures. That is, open your arms or cup one hand in a gathering motion to emphasize your desire for input. If no one responds after an uncomfortable length of time, then restate the question and/or call on specific people by name.

4.9.2.3   Controlling Discussion

Note that much like a traffic cop, you can use delivery techniques to turn the action off as well as on. For example, if several speakers are competing for attention, raise your hand as you would to stop a moving vehicle at a crosswalk. Continue to look directly at the offenders without speaking. As you will discover if you have occasion to use this technique, silence is a much-overlooked tool in controlling groups.

If everyone seems to want to talk at once, you can manage their participation by providing clear directions for the order of speakers and by reassuring participants that everyone will be given a turn. Of course, at some point, you must draw the discussion to a close and should give participants some warning that you are about to move on to another topic. You might close with something along the following lines: *We are running out of time, so I'll just take two more comments—first from you, Tam, and then from Agnes.*

If participants are particularly long winded and/or monopolizing the conversation, you may need to exert control while they are still talking. In such cases, attempt to maintain both your own composure and the dignity of participants. One of the easiest ways to stop speakers prematurely is to call them by name while making direct eye contact. As soon as a speaker pauses, compliment his or her contribution, note the problem of time or the need for others to have an opportunity for input, and then remove eye contact. You can then call on the next person in line to speak, or you can sum up and move to the next agenda item.

4.10   SUMMING UP

At the beginning of the twenty-first century, engineers face increasing demands on their interpersonal and communication skills. In your career, you will encounter increasingly complex, multidisciplinary, and international projects, a diverse workforce, globalization, and virtual work environments. Consequently, you must be a team player and a team leader—a good listener who is flexible and open minded, prepared to accommodate many different perspectives, respectful of the opinions of others, and willing to negotiate consensus. You must also be prepared to write and read more than any generation of engineers before you. Further, you must develop powers of persuasion strong enough to influence people from a wide range of backgrounds with many points of view.

In a world undergoing enormous change, your success as an engineer may depend less on your technical know how than on how well prepared you are to face the interpersonal and communication challenges of the new millennium. One way that you can prepare yourself is by starting small, by turning routine encounters with colleagues and opportunities to chair meetings into occasions to practice and polish your interpersonal and communication skills. The following checklists and exercises also provide guidance for improving your ability to work and write as part of a team, to engage in routine office communications, and to chair effective meetings.

4.11   TEAM-WRITING HEURISTIC

The following set of questions is aimed at helping you clarify some of the key issues that writing teams must address. Whenever two or more people are responsible for producing documentation or writing a report, we suggest that writing partners collectively answer these questions. Finding and agreeing on answers will enhance efficiency and help ensure overall success.

This list of questions is by no means exhaustive; it provides you with a starting point for analyzing and improving the dynamics of a writing team. Add questions to reflect the challenges and constraints of a particular rhetorical situation.

Assigning Roles and Tasks

- What sort of organizational structure will your team follow?
- How will you determine what role each team member will undertake?
- Which activities will be collaborative and which will be individual?
- What criteria will you use for assigning these tasks in an equitable and productive way?
- What issues relating to computing platforms and software must be considered and agreed on when assigning tasks?

Scheduling

- When establishing deadlines, what are the potential problems that must be accounted for or addressed to ensure tasks can be completed on time and in an effective and efficient manner?

Communicating

- What potential communication problems do you foresee?
- What lines of communication will you use to ensure that everyone is clear about and agrees with the overall approach, task division, and schedule?
- How will you handle conflicts among team members?
- What might you do to ensure that alternative viewpoints are not discounted or dismissed without consideration?

4.12   TEAMWORK CHECKLIST

We provide the following checklist for you to use while working as part of a team. We suggest that you use this checklist by considering the points that it raises *while* you are working and writing as a member of a team. By noting those elements that are functioning well or functioning poorly in your team, you can determine what changes to make in order to help your team work together more cffcctively and efficiently.

4.12.1   Listening Skills

Do you listen actively? _____

Do you concentrate fully and tune out all distractions? _____

Are you interested in what the speaker is saying? _____

Do you listen with an open mind? _____

Do you listen to the speaker's feelings as well as to the content of what is being said? _____

Are you aware of your biases and paying attention to how you react to certain topics? _____

Do you pay attention to nonverbal communication such as body language? _____

4.12.2   Team Dynamics

Has your team outlined its professional objectives and have team members articulated their personal goals? _____

Are team members interacting frequently? _____

Are team meetings well organized and well run? _____

Are you and other team members fulfilling all necessary task-oriented and social roles? _____

Are tasks divided equitably among team members both in terms of how long they take to accomplish and in terms of how interesting they are? _____

Does your team fairly share responsibility for failures and successes? _____

Do team members avoid blaming one another when problems arise? _____

Is conflict among team members handled appropriately? _____

Has your team avoided groupthink? _____

4.12.3   Writing as Part of a Team

Team Planning

Has the team clarified its goals?   _____

Have team members analyzed the audience?   _____

Has the team collaborated on a topic outline?   _____

Has the team agreed on how to divide the work?   _____

Have team members agreed to deadlines?   _____

Has everyone agreed on the style and format of the document?   _____

Team Dynamics

Are tasks divided equitably?   _____

Is everyone clear on tasks and deadlines?   _____

Does everyone agree to tasks and deadlines?   _____

Have leadership roles been assigned?   _____

Team Revising

Has a team meeting to discuss revisions been scheduled?   _____

Has everyone read and commented on the completed draft before the revision meeting?   _____

Is criticism of other people's sections of the document constructive?   _____

Have team members been assigned to play the role of the intended readers during a team revision meeting?   _____

Has everyone read an all-but-final draft and had an opportunity to draw attention to remaining points of concern or editing issues?   _____

4.12.4   Routine Workplace Communication

Are you practicing the traits of a good conversationalist by helping others talk about themselves and adapting to their style?   _____

Are you paying attention to how you sound on the phone, carefully preparing your voice mail messages, and leaving brief, clear messages for others?   _____

Are your informal office communications both collegial and respectful of others' time and responsibilities?   _____

Do you plan meetings to maximize the use of time and resources?   _____

4.12.5   Chairing Meetings

Are you following the five steps to an effective meeting?   _____

Have you clarified the purpose of the meeting?   _____

Have you identified who should attend, chosen a suitable location, and arranged for visual aids?   _____

Have you prepared and distributed an agenda?   _____

Have you demonstrated leadership by establishing an appropriate
level of formality, by playing host, by arranging for a minute taker,
and by promoting participation?   _____

Have you followed up by ensuring accurate minutes are distributed
and by dealing with any outstanding issues?   _____

4.13   EXERCISES

Ideally, you should apply the concepts, principles, and advice in this chapter to team activities that you already do for school or work. We provide the Teamwork Checklist on pages 144–146 of this chapter for that purpose. If you do not have opportunities to apply that checklist at school or work, then the following exercises may help you further understand some of the issues we raise.

1. Listening Skills
Get together with a friend or colleague and take turns describing something that is of interest to you (why you want to be an engineer, why you like cooking, the joys of surfing, etc.). While the other person is speaking, try to make it very obvious that you are listening to him or her through body language and spoken cues. After two or three minutes, then try to make it very obvious that you are not listening. Next, ask him or her to describe what you did that encouraged conversation and what you did that discouraged it. Finally, reverse the roles.

2. Team Dynamics
Think back to the last time you worked as part of a team at school or work. What things made the experience a good one or a bad one? List these points and then review pages 117–122 of this chapter. What strategies could you (or the team leader) have applied to make the outcome better? Discuss the situation with friends or colleagues and ask them for their opinions about your strategies.

3. Cross Functional Teams
Get together with a friend or colleague from a different cultural, professional, or academic group. Through discussion, determine whether you and your partner have different understandings and experiences when dealing with other people in both social and professional situations. After determining any differences that exist, brainstorm ways you would compensate for these differences if you worked together on a project.

4. Team Writing
Working in a group of four or five people and using the Team-Writing Heuristic on pages 143–144 of this chapter, plan a hypothetical team-writing assignment of about 100 pages that might be required in your field of engineering. Once you have completed your plan, discuss as a group what sort of problems you foresee arising while working on the task.

5. Telephone Communications
Working with a friend or colleague on the telephone, practice leaving messages on an answering machine (refer to pages 133–135 of this chapter for some guidelines about effective messages). Ask your friend to critique your messages. Then switch roles.

6. Chairing Meetings
Imagine you have been asked to chair a meeting with a rowdy group of fellow students or colleagues. The main purpose of the meeting is for a presentation by a guest speaker about how to run effective meetings. Using the information about how to run effective meetings in this chapter (pages 137–143), undertake the following. First, prepare an agenda for the meeting. Second, think about how you would get everyone's attention at the start of the meeting. Third, introduce the speaker. You could role play this exercise with a group of your friends (if you do so, ask them to be especially difficult).

Chapter 5

Oral Presentation Strategies

5.1 SUMMARY

As an engineer, you will be frequently called on to present. During a meeting, your boss may unexpectedly ask you to brief others on your work. At other times, you may feel compelled to take the floor to argue a point or to provide information. You can also expect to make scheduled presentations to update superiors or clients on the progress of a project and may be given as little as three minutes to present the accomplishments of your team and to make the case for increased resources. You may present alone or as part of a team, but one thing for certain is that you will be called on to make *many* presentations.

This chapter focuses on formal oral presentations, but also provides advice that applies to less formal speaking occasions. Specific topics relating to performance both as an individual and as part of a team include the following:

1. **Controlling Anxiety**
   A little nervousness can be a good thing, but if you are overly anxious, you should follow the advice offered on pages 150–152. The most important points to remember are to allow sufficient time for rehearsals and to take every opportunity to practice presenting.

2. **Formal Presentations**
   Oral presentations, like all other forms of communication, must respond to a specific rhetorical situation, so the foundation for an effective presentation is attention to the audience, subject matter, context, and persuasive or informative aim as well as to developing presentation skills. When you present, also demonstrate your enthusiasm for your topic to your audience and show your respect for those attending your presentation.

   a. **Planning a Presentation**
   To plan a presentation effectively, analyze your audience so that you can adapt your presentation to their needs and level of expertise. Also clarify your objective, limit your talk to no more than five main points, choose an organizational pattern that will be obvious to your audience, create links between

points, prepare visual aids that contain no unnecessary detail, compose speaking notes if required, and plan for the question period.

b. Rehearsing a Presentation

Whenever possible, allow enough time for several rehearsals so that you can concentrate on different aspects of your presentation and work on aspects that need particular attention. If you have only short periods of free time, rehearse your presentation in sections, beginning and ending with transitions between points. Also rehearse with a practice audience and ensure that you can present in a little less than the time allowed. You can also prepare psychologically by taking a few minutes before you present to imagine yourself handling everything smoothly and professionally.

c. Delivering a Presentation

The best way to improve your delivery is by making presentations. The more you present, the more confident and controlled you will become. However, a number of issues relating to presentations also deserve attention. We provide advice and strategies for improving body language, voice, and speaking style.

3. Team Presentations

When presenting as part of a team, you can apply much of the advice for presenting alone, but you must also engage in a number of group activities. For example, all members of your presentation team must agree on the purpose of your presentation, and play a role in determining content and organization, in selecting visual aids, and in deciding on a delivery style and an approach for dealing with the question period. Members of the team should also be prepared to act as coaches for one another during team rehearsals. We identify common problems with body language, voice, and speaking style and provide advice for helping team members overcome these problems. Coaches must make others aware of such problems with delivery, but should avoid addressing too many problems at once and should take teammates' feelings into account when deciding how to offer advice.

Our discussion of formal presentations ends with a reminder that outstanding speakers are not born; they develop their skill through repeated opportunities to present. You can accelerate your improvement by joining a Toastmaster's club or engaging in another related activity.

This chapter also includes a checklist for oral presentations and a heuristic (a list of questions) you can use to help improve your delivery. The exercises in the final section reinforce the point that the most important thing you can do to improve your presentations is to practice, practice, and then practice some more.

5.2 ORAL COMMUNICATION: A KEY TO SUCCESS

Being a skillful oral communicator is one of the keys to a successful engineering career. Engineers are frequently called on to share ideas and to present information orally. As an engineer, you will address subordinates, peers, supervisors, and nonengineering audiences, engage in brainstorming sessions, speak informally at meetings, interact with team members and clients, and make formal presentations. You may also network at trade shows and conventions, participate on industry panels, and present at conferences.

The more successful your career, the more often you will be called on to communicate orally for a wide variety of situations and audiences. In Chapter 4 (pages 132–143), we addressed a number of issues relating to informal office communications and to organizing and leading meetings. In this chapter, we focus on presentation skills and on individual and team presentations. While you will likely be called on most often to make technical presentations, much of the following advice applies equally well to other formal speaking occasions.

5.2.1   Occasions for Oral Presentations

The occasions for individual oral presentations range from unanticipated requests to present on a moment's notice to carefully prepared papers that are read at conferences. *Impromptu presentations* are common in most workplaces. When the occasion arises in a meeting, you will be called on to share your expertise or specialized knowledge with colleagues, superiors, or clients. Or perhaps you will decide that you need to clarify a point or provide information or background not known to the majority of participants. As far as possible, you can plan for the unexpected by coming prepared to meetings. For such impromptu presentations, you will generally know your audience well and have a very clear grasp of your purpose, so the best use of the few seconds you have before beginning to speak is to spend them organizing how you will present information or attempt to persuade your audience. Long-term preparation for this sort of presentation involves developing your ability to organize information, sharpening your powers of persuasion, and practicing the delivery skills discussed later in this chapter.

Read papers are most common at conferences and you are most likely to read a prepared paper if you become a researcher or academic. If you are ever called on to write a paper to be read, keep your sentences relatively simple and short. Also read your paper out loud (preferably before an audience) and revise any sentences you stumble over (or that your listeners find problematic). While conference participants routinely write papers for publication in conference proceedings, in many fields of engineering, participants present their results without reading from a prepared text.

Formal presentations—the kind that are carefully prepared but delivered with minimal reliance on notes and significant use of visual aids—are the focus of this chapter. You may present to convey information also contained in a report, or the presentation may be the sole means of providing information or of attempting to persuade an audience. Formal presentations are popular in business and industry because they are generally relatively short and focused only on those aspects of a subject relevant to the audience. Attending an oral presentation may save listeners from needing to read a lengthy report or it may motivate them to learn more about the subject. Presentations also provide opportunities for attendees to ask questions, to gain clarification, to offer suggestions, and to present opposing viewpoints. While documentation is often necessary to create a record of activities and to provide information to those distant in time and space, presentations are essential to the present—to the efficient functioning of projects and project teams and to maintaining productive relationships with superiors and clients.

5.2.2   The Challenges of Oral Communication

Many of the challenges of oral communication result from a presenter's need to address what is happening in the here and now. A fundamental difference between written and oral communication lies in the immediacy of the audience. When you write, you face the challenges of addressing readers who are absent, but you have time to consider your thoughts carefully before transmitting them to a reader. When you speak, you face the challenges of addressing listeners who are immediately before you.

Given this immediacy, you must constantly observe and react to listeners' needs and be prepared to adjust the complexity, style, and content of your presentation to meet those needs. When members of your audience speak, whether in words or nonverbally (i.e., through facial expression, grunts, or changes in body language), you must listen and observe. For example, you must watch for signs that the audience wants to interact, to ask questions, or even to end your presentation.

Capturing and maintaining the audience's attention are major challenges for presenters. While the reader of a written report can review previous material and stop to contemplate the intended meaning of words and phrases, listeners cannot replay what they miss when their attention wanders. Because attention will inevitably wander—at least momentarily—in even the most gripping presentation, information must be packaged for listening and delivered with listeners in mind. That is, the structure and organization of an oral presentation must be clear to listeners even before the content is presented, and major points must be emphasized and repeated in appropriate places.

Skilled presenters will further emphasize the structure of their presentations through pauses, through interactions with the audience, through changes in delivery techniques, and through visual aids. Visual aids play a major role in presentations because they emphasize critical information, draw attention to areas you want to open for discussion, promote interest in your topic, and generally help the audience remember what you are saying. To become a skilled presenter, you must learn to emphasize organization and repeat key points to ensure that momentary distractions will not disorient your listeners. You must also create visual aids that hold attention and promote understanding.

One added challenge is that listeners respond not only to your words and visual aids, but also to a range of other visual and auditory signals. For example, when you present, your audience will be affected by how you are dressed, stand, move, and gesture as well as by the pace, volume, tone, and quality of your voice. They will also react to your demonstrated passion for your subject and commitment to your ideas and recommendations.

Your message, your visual aids, and your delivery are parts of a performance package. This notion of performing can trigger stage fright. This anxiety may intensify when members of the audience interrupt, disagree, or otherwise distract your attention from a rehearsed train of thought or move the conversation away from planned remarks. Managing performance anxiety and accommodating the unexpected are, therefore, essential oral communication skills.

5.2.3   Controlling Anxiety

When you are in front of an audience, you should expect to feel at least a little nervous. Without a little nervous excitement, you will likely make a dull, lifeless presentation. But excessive nervousness can inhibit spontaneity and reduce your ability to focus on the audience. If your nervousness is obvious in shaky hands or a thin, cracking voice, some members of the audience will also feel uncomfortable and will focus on the symptoms of your nervousness rather than on your message.

If you suffer from nerves when addressing an audience, developing the following habits will help you gain control. Even confident speakers should develop the first and last habits listed below:

- Make time to rehearse
- Warm up with tension/relaxation exercises
- Begin with a smile
- Breathe deeply

- Maintain eye contact
- Drink water
- Practice, practice, practice

Note that while performance anxiety is most often a problem when making a formal presentation, you can also use these techniques to overcome anxiety in less formal situations.

5.2.3.1   Make Time to Rehearse

You are most likely to feel panic if you face an audience knowing you are not fully prepared or if you check the clock and realize you are running out of time with much more to say. You can avoid these reasons for panic by beginning work early so that you leave adequate time to plan and to organize your points, to create visual aids, and, most importantly, to rehearse.

Allow time for several dry runs. On your own, work on timing and becoming comfortable talking from your aids rather than reading from notes. Then perform for colleagues willing to serve as a practice audience. After presenting to your peers, you should be more confident presenting to your final audience.

5.2.3.2   Warm Up with Tension/Relaxation Exercises

Your brain prepares for the stress of performance by releasing hormones, creating what we generally refer to as an adrenaline rush. Before you even step up to the podium, you may be a little shaky, feel your muscles tense, or suffer from excessive perspiration, shallow breathing, burning ears, and a red face. Brisk exercise before your presentation will naturally lower your adrenaline level and reduce these symptoms. But even if you cannot find time to exercise, you can practice tension/relaxation exercises immediately before performing.

Note that muscles relax more completely once they have been fully contracted. So if you tense the muscles of your neck, shoulders, arms, and hands and then fully relax them, you are less likely to cramp or shake. When these tension/relaxation exercises help your muscles relax, other symptoms of nerves will likely subside as well.

5.2.3.3   Begin with a Smile

By starting with a smile, you offer a warm welcome to your audience that helps them and you relax. Facing an audience that appears friendly and relaxed will help keep you calm. Smiling is also more relaxing than frowning.

5.2.3.4   Breathe Deeply

Calm yourself further by taking a couple of deep breaths just before beginning to speak. If you take a deep breath before launching into your introduction, your first words will sound louder and more confident. Sounding confident actually helps build confidence. The extra oxygen will also help you concentrate.

5.2.3.5   Make Eye Contact

To reduce the visual stimulus your brain receives from looking at a mass of faces, limit your gaze, making eye contact with one person for a few seconds before moving on to the next person. By reducing stimulus, you reduce panic. If you happen to focus on a scowling listener, quickly move on to a friendlier face. Return your gaze to apparently less supportive listeners when you are well into your presentation and feeling more confident. When you realize that your audience is generally interested in what you are saying, you will feel even more confident.

5.2.3.6   Drink Water

Plan ahead to ensure you have a glass of water at hand during your presentation. Then, if nerves make your voice thin and squeaky or if they dry out your mouth, you can pause to take a sip of water. The pause will help you regain your composure as will drinking the water because swallowing is a tension/relaxation exercise.

5.2.3.7   Practice, Practice, Practice

While developing the above habits will increase your confidence, repeated opportunities to present are crucial. The more you present, the more confident and skilled you will become. So the less confident you feel, the more you need to stand in front of an audience and present. Seeking opportunities to practice public speaking is perhaps the most important way to control anxiety.

5.3   FORMAL PRESENTATIONS

Practicing engineers make numerous formal presentations to brief co-workers, managers, or clients. Such presentations may provide technical information or may encourage potential clients to buy a product or service, convince potential investors of the future success of a company or proposal, or persuade managers to change a plan or to deal with a safety issue. The following advice focuses on the attitudes and delivery skills required for informative and persuasive technical presentations.

5.3.1   The Importance of Attitude

A positive attitude is essential. Listeners recognize when you are enthusiastic, feel prepared, are confident in your material, and are comfortable presenting it. Your attitude also tells your listeners much about the nature of your preparation, the adequacy of your practice, and the degree of skill you have acquired.

Because the key to connecting with your audience is respect, your efforts should come from a genuine desire to build rapport, increase understanding, and establish agreement. How you express your attitude to an audience should also reflect your rhetorical and physical situation. For example, your level of energy and excitement may be appropriate to your purpose and surroundings in one context but could be perceived as an overly dramatic style of delivery in another. For example, in a small area with only a few listeners, you may need to use smaller movements and less vocal variety to convey your enthusiasm than would be appropriate in a large room with more listeners.

You also exhibit the appropriate attitude when you think in terms of serving your audience—of giving them what they need and keeping everyone focused to avoid wasting time and resources. Take time before the presentation to find out who will attend so that you can adjust your material to their interests and experience. You can also communicate your concern for, interest in, and connection with the members of your audience during your presentation by doing the following:

- Making eye contact
- Using people's names and referring to their companies, products, projects, or expertise
- Providing sufficient background and context
- Using vocabulary and examples suited to the knowledge and experience of the audience
- Expressing appreciation to those who have provided you with information or assistance

Audiences also appreciate speakers who convey a positive attitude toward their subject, themselves, and life in general. Avoid the temptation to apologize for being nervous or for the quality of a visual aid or to otherwise point to a fault or to make an excuse for some aspect of your presentation. Beginning with an apology leaves listeners wondering why you failed to take the time to prepare properly, why you are wasting their time, or why you are being so self-effacing or insecure. They do not want to hear excuses, and they generally will not admire your honesty. Being well prepared is clearly the best defense.

Imagine that your computer crashes five minutes into a computer-assisted presentation, thus making your planned demonstration impossible. How you handle yourself at such a moment is crucial. When your computer crashes, you might as well laugh at the problems of technology and then tell your audience how you will proceed. They may not get the great presentation you planned, but they will certainly be sold on your positive approach and ability to deliver when faced with a major challenge. In such situations, careful planning is crucial.

5.3.2   Planning a Presentation

While you may sometimes be called on to speak at a moment's notice, time to prepare is usually available and should be used efficiently. If you master techniques for effective preparation, even a half-hour's effort can improve the quality of a presentation. Of course, the earlier you can start and the more time you allow for preparation, the better you are likely to perform. Far too many presenters leave preparation until the last minute. Some scribble out detailed notes focusing entirely on content and then, lacking rehearsal time, muddle through by reading to the audience. Others create detailed computer slides but stumble over what to say about them because they allowed no time to rehearse. Some obviously speak without any preparation.

No audience is oblivious to such tactics and will likely perceive the ensuing presentation as a waste of time. To ensure your presentations are productive for both you and your audience, follow the advice in the next sections.

5.3.2.1   Learn about Your Audience and Location

Your first task as a speaker is to learn as much as you can about your audience. In some situations, you may know very little about the people you will address. If so, then you must find sources of information about them. One obvious source is the person who organized the meeting or event at which you are presenting. If appropriate, consider contacting a small sample of participants to assess their interests, knowledge, points of contention or concerns, and to determine how your presentation fits their agenda.

If you will be speaking to representatives of a company, you can often find a wealth of background information on the World Wide Web. You can also seek out others who have recently addressed or dealt with the group in question. Recent presenters will have first-hand knowledge of your audience and can provide useful information on the room you will present in and on the equipment available.

While such effort may be extreme for routine presentations, to ensure the success of a significant presentation, you must be thoroughly familiar with your audience and with the facility where you will speak. Plan to arrive well ahead of time so your setup will be complete when the audience arrives. Then you can spend time visiting or networking in order to determine if any last-minute adjustments are required. The time it takes for people to congregate also provides an opportunity to learn a few names, discover the roles of various audience members, and build rapport.

5.3.2.2 Adapt to Your Audience

Many presentations are edited versions of lengthy reports or summations of ongoing projects about which you know far more than you can discuss in the time allowed. For example, you may be given ten minutes to report on a project you spent four months working on. In such cases, you must think carefully about what content to include. Your audience is your best guide to selecting content.

The process of selecting appropriate content is one of adapting to the needs and attitudes of your audience. Begin by considering the general makeup, expertise, and attitudes of the group. Are you addressing co-workers who understand the big picture but are unfamiliar with your part of the project? Will your audience comprise a parallel team whose members are concerned that your group is not working as hard as theirs? Will your listeners already think they know what you have to say? Are they interested in obtaining a general understanding of your project? Are they likely to feel nervous and to be unsure they can follow the technical detail? Are they pressured for time and attending only because they have been told to come? Are they stakeholders who will be receptive to your ideas?

The answers to such questions can be complex given that audience members usually represent a wide range of knowledge, skills, interests, capabilities, and attitudes. Sometimes you must focus more on some members of the audience than others. But normally, you should seek compromises that make your presentation as useful and effective as possible for all members of your audience.

A number of factors should influence your approach to a particular audience:

- Motivation
- Attitude
- Group dynamics
- Linguistic and cultural background
- Demographics
- Physical environment

Motivation   How motivated your audience is to hear your presentation should influence your approach. For example, if the majority of a group has been sent on their noon hour to hear your briefing, acknowledge their sacrifice and encourage them to eat their lunches while you speak. In this case, keep your remarks as brief as possible and consider adding elements of humor to win and maintain attention. By contrast, if a group has specifically asked for your presentation or traveled some distance to hear you, too brief a presentation may disappoint or even annoy them. In this case, being entertaining is less of an issue.

Attitude   While an audience may be motivated to hear what you have to say, they may also hold an opposing view of the action or outcome you recommend. When a topic could be controversial, you are well advised to determine audience members' attitudes and to decide whether they are likely to be openly hostile, quietly argumentative, or simply skeptical. The more you know in advance about the audience's attitudes and potential reactions to your topic, the more likely you will adequately address their concerns and gain their respect, if not their agreement.

Acknowledging and demonstrating respect for conflicting opinions and viewpoints is an important step in helping a potentially skeptical or argumentative audience appreciate the value of your recommendations. Such acknowledgments also improve the likelihood of a productive discussion.

Knowledge and Expertise   Your audience's level of knowledge about your subject and their areas of expertise should strongly influence the content of your presentation and the degree of detail you provide. For example, an audience of entrepreneurs or investors will appreciate and readily comprehend the commercial or practical value of your technology. But if they need to understand how the technology works, they will require more background, more examples, more definitions, and less specialized language than a technically oriented audience. Generally, however, this audience will be less concerned with technical details than a group of engineers would be.

Group Dynamics   You should mentally prepare yourself for the way a particular audience is likely to behave and plan your presentation accordingly. For example, groups who know each other well are more likely to interrupt you to ask questions and indulge in side discussions seemingly unrelated to your presentation. Because such a group would likely be frustrated if you allowed no interruptions, you should allow for questions and comments during your presentation but also plan how you will maintain control. On the other hand, audiences comprised of strangers (or near strangers) tend to be reserved. Rather than controlling participation, you will need to find ways of drawing this kind of audience into a discussion.

Linguistic and Cultural Background   Whether you present locally or in a foreign country, you can expect at least some members of the audience to have linguistic or cultural backgrounds different from your own. Those struggling with English or with your accent will appreciate relatively slow speech and visual aids designed to assist their understanding. You must also adjust your style to accommodate cultural differences. For example, people from a particular culture may be uncomfortable asking questions or may not indicate their feelings through facial expressions. If you know in advance how your audience is likely to react, you are more likely to structure your presentation and adapt your style to increase their comfort level and less likely to be distressed by their reactions.

Demographics   Learn what you can about the size, gender distribution, age range, and other physical attributes of your audience. For example, will you be addressing 30 young women who are touring your facility as part of a program to encourage them to continue taking science and math courses? Or will you be addressing eight decision makers with an average age of 50 from a venture capital corporation who are considering funding your next project? Will you present for an audience of 400 scientists and engineers or for only a handful of undergraduate engineering students? Your language, visuals, and style should reflect the demographics of your audience.

Physical Environment   Also pay attention to the physical conditions confronting your audience such as the level of comfort afforded by seating, room temperature, and lighting. Has your audience been seated for an hour on hard metal chairs with an enthusiastic air conditioner blasting them from above? Will intense overhead lighting render your Power-Point presentation ineffective and threaten viewers with eyestrain? If you recognize such potential problems early enough, you can ask for alternative arrangements or adjust your materials, visual aids, and style to accommodate the situation.

5.3.2.3   Clarify Objectives

A clear picture of your audience must be accompanied by a clear sense of the objectives you hope to achieve. To be useful, these objectives must reflect the audience's perspective. To this end, many speakers find completing the following sentence helpful:

> *If I am successful, my listeners will . . . .*

You will select content, structure, and visual support most easily once you know what you are trying to accomplish. Suppose, for instance, that you must brief an audience on the development of a hand-held diameter gauge in order to demonstrate the range of issues and challenges involved in designing precision tools. Your sentence might read: *If I am successful, the audience will appreciate the complexities of creating precision tools.*

For a long presentation, you will need several objectives—one for each major section. Assuming that your audience is engineering students, another objective of the diameter gauge presentation could read as follows: *If I am successful, students will see the practical applications for their lab assignment.* If the presentation has a persuasive aim, the objective will reflect this emphasis: *If I am successful, students will be interested in seeking internships and job opportunities with our firm.* Different objectives direct you toward different approaches, content, and delivery styles. The more explicit the objective statement, the more apparent what details to include, what visual aids to use, what style to adopt, and so on.

5.3.2.4   Limit and Organize Content

Structure your presentation in terms of three or four major sections: an opening, a body, a closing, and if appropriate, a question period. In the opening, start with the big picture (the background and context) and then provide your audience with an overview (outline) before providing the details for each point in the body of your presentation. Limit the body of your presentation to between three and five main points. (For a lengthy presentation, each of these main points may be divided into three to five subpoints.) In the closing, summarize your main points and tell the audience what you want them to do with the information or ideas you have presented, or remind them of what they should have learned. You can accept questions throughout your presentation or contain them at the end. Figure 5.1 contrasts the two structures depending on how questions are handled.

Opening   Begin your presentation with either a *greeting* or a *grabber*. A greeting could be as simple as saying *Good afternoon. Thank you for joining me to discuss the fuel gauge problem.* A grabber is a startling statement, a particularly interesting or illustrative example,

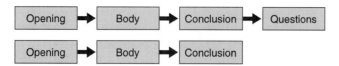

Figure 5.1 Structure of Presentation Depending on Where Questions Are Asked

or a dramatic visual. For example, consider how one engineer might grab audience attention by beginning a presentation about fiber optics by describing an early morning call from her father in England. She describes how he called on his videophone on Christmas Day to thank her for his present. She describes her pleasure at seeing him hold up the sweater she had sent him and then pull it over his head and model it for her. She then tells the audience this scenario represents only one of the exciting possibilities of fiber optics in the world of telecommunications.

In the opening, you should also outline the presentation, telling the audience what topics you will cover or what points you will make. For example, in the presentation on fiber optics, the speaker could say that she will first explain what fiber optics is, provide a progress report on the installation system in North America, and then introduce some of the telecommunications services we will enjoy once the system is activated.

The outline should include or be followed by a *value statement*, explaining why the audience should listen to you. Again, the fiber optics speaker indicates the value of the information she is presenting by telling us that her presentation will help us plan and make better business decisions regarding communications technology.

In a lengthy presentation, your opening should also include *housekeeping information*, explaining how and when you want members of the audience to ask questions, mentioning any print material available to them, giving details of timing, and so on. Our fiber-optics speaker might recommend that we ask questions throughout her session, and she could offer to remain after the presentation if anyone has specific questions they were reluctant to ask during the presentation. Such a well-structured opening and thoughtful behavior would set the stage for a cooperative audience response.

Body and Organizational Patterns   Ensure that the three to five points in the body of your presentation are logically arranged to assist listeners in understanding and recalling your information. Sometimes the parts of a presentation virtually arrange themselves, but other times, the order or structure is much less obvious, and a conventional organizational pattern may lend guidance.

One of the following organizational patterns should help you organize your presentation:

- Time sequence
- Selling
- Problem solving
- Spatial relationships
- PREP (Point, Reason, Example, Point)

Note that these patterns are also useful for organizing sections within a longer presentation.

Time Sequence   A time-sequenced approach is appropriate when your purpose is to provide an update on a project, to describe your company's history and future outlook, or to lead the audience through a process. The basic pattern is *first, next,*

Figure 5.2 *Time-Sequence* Pattern for Organizing the Body of a Presentation

last. Note how a student could use this approach to structure part of the opening to his presentation:

> *I spent my industrial internship working for Biochip International. The company was formed in 1990 to produce. . . . Today, they are primarily engaged in. . . . And the company is now expanding rapidly and is about to move into . . .*

Figure 5.2 illustrates the *time sequence* pattern.

Selling   Use a selling approach if your objective is to win the support of your audience. For this organizational pattern, describe the problem you are facing, explain how it affects or potentially impacts your audience, then offer a solution. Be specific about the benefits of the solution you are suggesting. The basic pattern is *problem, impact, solution, benefits*, as demonstrated by the following example:

If the president of our fictional company, Biochip International, is speaking to investors, she could encourage financial participation with the following opening:

> *Current uncertainties in world markets create enormous risks for investors. Your challenge is to invest in companies with an opportunity for extraordinary growth without extraordinary risk. Our firm offers you that growth and a high degree of security because we are the only firm globally offering a chip that safely increases milk production in cows and goats. The alternatives are hormonal therapies that, quite frankly, frighten consumers. Our product has increased milk production by as much as fifty percent without shortening the animal's life. It has passed preliminary health standards without difficulty. If you participate in this amazing technological development, your investment will yield exceptional returns.*

Figure 5.3 illustrates the pattern for the *selling* approach.

Problem Solving   A problem-solving approach is useful to invite participants to help solve a problem or to clarify why you have chosen a particular solution. To explain your choice of solutions, you could lead a group through the thought processes you used to solve the problem. To work out a solution, you might state the problem, then discuss one or two possible solutions (and their drawbacks, if appropriate), and then invite members of the audience to contribute their thoughts on a solution. The basic problem-solving pattern is *problem, possible solutions, discussion, decision*.

In the following example, the Biochip intern uses the problem-solving approach to lead his audience through the steps taken to reach a solution:

> *During my internship at Biochip, I had an opportunity to learn how milk production can be enhanced. Because milk production increases following births, farmers have historically attempted to increase production by developing better feeds, through careful breeding, and by breeding cattle more frequently. But these solutions are no longer sufficient. And so other methods are being sought. One solution is to use hormones that make dairy cattle perform as if they have recently calved. But these hormones shorten the animal's life, and some consumers fear side effects from consuming milk products with hormonal additives. At Biochip we have developed a better solution . . .*

Figure 5.4 illustrates the *problem-solving* pattern.

Figure 5.3 *Selling Pattern* for Organizing the Body of a Presentation

Figure 5.4 *Problem-Solving* Pattern for Organizing the Body of a Presentation

Spatial Relationships A spatial-relationship approach is often appropriate when providing technical detail. First explain the practical applications of the technology and then provide an overview of a product or process and discuss the parts. Clarifying the value or utility of a project or process up front ensures that the audience understands why you are providing them with various details and explanations later on. A failure to explain what a device, process, or system is used for frustrates and confuses those members of your audience who are unfamiliar with the one under discussion. These may be the very people you most want to inform and impress.

The basic pattern is *application, overview, description of parts*, as demonstrated by the following abbreviated example:

> *An optical character recognition system transfers information from hardcopy to softcopy, translating books, reports, articles, etc. into text files. The system requires a scanner and specialized software to convert the scanned image into a text file. The scanner. . . . The software . . .*

The presenter would continue by describing the scanner and the software for the system. The pattern for the *spatial relationship* approach is illustrated in Figure 5.5.

PREP Approach A PREP approach is useful for persuasive presentations. It involves stating your position or point of view, explaining the reason you have taken that position, providing a story, fact, or analogy as an example, and concluding by restating your position or point. The basic pattern for this approach explains the acronym P*oint*, R*eason*, E*xample*, P*oint*.

In the following example of the PREP approach, a student leader is encouraging fellow engineering students to participate in a fundraiser that involves an engineering student or professor jumping into the icy winter water of the campus pond for every $20 collected.

> *I am here today to ask you to participate in this year's Polar Plunge, which will be held next Friday at noon. I know you will have a great time and feel really good about helping us raise money for a new Variety Club van. Last year the event raised more than two thousand dollars. How many of you plunged last year? (A show of hands.) And how many of you will do it again? (Much laughter, many hands up). It's lots of fun and for a worthwhile cause. So please, collect money to dunk your favorite prof or lab partner and join me next Friday.*

Figure 5.6 illustrates the *PREP* approach.

Closing One way to end a presentation is to review the main points. If a presentation aims to persuade the audience, then the closing should emphasize information critical to the decision-making process. For example, a presenter might provide the following review:

> *I've provided you a brief introduction to the world of fiber optics and given you some idea of when it will be available in your area. Perhaps the most exciting part of my presentation, judging by your many questions and our lively discussion, was the potential applications of this technology in your workplace. My company is well situated to help you plan for that future.*

Note how this example also provides a subtle reminder of the practical value of the presentation.

Figure 5.5 *Spatial Relationship* Pattern for Organizing the Body of a Presentation

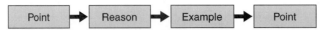

Figure 5.6 *PREP* Pattern for Organizing the Body of a Presentation

If a presentation on fiber optics aimed to inform, the closing might remind the audience of new services about to come on-line. Whether your emphasis is on informing or persuading, at the end of your presentation, remind the audience of your purpose. Tell them what you want them to do, think, or remember based on your presentation. For example, the talk on fiber optics might end as follows:

> *Through my presentation, you've begun to understand the exciting telecommunications opportunities of the next two decades. I urge you to study this topic further. You are welcome to contact our firm if you'd like additional information. I can also stay to talk to you when this session wraps up.*

5.3.2.5 Create Links

When your presentation is organized, plan how you will bridge or link points so that you move smoothly from one part of your presentation to the next. For example, the following statement creates a bridge from a definition of fiber optics to a discussion of installations in North America: *Now that you have a basic understanding of fiber optics, I'm sure you'll want to know when it will be available in your area.*

You can also use your linking statements to reinforce the value of each point to the listener. For example, the point about the installation of fiber optics can be linked to new telecommunication services with a statement that emphasizes their value from a business perspective: *The next part of my presentation will help you plan your telecommunications strategies.*

5.3.2.6 Select and Prepare Visual Aids

Another essential aspect of your presentation is your visual aids. We live in a visual society and many people understand and retain information better if it is presented visually. Visual aids can increase interest, illustrate key points, and signal transitions from one part of your presentation to the next. They also significantly increase the impact of your message, help listeners retain important information, and help you present ideas without depending on notes. Further, for those not listening in their first language or who are unfamiliar with your accent, visual aids can turn the incomprehensible into something understandable.

Deciding what aids to use is an ongoing process that can occur at any and all points of the planning process. As you organize, prepare rough drafts or sketches of your visual aids. Creating final versions is best left until you have a firm idea of your outline and have begun rehearsing. However, note any lengthy production times early in the planning process.

As you plan what aids to include, also consider how you will use them to ensure a smooth, professional performance. Whether you are using a simple object or interfacing to on-line resources, plan ahead. The following sections discuss various visual aids you can use in your presentations.

Overhead Transparencies and Computer-Projected Images   Overhead projectors remain standard presentation equipment because they are relatively inexpensive, can be used in a lighted room, and a speaker can write on them while facing the audience. Overhead transparencies are also relatively inexpensive to prepare and can be modified during the presentation and then cleaned for later use with another group. Increasingly, however,

The major considerations for review are as follows:

1. **Teamwork**

   - Emphasize team member participation in planning and managing each project.
   - Encourage team building.
   - Increase lateral communications between functional departments.

2. **Coordination**

   - Develop balanced resource planning.
   - Require marketing to solidify commercial specifications before commitment to build.
   - Increase the involvement of senior management in the project process and the coordination of the functional departments.

3. **Training**

   - Provide formal project management training to all members of the project group.
   - Develop an effective bid-response system.

Figure 5.7 Example of a Bulleted List in a Report

computers and more expensive projection equipment are being used, and visual aids are being created using presentation software such as Microsoft's *PowerPoint*.

Presentation software can help you create full-color slides that are easily edited and that support interesting on-screen features such as text fading in, flying out, and dissolving. These features are fun to use and can add excitement and interest to your presentations. However, too many presenters—especially those just learning to use presentation software—become so focused on creating visual effects that they ignore content. When you use presentation software, be sure that your form supports your content rather than substituting for it.

Whether traditional transparencies or computer-projected images are used, the general guidelines remain the same: limit content and if an image is primarily words, follow the 5×5 rule: strive for no more than five lines per slide and no more than five words per line. Figures 5.7 and 5.8 provide examples of what a bulleted list might look like in a document compared to how it should appear on an overhead transparency. Note that much of the detail has been eliminated in Figure 5.8.

Recommendations

1. **Teamwork**

   - Team building
   - Lateral communications between departments
   - Team participation in planning and management

2. **Coordination**

   - Balanced resource planning
   - Increased involvement of senior management
   - Marketing must solidify commercial specifications

3. **Training**

   - Project management
   - Effective bid-response system

Figure 5.8 Example of a Bulleted List on an Overhead

Table 12 presents the *Moh's Hardness Scale* for minerals:

Table 12: Mohs Hardness Scale for Minerals

Hardness	Common name	Chemical formula	Sp. gravity
1	Talc	$Mg_3Si_4O_{10}(OH)_2$	2.7–2.8
2	Gypsum	$CaSO_4 \cdot 2H_2O$	2.3–2.4
3	Calcite	$CaCO_3$	2.7
4	Fluorite	CaF_2	3.0–3.2
5	Apatite	$Ca_5(PO_4)_3(F,Cl,OH)$	3.1–3.2
6	Feldspar[†]	$KAlSi_3O_8$	2.5–2.6
7	Quartz	SiO_2	2.7
8	Topaz	$Al_2SiO_4(F,OH)_2$	3.4–3.6
9	Corundum	Al_2O_3	3.9–4.1
10	Diamond	C	3.5

Source: Chesterson, C.W. and K.E. Lowe, *The Audubon Society Field Guide to North American Rocks and Minerals*, 1978, (New York: Alfred A. Knopf).

[†]Figures are for Orthoclase Feldspar. The hardness of Plagioclase Feldspar ($NaAlSi_3O_3$) is the same, but the specific gravity is greater (2.62–2.76).

Note that there is not a simple relationship between the hardness of a mineral and its specific gravity. Also note that you can easily remember the order of hardness here by memorizing the following sentence and associating the first letter of the minerals with the first letter of the words in the sentence: *The Geologists Can Find All Friends Quickly Through Clever Devices.*

Figure 5.9 Example of a Table in a Report

As Figure 5.8 suggests, you should provide viewers with the bare bones. Add detail as you speak.

Also keep graphics and diagrams simple and remove unnecessary detail. Figures 5.9 and 5.10 present examples of a table as it might appear in a document and as it would appear on an overhead transparency. Again note that excess detail has been eliminated.

Moh's Hardness Scale for Minerals

Hardness	Common name	Sp. gravity
1	Talc	2.7–2.8
2	Gypsum	2.3–2.4
3	Calcite	2.7
4	Fluorite	3.0–3.2
5	Apatite	3.1–3.2
6	Feldspar	2.5–2.6
7	Quartz	2.7
8	Topaz	3.4–3.6
9	Corundum	3.9–4.1
10	Diamond	3.5

- **Increase in Hardness $\neq$ Increase in Specific Gravity**
- **The Geologists Can Find All Friends Quickly Through Clever Devices**

Figure 5.10 Example of a Table on an Overhead

RECOMMENDATIONS

1. TEAMWORK

- TEAM BUILDING
- LATERAL COMMUNICATIONS BETWEEN DEPARTMENTS
- TEAM PARTICIPATION IN PLANNING AND MANAGEMENT

2. COORDINATION

- BALANCED RESOURCE PLANNING
- INCREASED INVOLVEMENT OF SENIOR MANAGEMENT
- MARKETING MUST SOLIDIFY COMMERCIAL SPECIFICATIONS

3. TRAINING

- PROJECT MANAGEMENT
- EFFECTIVE BID-RESPONSE SYSTEM

Figure 5.11 Overhead Using Kaufmann Uppercase Font

Also keep your transparencies and slides uncluttered so that your audience attends to the topic-related content of your presentation rather than admiring your corporate art-work. Each slide should be titled, but the title should be brief. If a corporate logo must appear, put it in the lower left-hand corner where it will be least distracting. Also avoid the temptation to add multiple colors and exciting backdrops to your slides. Limit yourself to two or three main colors and keep the background simple. Various shades of blue are effective background colors.

Also pay attention to font type and size. Avoid decorative fonts such as *Kaufmann* because they are too hard to read (not to mention inappropriate in an engineering context). Instead use a less-ornate font such as *Arial*, and keep the use of uppercase to a minimum because it is harder to read than lowercase. Work in a minimum 24-point size so that your viewers will not need to strain to see what you have written. Figures 5.11 and 5.12 contrast the readability of *Kaufmann* (all uppercase) with the readability of *Arial* (mixed upper- and lowercase) to indicate just how dramatically the choices you make can affect readability.

Note that if you are in a room with participants at the same level as you or at a lower level than you are and with the projector at the front of the room, the projector may

---

Recommendations

1. **Teamwork**

- Team building
- Lateral communications between departments
- Team participation in planning and management

2. **Coordination**

- Balanced resource planning
- Increased involvement of senior management
- Marketing must solidify commercial specifications

3. **Training**

- Project management
- Effective bid-response system

Figure 5.12 Overhead Using Arial Mixed Upper/Lowercase Font

obscure the lower portion of the image. To accommodate this common setup, keep as much of your text or image as possible in the top ⅔ of your slide. When using presentation software, you can avoid using the lower portion of the screen by choosing a landscape orientation. Note that Figure 5.12 presented an example of an overhead with the text in the top ⅔ of the screen. See the information about format in Chapter 7 (pages 254–273) for more information on how to present information effectively.

Figure 5.13 presents the six slides used to support a 10-minute formal presentation using the various organizational and formatting principles outlined above.[1]

White Boards or Flip Charts Anytime you cannot anticipate all the visual information your audience may require, a white (or black) board is useful. In preparing for your presentation, clean the board and check the condition of markers (or make sure long-enough pieces of chalk are on hand). When using a white board, note that black or blue markers are easier to see than yellow or orange, although light colors are useful for highlighting points.

For a discussion or brainstorming session, also arrange ahead of time for another participant to record ideas so that you are free to continue a discussion or to comment while points are being recorded. Because you should avoid speaking while turned to the board, if you are writing or drawing yourself, stop talking while you work and then turn back to face your audience before commenting.

Computer Interfaces to On-line Resources You can also connect your computer to on-line programs or the Internet, linking to a database or your home page in order to create real-time demonstrations or to take advantage of various Internet resources. But proceed with caution. Many existing programs are not designed for group viewing, and some projection equipment will not produce an image of sufficient quality for the audience to see clearly what is happening on the screen. Before including an on-line demonstration in your presentation, also make sure you are sufficiently skilled at manipulating the program. Otherwise, you will project errors, waste time, and confuse or irritate your audience. Rehearsal and a final test run in the meeting room are essential to successful on-line demonstrations.

Even the best-prepared on-line presentation is susceptible to equipment failure or the sudden loss of an Internet connection. Plan for what you will do if your Internet link goes down just as you start to demonstrate your web site. In such moments, good humor is a great ally, but so is planning. If you can produce a set of overheads and carry on with a lower level of technology, your audience will be both relieved and impressed. The more significant your presentation, the more you need a backup plan.

Video Short video clips are useful when you want to portray the operation of a process or device, to showcase a facility, or to demonstrate a working prototype. Ensure the images you project are large enough so everyone in the audience can see them easily. Before your audience arrives, also set your videotape to the precise starting point. Before running the tape, explain to the audience what they will be watching and why you consider this video clip of value to them. Your audience will be more attentive and the video more effective if the context and purpose are clear before the viewing begins.

---

[1]We would like to thank Yann Le Du for his permission to use the overheads he prepared to organize and illustrate his presentation.

Yann Le Du,
Research Assistant

**Development of a Non-Contact
Diameter Gauge**

CREO Products Inc.

Burnaby, BC

Overview

• **Introduction and background**

• **Motivations and objectives**

• **How the gauge works**

• **Test results**

• **Conclusions**

Figure 5.13 Sample Overhead Transparencies for a 10-Minute Formal Presentation

Objectives of the Project

To build a gauge that is

• **Accurate to 3μm in 300mm**

• **Suitable for a range of diameters**

• **Hand held**

• **Unaffected by surface irregularities caused by machining**

How the Diameter Gauge Works

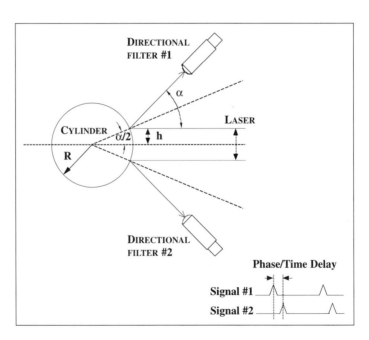

$$\text{Sin} \ (\alpha/2) = h/R$$

Figure 5.13 *(Continued)*

How the Directional Filter Works

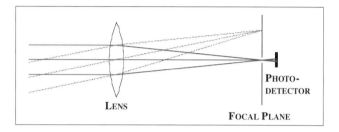

PHOTO-
DETECTOR

LENS

FOCAL PLANE

Test Results

• Accuracy of 1:100 000 achieved if incident rays limited to center of directional filter lens.

• To restrict light, gauge must be rigidly supported.

• Supported gauge must be calibrated to measure cylinders of only one nominal size.

Figure 5.13 (*Continued*)

Objects   If you use objects to illustrate a presentation, be sure they are large enough so that you can avoid passing them around and thus avoid distracting your listeners. Plan to hold an object so that everyone can see it easily and to display it only as long as it is relevant to your listeners' understanding of the material. Plan what you will do with it once you have finished using it. Preferably, set it aside or even out of sight so it will not continue to draw attention. If you sense that some members of your audience would like a closer look, you can invite them to view the object after you conclude.

35mm Slides   Presentation software and digital cameras are displacing 35mm slides as a presentation medium. But if you should use 35mm slides, prepare them following the guidelines for overhead transparencies and computer-projected slides. Remember the 5×5 guideline: no more than five points and no more than five words per point. Limit the slide segment of your presentation to no more than 10 minutes. Because most 35mm projectors require a darkened room, you lose contact with the audience and may not realize if people are beginning to fall asleep. If you have more than 10 minutes worth of slide material, plan a break, taking time out for questions or to summarize what has been discussed. With any technology that requires dimming the lights, brighten and dim the lights gradually if possible or warn the audience that a sudden change in light level is coming.

5.3.3   Preparing and Using Speaker's Notes

Ideally, you will speak to your audience without constant reference to notes, but you should prepare notes and have them available in case you need them. Some people prefer small index cards, thinking their notes will be inconspicuous. We see no reason to attempt hiding your notes. If you find notes useful, then the format should be large enough so you can read them at a glance.

When rehearsing, note any information you should include. For example, you may include information that corresponds to a diagram, reminding you of the points you plan to make. You may also include details and facts you may not remember.

Use at least a 14-point serif font, avoid crowding information on the page, and fill only the upper half to two-thirds of each page in order to avoid the need to dip your head to scan the lower part of a page. Leave pages loose so that as you finish reading each page, you can slide it off the pile, minimizing the audience's attention to your notes.

To avoid looking down at your notes and leaving the audience to view the top of your head, arrange for a lectern to place your notes on. Alternatively, you can hold your notes up in one hand, but ensure the audience has a clear view of your face.

5.3.4   Preparing for Questions

A question period following or questions asked throughout your presentation benefit both you and your listeners. Questions provide you with feedback, indicating the extent to which your listeners accept your ideas. Through your answers, you can address points that confuse your listeners as well as reinforce your message. You should, therefore, plan for questions and you should develop strategies to deal with difficult questions and questioners.

5.3.4.1   Preparing for Questions

While preparing a presentation, anticipate questions. Make a list of ones you think your audience will ask and plan for how you will answer them. If specific details are required to answer them, list the necessary facts in your notes for easy reference. When you rehearse before a practice audience, ask your listeners to note any questions they think you will be asked.

Also decide when you want the audience to ask questions. Is your presentation best suited to questions at any point, after each section, or at the end? Also tell your audience how you plan to handle questions. When you reach a point at which you will accept questions, invite responses verbally: *I will answer any questions now.* If the audience seems reluctant to respond to your verbal invitation, use nonverbal signals as well. The most direct approach is to simply look at them and remain silent. In a matter of seconds, someone will inevitably ask a question or comment on some aspect of your presentation. The trick is to remain silent long enough. At first, you may feel like you are waiting for several minutes, but if you count off the seconds, you will be surprised by how little time passes before someone addresses you.

5.3.4.2   Answering Questions

Always treat questions as compliments, as signs that the audience is interested in what you have to say. In this way, you will avoid being defensive and reassure listeners that you welcome their questions. Also assume that all members of the audience are interested in the answers. Repeat a question if everyone may not have heard it and paraphrase a complex or poorly worded question to clarify what is being asked. Also identify the components of a complex, multifaceted question and answer each in turn. If you are uncertain as to what is being asked, repeat what you think you heard or rephrase the question, then ask if your interpretation is correct.

If you do not know the answer to a question, admit it. Another member of the audience may offer an answer. If not, you can always offer to provide the requested information after the presentation. Occasionally, the person who asked the question actually knows the answer, but was just checking to see if you did as well. If you admit not knowing the answer in this situation, be assured the questioner will provide it. In any case, admitting your ignorance is preferable to providing incorrect information.

Think before you speak. Organize your thoughts so that your answers support the points made in your presentation. If you need a moment to gather your thoughts, look down briefly; your audience will wait for you and will appreciate your thoughtful response.

5.3.4.3   Dealing with Difficult Questions

Infrequently, a questioner will try to trip up a presenter with a query based on false premises or irrelevant assumptions. In such cases, your best defense is to maintain a positive attitude and be polite, but firm. Avoid any comment or actions indicating you are annoyed and, as soon as possible, direct your attention to other participants and encourage their questions. If you hold your ground, you can usually disarm such questioners by asking them to clarify their question and to share the source of their information.

Also be prepared to divert inappropriate questions. Your audience will appreciate your ability to avoid wasting time on questions that are off topic. Even if you know the answer to an apparently irrelevant question, a good strategy is to politely ask the person to clarify how the question relates to your presentation. In most cases, the answer you receive will allow you to move on. In other cases, the answer will clarify that the question is indeed relevant—which is one good reason for always being polite and respectful.

You may also encounter individuals who try to monopolize the question period. If someone keeps asking questions while others are obviously waiting to ask theirs, raise your open hand as if to stop traffic. If necessary, you can make a pointed comment: *Excuse me, I'd like to give everyone a chance to ask questions.* Then direct your attention to someone else.

5.3.4.4   Ending the Question Period and Presentation

Plan how to end your presentation and schedule a few minutes for a brief summary, for a reminder of the points you want the audience to remember, or for some other upbeat ending. As you answer questions, watch the time so that you can tell the audience when you have time left for only one or two more questions. You can then control the way your presentation ends.

5.3.5   Rehearsing Your Presentation

Many presenters spend too little time rehearsing, often because they attempt to finalize the content and polish their visuals before beginning to rehearse. When they are finally satisfied, little time is left for practice. Moving rehearsals earlier in the process allows for a more accurate estimate of the time required to present the planned information, to revise or eliminate visuals before too much time is spent polishing them, and to develop a natural, relaxed, professional presentation. If a presentation is well rehearsed, you should be able to speak in a conversational voice, to avoid excessive reference to notes, and to handle disruptions without obvious concern or confusion.

If time permits, we recommend the following sequence of events:

1. Begin with a walkthrough rehearsal.
2. Rehearse in sections.
3. Rehearse using visual aids.
4. Rehearse the entire presentation.
5. Rehearse with an audience.
6. Perfect your timing.
7. Prepare psychologically.

Try following this sequence of events when preparing for your next presentation.

5.3.5.1   Begin with a Walkthrough Rehearsal

For a walkthrough rehearsal, you simply describe the outline of the presentation, referring to details but not actually delivering them. This type of rehearsal takes very little time and is particularly useful for developing a good sense of the flow of your presentation. A walkthrough is also a good time for reorganizing content and deciding where additional transitions or visual aids are needed.

5.3.5.2   Rehearse in Sections

When you begin a full rehearsal, consider breaking your presentation into sections, rehearsing each part until you are comfortable with it before moving on to the next part. For example, rehearse your opening and transition to the body of your presentation until you are comfortable with that material, then begin with the transition and practice the first point in the body and the transition to the next point. Start and end each section you practice with a transition between points until you reach the end. This technique is particularly useful when practice time must be broken into short periods, but it is also a good way to reduce the frustration of trying to remember the entire presentation at an early rehearsal.

5.3.5.3   Rehearse Using Visual Aids

When you are comfortable with your content, focus attention on how you are presenting your visual aids. For example, to manage the audience's attention, use some sort of pointer to indicate what part of a slide you are addressing. For an overhead transparency, you can use a pen or pointer placed on the glass surface of the projector, or you can step back flush with the screen and use a telescoping pointer. You can even use your hand if you can hold it still and can reach high enough. When using presentation software such as PowerPoint, a laser pointer is particularly useful. But note that laser pointers must be held very still because any hand movement is greatly exaggerated by the time the light strikes the image on the screen.

Some presenters are fond of revelation techniques such as covering those points or areas of a screen they do not want the audience to read until they are ready to discuss them. This technique is a source of irritation for many viewers, so use it sparingly, if at all. If you are using presentation software, you can achieve the same effect by bringing each image or line of text onto the screen as you are ready to discuss it. Viewers seem more accepting of this technique than they are when a presenter covers areas of a transparency.

5.3.5.4   Rehearse the Entire Presentation

Make at least one dry run of your entire presentation complete with visual aids and note how long it takes. Deliver your remarks as if the audience were present: project your voice, gesture appropriately, and imagine each pair of eyes before you as you speak. If stage fright slows down your speech so that you lose fluency, consider another dry run in which you deliver your presentation at enormous speed, not worrying about diction, expression, or sentence structure. The goal of this type of rehearsal is to increase your comfort with the flow of the presentation.

If nerves make you speak too quickly and ignore pauses, then you should rehearse a slowed-down version of your presentation, particularly of your opening. In this case, exaggerate the pauses, enunciate every word carefully, and stress key words. Of course, you won't deliver your presentation in this exaggerated form, but this sort of practice can improve the pace at which you deliver.

If your body language feels awkward or you are unsure how you appear to your audience, rehearse in front of a large mirror, or videotape a practice session, so that you can see your posture, movements, and gestures. If you are worried about how you sound, an audio recording will alert you to such problems as a monotone voice, excessive use of *um* or *ah*, and parts of your presentation that sound unnatural. Later in this chapter, we discuss issues relating to movement and voice in more detail.

5.3.5.5   Rehearse with an Audience

When you are reasonably comfortable with your presentation, invite at least one co-worker, family member, or peer to observe a full rehearsal. If you suffer from stage fright, rehearsing in front of a small, friendly audience will help build confidence. If you use notes, a dress rehearsal is particularly useful for practice maintaining eye contact with the members of your audience. You should also practice pausing between points. If you use notes, you can use these pauses to look at your notes before you begin your next point.

If, like many presenters, you tend to add information or change your style of delivery when facing an audience, a dress rehearsal will help you gain a better sense of the actual time required to complete your presentation. You can also ask your practice audience to provide feedback based on the items in the heuristic for improving delivery later in this chapter.

5.3.5.6   Perfect Your Timing

Good timing is crucial and you should rehearse a formal presentation until it is a little under time because staying within time limits is a mark of courtesy and professionalism. Engineers are often called on to make significant presentations in limited amounts of time. For instance, engineers in one large company tell us that project managers are given only three minutes to make annual progress reports to company executives. In this very short time, they must inform executives of their teams' accomplishments, convince the audience that they need a bigger budget or more personnel, and more—all within *three minutes.*

The shorter the time allowed, the more difficult the task. The more difficult the task, the greater the need not just for careful planning and effective visual aids but also for diligent rehearsal.

5.3.5.7   Prepare Psychologically

Aside from physical rehearsals, you can also rehearse in your mind just before your presentation. This sort of practice is particularly useful if you suffer from anxiety. In a quiet place, close your eyes and imagine yourself in front of your audience, handling everything smoothly and professionally while the audience responds enthusiastically. Because our minds seem not to distinguish mental practice from real experience, positive images can count as a successful speaking experience. Following mental practice, you should feel more relaxed, more confident, and ready for your presentation.

5.3.6   Other Preparation Tips

A true professional will go beyond rehearsing to look after a number of other preparation details. When you prepare for a presentation, assemble all your visual aids and check that they are complete and, if used previously, that they are up-to-date and in good condition. Leave sufficient time to make changes or to recreate a missing visual. Reserve equipment such as microphones, computers, or projectors well ahead of time and prepare a contingency plan in case equipment fails to operate or does not arrive. Also gather pens, pointers, or any other items you may need. Determine whether water will be available and, if not, make appropriate arrangements to have some on hand.

Beyond what you are expected to bring yourself, ask for what you need. Presenters often work in poor conditions because they failed to ask for the right things. For example, a remote lapel microphone is preferable to a hand-held microphone or to one with an electrical cord, and most conference facilities and many companies can supply them. Yet too many speakers struggle with hand-held microphones and dangling cords, or remain bent over fixed microphones simply because they did not ask if other, more appropriate, equipment was available.

If you will be presenting for the first time in a particular room, find out what equipment is in stock and what can be ordered. Also arrive early to make sure the room is set up properly with sufficient seating. Learn how to adjust the lighting and set up and test the audio and visual systems so that your presentation can proceed smoothly. If you are presenting away from your immediate area, stay in close contact with the meeting planners, hotel staff, or other administrative personnel who are coordinating your meeting or event. Build a healthy relationship with them, and you should receive all the help you require.

5.4   DELIVERING YOUR PRESENTATION

No one becomes a skilled presenter overnight; everyone needs time to develop delivery skills. The more times you present, the more likely you are to become a confident and controlled presenter. This section explores a number of issues relating to three key aspects of delivery: body language, voice, and speaking style.

5.4.1   Body Language

Body language is critical in oral communication because we see speakers before they start talking and immediately begin forming opinions of their abilities and attitudes based on nonverbal signals. Body language includes all aspects of appearance, including what you wear, how you stand, eye contact, movement, facial expression, use of space, and physical habits such as constantly licking your lips, jingling the change in your pocket, or playing with your hair. Eye contact, use of space (or movement), and gesture are key concerns.

5.4.1.1   Eye Contact

Looking directly at listeners builds rapport. Sustained eye contact tells your audience that you *believe in this message, that you want them to pay attention*, and that you are *giving them the honest goods.* However, avoid making anyone uncomfortable by appearing to stare at them. Hold eye contact for three to five seconds and then move on to someone else. Alternatively, you may glance at several members of the audience long enough to register their reactions before selecting an individual for sustained—three- to five-second—contact. Eye contact is especially important as you start your presentation and as you begin each new section. When you need to break eye contact in order to refer to your notes or collect your thoughts, looking down briefly is less distracting than looking at the ceiling or off to one side.

When using visual aids, you may be tempted to talk while looking at the projected image. Certainly, take a moment to ensure that the image is focused, properly oriented, or projected squarely on the screen. But then turn back to your audience and continue to make regular eye contact while you discuss the content of your visual aid. Otherwise, you may literally turn your back on your audience and undermine the rapport you have been establishing.

5.4.1.2   Use of Space and Movement

In our society, space is used to signal power and status. For instance, a corporate CEO has a larger office than more junior employees, the chairperson is given extra space at the head of a table, and you will most often be encouraged to speak from an open space in front of your audience. You should use the psychological power of this space to advantage.

Size your gesture and movement to the space available. For instance, when standing with generous space to maneuver in, move more boldly than usual and increase the size of your gestures. But even when seated at a table where you must use relatively small gestures, you can still reach out over the surface of the table to subtly extend your territory or space.

While making full use of the space available expresses control and authority, moving closer to your audience is useful to invite discussion, to express agreement, or to emphasize a point. Be careful, however, not to retreat from this intimate position as you make a key point or conclude your presentation. If you do, you encourage the audience to downplay the importance of your point.

Dynamic, forceful, influential speakers understand the importance of space and use it to advantage. You can learn a great deal by observing them at work and by noting how others use space to their advantage or disadvantage.

5.4.1.3   Gesture

Gestures, including posture and facial expressions, project attitudes and are effective means of holding an audience's attention. Upright posture and emphatic movement express confidence, enthusiasm, and control. Facial expressions can be used to emphasize your message, to express enthusiasm, and, in the case of a smile, to indicate that you are friendly and approachable.

Arm and hand gestures can also enhance your message while adding color and emphasis to key points. Bold movements and sustained gestures let the audience know you are confident and expect their attention; sustained gesture is also effective as you complete each main point. Note, however, that overly dramatic gestures distract your listeners.

Be sure to face your audience as you speak to signal that you are paying attention to them and that you expect them to do likewise. As much as possible, use open body language (keeping the palms of your hands open to the audience) to signal your confidence. If you find this orientation strange or awkward, try rehearsing in front of a mirror. (You can also practice in front of a window when it is dark outside and light inside.)

Stiff, unnatural positions such as the *fig leaf* (hands crossed below the waist with the back of the hand toward the audience) send the message that you are unsure of yourself and your message. Instead, lower you hands to your sides in a natural, relaxed resting posture. For variety, you can rest one hand in your pocket or on a table, but keep the duration of such gestures brief and refrain from using them too often so that you look relaxed but not lazy. Relaxing your posture slightly will also encourage questions and discussion. Figures 5.14 and 5.15 contrast the fig leaf position with a more open posture.

Figure 5.14 Fig Leaf Position   **Figure 5.15** Open Posture

5.4.1.4   Other Ways to Improve Body Language

If possible, videotape a segment of your presentation, and then, as you play it back, identify one aspect of body language you want to improve. After further rehearsal, record the segment again to note improvement and to check whether any other aspects require work. Practicing in front of a mirror is also useful to improve facial expression while practicing in front of a large window at night is useful to improve posture and gesture because only these aspects of your performance are reflected.

Practicing in front of an audience, even of one friendly face, is a test of your readiness. Once you have worked through your entire presentation and are comfortable with your movements and gestures, invite a friend to watch your performance and to note what aspects of your body language are particularly strong or need improving.

5.4.2   Voice

The clarity and qualities of your voice impact the way an audience receives your presentation. While some aspects of voice are difficult to alter or take time to develop, others largely depend on your delivery. For instance, whether someone is reading from notes or speaking naturally is generally evident to an audience. Because reading voices tend to be more monotone and less interesting than speaking voices, you are generally better received when you make a presentation without reading from a script. Other issues relating to voice are projection, pitch, and pace.

5.4.2.1   Projection

Note that loudness and projection are different aspects of voice: your voice must always project but need not always be loud. Stage actors are trained so that they can speak in a whisper and yet be heard in the back row of a theater. This ability to project sound is essential to presenting without a microphone. If you master projection, then you can lower or raise your voice to hold your audience's attention. For example, if you begin in a loud voice, you will command attention; shifting to a softer, quieter voice—even to a whisper—when appropriate will hold that attention. Note that you generally want a louder voice when emphasizing key points and when beginning a new section.

To make your voice project (i.e., travel throughout the room), breathe deeply. You need air in your lungs to project. Too often, presenters let their voices fade out at the end of a thought, just when they are delivering critical information. Note that your voice need not be loud at the end of a point, but it must project.

5.4.2.2   Pitch

Pitch is also influenced by your air supply: if you run out of air, you cannot control the pitch of your voice. Pitch is also affected by the stress you place on your vocal cords. Like the strings of a guitar, if you tense the vocal cords, a higher pitch results, and if you relax the vocal cords, a lower pitch results. To avoid a high-pitched voice, keep your chin level so that the area around the vocal cords is relaxed. Tilting your chin up or down reduces your ability to control pitch.

Use a variety of pitches to hold your listeners' attention. A low-pitched voice is pleasant, projects an aura of authority, and is most appropriate when delivering key points. A higher-pitched voice is associated with excitement, humor, and friendliness and is appropriate when providing examples. However, avoid raising the pitch of your voice as you end a sentence or statement. Often called *pitching up*, this vocal pattern makes your remarks sound tentative or unfinished.

5.4.2.3   Pace

Speakers are often cautioned to speak slowly, and yet when they do they often sound dull and boring. Our advice is to vary your speaking pace in order to enhance your message and to use pauses to create emphasis. A rushed delivery with no pauses suggests panic, while a too-slow delivery is simply boring and encourages listeners to engage in other activities. A well-paced, varied message suggests enthusiasm, self-assurance, and awareness of audience.

A varied pace and adequate pauses maintain an audience's attention. Speed up when you are telling a story or reviewing known ideas, but slow down when you are giving new and difficult information. Pause after key statements, between main ideas, and just before you conclude.

5.4.3   Speaking Style

Speaking style is another crucial aspect of delivery. Listeners generally require shorter sentences, simpler words, and more repetition than readers do. When you speak, your words are transitory, and so being clear and easy to understand is particularly important. If your words confuse your listeners, they will quickly become irritated and alienated. The following strategies will help keep your audience alert and responsive.

5.4.3.1   Personalize Your Presentation

Address your audience directly, using participants' names when you know them. Those addressed will appreciate the recognition and tend to pay closer attention after being acknowledged. Also develop a habit of using personal pronouns such as *you, your, we, our*, and *us* in your presentations. For example, *some people around here have noticed a decline in morale among the office staff* can be phrased more personally as *you have likely noticed a decline in morale among the office staff*. Or *a solution must be found* can become *we will all benefit by working together to solve this problem*.

5.4.3.2   Use Clear, Simple Language

Because your audience cannot check a dictionary or a thesaurus or review a passage, your language must be as clear and simple as possible. Ask your practice audience to listen for passages that lack clarity and simplicity. You can also review an audio recording of your presentation in search of problematic words, phrases, and sentence. For example, you can revise a sentence such as *Due to parental ignorance and apathy, pertussis is once again a threat to juvenile patients* to make the point more clearly: *Because many parents are not aware of the danger, we are again treating children for whooping cough.*

5.4.3.3   Avoid Jargon

The difference between alienating jargon and useful specialized language is in the ears (and experience) of the audience. The language you use productively with other members of your project team may be incomprehensible to someone outside your group or company. Choose a practice audience with no specialized knowledge of your topic, and ask participants to make a list of unfamiliar terms. Then find everyday alternatives, offer explanations, or provide analogies and examples familiar to everyone likely to attend your presentation. Acronyms should also be used sparingly and explained when first used. Providing the word for each letter on a visual aid will help members of the audience remember the meaning of the acronym. See Chapter 6 (pages 215–218) for more information about jargon.

5.4.3.4   Minimize Filler Sounds and Words

Pauses should be moments of silence. If you have developed a habit of filling pauses with sounds such as *um* or *ah*, you will distract some members of your audience and annoy others. Some may also judge you as being uncomfortable speaking or uncertain of your material.

Words, phrases, and questions such as *like, basically, right, you know*, and *okay*? are also common fillers. If your practice audience points out this tendency to you, then practice pausing *silently*. Do not be overly worried by the odd filler. But the fewer fillers you use, the more confident and certain of your material you will sound.

5.4.3.5   Other Ways to Improve Voice and Speaking Style

Audiotape a section of your presentation or listen to the sound of a video recording without watching the picture to avoid being distracted. As with body language, work on only one aspect of voice or speaking style at a time. Then record that segment again to gauge your progress. You can also improve voice and speaking style with other kinds of practice such as reading aloud to become more familiar with hearing yourself perform. Try reading children's stories, giving each character a unique way of speaking, to develop vocal variety. Reciting tongue twisters such as *Peter Piper picked a peck of pickled peppers* also improves diction. The advice on evaluating delivery offered for team rehearsals is also useful for evaluating individual presentations.

5.5   TEAM PRESENTATIONS

Because engineers frequently work in teams, you can expect to be part of numerous team presentations during your engineering career. All of the advice offered in the previous sections applies to your part of a team presentation, but when presenting as a team, you must also consider additional issues relating to team planning, group delivery, team rehearsals, and group question periods.

5.5.1   Team Planning

All team members should be familiar with the content of the entire presentation. Therefore, begin with a group planning session to clarify objectives, discuss audience considerations, identify appropriate content, and choose an organizational strategy. A white board or flip chart and some colored pens are ideal aids for this planning session.

5.5.1.1   Audience and Purpose

As for an individual presentation, begin group planning with a discussion of purpose to ensure that everyone agrees on what the team expects to accomplish. A written statement of purpose will guide the process of preparing the presentation. Also analyze the audience, identifying their needs, reasons for attending, knowledge and expertise, group dynamics, linguistic and cultural background, demographics, and physical environment (see pages 154–156 of this chapter). If your presentation is persuasive, pay particular attention to the audience's attitude toward your proposal or request and to their values, fears, and expectations in relation to your topic (pages 82–84 of Chapter 3).

5.5.1.2   Content

Brainstorm to plan content. As a group, begin by writing down any ideas that come to mind, regardless of their apparent usefulness. When you have exhausted your explorations, take a brief break and then analyze the ideas you collected during the brainstorming session. Discard any that prove unfeasible or impractical and organize the rest into three to five main points. You might try using a colored pen to circle your main points. Next indicate which secondary points belong with each main point, and then decide how to order the main points. (See pages 15–25 in Chapter 1 for a full discussion of inventing strategies.) Alternatively, you may decide to use one of the organizational patterns discussed earlier: a time sequence, selling, problem solving, spatial relationship, or PREP approach. Working as a team, you should also decide on the basic content of the opening and closing sections. Once the content is organized, you can create a rough outline of the presentation, which should be distributed to all team members.

5.5.1.3   Visual Aids

When the content is determined, decide what visual aids are required for each section of the presentation. Also determine what media you will use for what part of the presentation (i.e., overhead transparencies or presentation software, white board or flip chart, video, objects, etc.) and what each screen or slide you prepare will look like. If possible, produce a rough sketch of appropriate visual aids while working as a group. If more than one person will produce the visuals, decide on formatting details at this stage to ensure consistency. (For a more detailed discussion of visual aids, see pages 160–168 of this chapter.)

5.5.1.4   Tasks

At some point in the planning process, you must decide who will take responsibility for each aspect of preparing the presentation. As far as possible, ensure that all team members are content with their responsibilities and that the work is equitably divided. Also ensure that everyone is clear on deadlines and is informed of who is responsible for any research or fact finding required, who will create the visual aids, and who will take responsibility for the various sections of the presentation.

5.5.2   Group Delivery Styles

When deciding how to deliver a presentation, the team could consider one of the styles listed below. Choose an approach (or combination of approaches) that suits your content and will hold the audience's attention. Time constraints and team members' talents should be kept in mind as your team selects an approach.

5.5.2.1   One Presentation, Several Presenters

For this approach, the presentation is organized in the same way as an individual presentation and then each member of the team takes responsibility for a section. For instance, one person might deliver the opening and the first point, another the second and third points, and so on. Because this type of presentation is modeled on the individual presentations that team members are most familiar with, it is a good choice when preparation time is short or rehearsal time severely limited.

5.5.2.2 Panel Presentation

For this approach, one team member serves as moderator while others act as panelists, representing different points of view or specialist roles. The moderator introduces the panelists and their specialties, creates the links between speakers, manages the audience's questions, and brings the presentation to a close. This approach could prove effective to demonstrate various aspects of an issue, to present the perspectives of various members of a cross-functional team, or to represent competing solutions. An entire presentation could be organized as a panel discussion, or this presentation style could be used to introduce a controversial or many-faceted topic.

5.5.2.3 Debate Format

The debate format is similar to the panel presentation except panelists demonstrate their disagreement with one another's positions. Again, one member serves as moderator, introducing the players and the issue under discussion, mediating the debate, ensuring the point the team wants to make is clear to the audience, managing the audience's questions, and ending the presentation. This approach can be beneficial for internal company presentations when the team wants to draw attention to opposing viewpoints but is rarely appropriate when presenting to a client.

5.5.2.4 Role Playing

Role playing can bring a team presentation to life—given sufficient acting ability. For this approach, team members perform like characters in a play, recreating an event or discussing an issue as if they were in the situation of the characters they are playing. A moderator may be required to introduce speakers and to facilitate discussion. At the end of the performance, the actors usually describe their roles, positions, and thinking and then members of the audience voice their opinions and ask questions. While this approach can be effective, with a technical audience, it should probably be employed for no more than a few minutes—just long enough to make a point.

5.5.3 Team Rehearsals

The advice that follows assumes that team members are not experienced presenters, are not used to presenting together, *or* are preparing for a presentation the team deems to be important enough to deserve a special effort. The latter type of presentation may be more common than you anticipate. When a team has one shot at convincing superiors to provide more resources or to accept a recommendation the team believes is crucial to their success, then expectations are high that all team members will perform to the best of their ability. A crucially important presentation merits rigorous rehearsals.

If parts of a presentation have been assigned to individuals, then team members should rehearse their sections individually before participating in a group rehearsal. For the first team rehearsal, team members should not worry about delivery style, but simply focus on ensuring that each section includes appropriate content and is presented within the available time. After any major issues with content, organization, timing, and delivery are identified and resolved, the team should rehearse again with team members coaching one another.

To coach effectively, you must think before you speak. In the role of a coach, take a moment to consider how you would react to the feedback you are about to give. An effective coach will state opinions in a positive manner, beginning and ending with an encouraging remark. To this end, clarify that your comments represent your opinions and are offered as suggestions. For example, use phrases such as, *I thought*, or *I notice* and follow up with suggestions for improvement: *Maybe if you tried breathing more deeply, the audience would be able to hear the ends of your sentences.* But avoid overloading your teammates with too many suggestions at one time. While you must take your teammates' feelings into account, you must also be honest. People need to know they have problems before they can improve.

Most novice presenters will encounter some problems with body language, voice, and/or speaking style. The following section describes common problems in these areas and offers language for talking about them with your colleagues. Everyone on your team should be familiar with the terminology in the following sections so that you can use it during coaching sessions. You should also find the following advice useful when reviewing video and audiotapes of your individual presentations.

5.5.3.1   Body Language

When providing feedback on team members' delivery, check their movement, posture, and gestures. What do they do well and what could they improve?

- Are their movements appropriate and do they use space effectively?
- Is their posture straight and natural?
- Are their gestures natural and appropriate to their context and situation?
- Do they make appropriate eye contact with members of the audience?
- Do they avoid repetitive movements or irritating habits?

Each question that receives a positive answer presents an opportunity to praise the speaker. If you answer any of the questions negatively, the speaker may suffer from one of the following problems.

Problems with Posture   The following problems with posture are all fairly simple to overcome once presenters are aware of what they are doing and understand why a habit is an issue.

Tilted posture is usually caused by putting weight mainly on one foot or hip rather than on both feet and is often combined with foot tapping, rocking, or swaying. This posture may distract members of the audience, drawing attention away from a presenter's message. Standing straight with weight evenly distributed on both feet should solve this problem.

Slumped posture results when shoulders fall forward, the chin thrusts out (or flops down), hips are forward, the back is arched, the belly protrudes, and knees are locked. Besides presenting a less-than-pretty sight, the audience may assume that the speaker is disinterested in the topic, bored, or lacking confidence. Concentrating on keeping the shoulders back and the knees relaxed should lead to a more upright posture.

Leaning away from the audience with weight back on the heels or on one leg makes speakers seem unfriendly, as if seeking to distance themselves from the audience. Keeping the knees slightly bent should resolve this problem.

Leaning on tables, lecterns, microphone stands, or ledges is often viewed as poor posture. Whether speakers lean forward or back for support, if they sustain these positions, they will soon appear tired, bored, uninterested, or dull. Leaning on something briefly to make a point is not a problem, but speakers who have a habit of leaning on things should try to position themselves away from inviting surfaces.

Closed gestures turn the back of the hands to the audience, signaling that the speaker lacks confidence or is holding back information and is possibly dishonest. Note that open gestures (with palms showing) seem friendlier, more open, and more honest than closed ones.

Fig Leaf posture is achieved by clasping hands in front of the waist or a little lower down. Members of the audience may decide that someone assuming this closed position looks scared, lacks confidence, is inexperienced, or lacks conviction. Dropping arms to the side creates a more positive image.

Problems with Gestures Alert team members if they use any of the following gestures. If appropriate, model their movements and then suggest alternatives.

Lecturing gestures are enacted by pointing a finger at the audience, by standing with arms crossed on the chest, or by standing with hands on hips. Because adults use these gestures to scold or control children, the audience may well sense that a speaker is lecturing or talking down to them.

Steepling describes the positioning of hands in front of the chest in the position of prayer (i.e., in the shape of a steeple). Sometimes this gesture is interpreted as begging, but if fingers point toward members of the audience, the speaker will appear to be lecturing or aggressive.

Too few or too small gestures are problematic in that gesture adds life to both voice and appearance. If speakers use few gestures or only very small ones, the presentation will lack energy and excitement. As a coach, you can model more appropriate gestures or identify another team member worth watching.

Too few sustained gestures distract the audience. While some gestures should be brief, many should be held while the speaker completes a point. Otherwise, arm movements appear not to match what is being said and the speaker seems nervous, desperate, or lacking conviction.

Other Problems *Lack of sustained eye contact* creates problems similar to lack of sustained gesture. If speakers merely glance at members of the audience, they appear not to care about the audience or not to believe their message. Team members with this problem should be encouraged to hold eye contact long enough to complete a sentence.

Repetitive movements and irritating habits such as fidgeting, pushing glasses up the nose, playing with hair, jingling the change in a pocket, rubbing hands together, or rising up on the toes also distract an audience, diverting attention away from what is being said onto actions unrelated to the presentation. Note that any gesture can divert attention away from what is being said if used too frequently.

5.5.3.2 Voice

Speakers have some control over how well listeners hear what they say and can influence listeners' interest in their topic through skillful use of their voices.

- Does the pitch remain level at the ends of sentences?
- Does the voice sound lively and natural?
- Is the voice consistently easy to hear?

- Does the speaker enunciate clearly?
- Is the speaker's pronunciation acceptable?

Again, positive answers provide opportunities for praise while negative answers suggest that one of the following issues may need attention.

Pitch You cannot expect speakers to change the natural pitch of their voices, but one common problem with pitch can be overcome.

Pitching up, or raising the pitch of the voice, generally signals the end of a question, while dropping the pitch indicates the end of a statement. If speakers raise their voices at the ends of statements, this rise in pitch makes them sound tentative. When coaching, draw this problem to a speaker's attention by repeating one or more of their sentences with and without the rise in pitch at the end. Working from an audiotape may prove particularly useful.

Pace Listen to the pace of a speaker's voice to determine whether he or she sounds unnatural or if words run together or are not fully enunciated.

Sounding stilted, or unnatural, occurs when the rhythm and expression of the voice do not match the accompanying words—as if the speaker were trying to remember a memorized passage. Reading from notes and putting emphasis on the same part of each sentence or phrase contribute to this problem. A stilted voice will sound boring, and the audience may assume that the speaker lacks sincerity or knowledge. Speaking from brief points on a screen is the best solution to this problem. Team members who sound stilted should also be encouraged to practice in front of an audience. If they are nervous, they should also be encouraged to employ the strategies to control anxiety presented earlier.

Not enunciating clearly often results from being nervous and speaking more quickly than usual so that words run together, word endings are omitted, and syllables are left out of words. A speaker can improve diction by pausing, breathing deeply, and slowing down. Otherwise, the audience will have difficulty following what is being said and will likely assume that the speaker is insecure and inexperienced.

Projection Even when presenters speak softly, you should be able to hear what they are saying. If not, they have a problem with projection and need to breathe more deeply so that the sound of their voices will carry to the back of the room.

Fading out results when the voice begins loud enough but becomes quiet and hard to hear just as a speaker reaches the important part of a message—usually at the end. A habit of fading out suggests a lack of confidence. If members of the audience must expend too much effort to hear the ends of sentences, they may simply stop listening.

Not projecting results when a speaker's voice is too quiet to hear properly because of insufficient breath in the lungs. Breathing more deeply or speaking in shorter, less-complicated sentences will solve this problem.

Other Problems Although some problems with voice may require significant effort on the part of a speaker to overcome, you can point to a problem and suggest possible solutions.

Sounding dull results when the voice lacks variety in tone, pitch, pace, volume, and expression. Without sufficient variety, the audience may assume a lack of enthusiasm for the subject under discussion or may be lulled to sleep. If speakers are enthusiastic about their topics, then this problem may result from nerves, and the strategies for controlling anxiety outlined earlier may help. People with this problem should also try listening to an audio recording of their voices, and they should pay careful attention to the variety in tone, pitch, pace, volume, and expression of particularly lively speakers.

Pronunciation problems are most likely to occur when a speaker is working in a second language or speaks a different dialect from that of the majority. If you note that a speaker has difficulty with specific sounds, you can note words containing those sounds that are repeated in the presentation, imitate how the speaker pronounces these words, and then model the standard pronunciation that the audience expects to hear. While hiring a speech coach is likely the best solution to serious pronunciation problems, you can help if the problem is confined to a few sounds. Pay careful attention to what *you* do with your tongue, teeth, and lips when you make a particular sound, and then slightly exaggerate the shape of your lips, position of your tongue, and so forth as you help your teammate reproduce that sound. Be warned that this sort of coaching leads to sore face muscles and is likely to be amusing to others, so you will probably want to find a private spot for this activity.

5.5.3.3 Speaking Style

Besides a distinctive voice, each presenter has an individual speaking style. Note whether a speaker has developed the following good habits:

- Does the speaker avoid jargon and use acronyms appropriately for the audience?
- Is the speaker concise?
- Are personal pronouns used to good effect?
- Does the speaker avoid fillers?

The most common problems with speaking style are listed below.

Using jargon or too many acronyms (strings of letters listeners are unfamiliar with such as XTB or JLSN) is a sign that a speaker has paid insufficient attention to the audience's needs. This problem results when speakers forget that what is useful language among specialists can be incomprehensible gibberish to nonspecialists. As coach, you should make a list of words and acronyms that the audience will be unfamiliar with and suggest using plain-English substitutes, providing definitions, and using the full form for uncommon acronyms.

Being wordy is exemplified by the following sentence: *My client, what she said was, that we should pay more attention to her.* Using unnecessary words to make a point, as this sentence does, irritates listeners. A concise speaker would rephrase the sentence: *My client demanded more attention.* To help someone overcome this problem, note wordy phrases, sentences, or passages in the presentation. Either write them down or make an audio recording so that you can review them with the speaker and suggest more concise alternatives. Much of the advice on style in Chapter 6 could prove helpful, particularly the advice on being concise on pages 224–234.

Lacking personal pronouns is a problem because personal pronouns help speakers connect with their audiences. For example, saying *most of us* is more personal and more likely to hold the audience's attention than using the phrase, *most people.* If a presentation is impersonal, suggest the speaker use *you, your, we, our,* and *us* to warm up a presentation and to make it more personal.

Using fillers—habitually repeating sounds, words, or questions—can be both annoying and distracting. Repeated use of *um* and *ah* sounds signals uncertainty. These sounds are best replaced with silence. Similarly, repeated words such as *actually* and *basically* as well as short questions (i.e., *you know*? and *okay*?) annoy listeners if used with any regularity. Coaches should also listen for the habitual use of *and* to join one sentence to the next, and the next, and so on. These fillers often become less frequent as speakers gain confidence, but you should point out the ones that occur with annoying frequency. In many cases, your best advice is to encourage speakers to pause silently.

5.5.4   Group Question Period

Because team presentations usually generate many questions, your team should plan for the question period. During rehearsals, members should compile a list of likely questions and decide who will come prepared to answer them. More generally, you should decide ahead of time how to manage the question period (with a moderator, for example) and who will field what types of questions. Throughout a question period, each member should remain alert and interested, paying attention to posture and eye contact. Maintaining a professional stance throughout this critical final part of the presentation ensures that the audience is left with a strong positive impression. The advice about answering questions on pages 168–170 of this chapter is applicable to team presenters.

5.6   A FINAL WORD

Most outstanding speakers have had a lot of practice, either because their jobs demand it or because they sought out instruction and practiced the techniques they learned. Your ability to make effective presentations and to deal with the other communication demands of the workplace will improve most dramatically with guided practice. One of the most productive ways to gain this practice is to join a Toastmasters' club. To locate the one nearest you, check the worldwide listing of clubs available from tminfo@toastmaster.org. For more general information and a link to this listing, visit the Toastmasters' web page at http://www.toastmasters.org.

Presentation skills courses are also available through many Community Colleges. Other useful activities include theater sports, singing or vocal coaching, or even auctioneering courses. If improving your oral communication and presentation skills seems like a daunting task, keep in mind that some of the most effective speakers you encounter probably began with no more innate ability than you possess. To maintain your motivation, also remember that learning to communicate effectively as a speaker is one of the most important things you can do to help ensure a successful career.

5.7   HEURISTIC FOR IMPROVING DELIVERY

The following set of questions is aimed at helping you assess and improve your delivery. The issues raised in these questions are discussed in more detail on pages 180–184 of this chapter. We suggest that you provide this list to the colleagues or friends in front of whom you will practice your presentations. Having a list will help them identify any issues you need to work on. But note that this list is by no means exhaustive; it addresses the most common issues relating to body language, voice, and speaking style. Customize it to focus on those issues most relevant to you and the contexts in which you routinely present.

5.7.1   Body Language

- Is your posture upright and relaxed and is your weight equally on both feet?
- Do you consistently face your audience?

- Are your gestures open, natural, and, when appropriate, sustained?
- Have you eliminated nervous habits?
- Are you using sustained eye contact and engaging individuals from all segments of your audience?

5.7.2 Voice

- Do you avoid pitching up and project your voice at the ends of sentences?
- Does your voice sound natural and relaxed?
- Is your voice loud enough and are your words spoken clearly?
- Does your voice express variety in tone, pitch, pace, volume, and expression?
- If you have problems with pronunciation, have you hired a voice coach or found some other means of improving?

5.7.3 Speaking Style

- Have you kept jargon and acronyms to a minimum?
- Is your presentation concise yet fully developed?
- Do you use personal pronouns such as *you, we,* and *us*?
- Do you avoid filler sounds, words, and questions?

5.8 ORAL PRESENTATION CHECKLIST

We provide the following checklists for you to use when making individual or team presentations. Answering the following questions will help you identify where you need to improve and where you are strong. Guided practice will help you achieve your goals.

5.8.1 Formal Presentations

A set of questions related to delivering your presentation is provided in the heuristic just prior to this section.

5.8.1.1 Controlling Anxiety

If you suffer from anxiety, are you practicing the strategies outlined below? _____

Do you make time to rehearse? _____

Do you warm up with tension/relaxation exercises? _____

Do you begin with a smile? _____

Do you remember to breathe deeply? _____

Do you limit your gaze to one person at a time? _____

Do you make sure you have water to drink when you need to regain your composure? _____

Do you take every opportunity to present? _____

5.8.1.2   Planning a Presentation

Have you analyzed your audience and adapted your presentation to
their needs?   _____

Have you clarified your objectives and written a statement of purpose?   _____

Have you limited your presentation to no more than five main points and
chosen an appropriate organizational pattern?   _____

Have you planned an effective opening and closing for your presentation
and linked the points of the body to create a coherent whole?   _____

Have you chosen visual aids that support your message and are easily seen
and comprehended by all members of the audience?   _____

Have you removed all unnecessary detail from your visual aids?   _____

Have you prepared for the question period?   _____

If appropriate, have you prepared speaker's notes?   _____

5.8.1.3   Rehearsing a Presentation

Have you allowed sufficient time for rehearsals?   _____

Have you found a practice audience willing to provide constructive
feedback on your presentation and to ask questions that will help you
prepare for the question period?   _____

Have you timed a rehearsal to ensure you can complete the presentation
and answer any questions in the allotted time?   _____

5.8.1.4   Team Presentations

Have you started your planning with a group brainstorming session?   _____

Have you agreed on a statement of purpose and a delivery style?   _____

Have you divided responsibilities equitably?   _____

Have you allowed sufficient time for group rehearsals?   _____

Are team members coaching one another?   _____

5.8.1.5   Ongoing Commitment

Have you made a habit of seeking opportunities to engage in public
speaking?   _____

Have you joined Toastmasters', taken a course, or engaged in other
activities that will improve your public speaking?   _____

5.9   EXERCISES

The best exercise for this chapter is to make a real presentation. We strongly advise you to seek out opportunities to present. Take a course that requires oral presentations, join a Toastmasters' club, or find some other activity that provides you with repeated practice making presentations. You should also seek opportunities to serve as a practice audience for your peers and to gain experience coaching them. The following exercises supplement these activities.

1. Planning an Individual Presentation

Plan a 10-minute presentation on a topic covered in one of your courses, on a project you have participated in, or on a hobby or subject of personal interest. Complete the following activities:

Analyze your Audience: Choose an appropriate audience for your topic and write a description of their motivation, attitudes, knowledge and expertise, group dynamics, linguistic and cultural background, demographics, and the physical environment in which you will present.

Clarify Your Objective: Write one or two sentences clarifying what you hope to achieve and what you want your audience to gain or to do as a result of your presentation.

Limit and Organize Your Content: Decide what points to cover in your presentation, limiting yourself to no more than five main points. Determine how you will organize your points and prepare an outline for your presentation, including an indication of how you plan to link your points.

Prepare Visual Aids: Decide what medium to use for your presentation and prepare rough drafts of your visual aids.

Prepare for Questions: Make a list of questions you expect your audience to ask and, if appropriate, prepare notes containing any information you will need to answer questions.

If feasible, rehearse your presentation, inviting one or more classmates or friends to act as your practice audience.

2. Planning a Team Presentation

Working with two to four other people, choose a topic for a presentation of no more than 20 minutes based on a topic covered in a course you have all taken, a project you have all participated in, or a hobby or skill you have in common. In a group planning session, clarify your objectives, choose and analyze your audience, brainstorm content, decide how you will organize your presentation, and choose a presentation medium. Produce an outline for your presentation and rough sketches of your visual aids. Also write a description of how you will handle questions in your presentation. Note that many of the instructions in exercise 1 also apply to group presentations.

3. Providing Feedback on Delivery

Find a presentation you can attend, and pretend you have been invited to provide feedback to the presenter. Addressing the speaker, write up to two pages, pointing out strengths and offering advice on how to improve the presentation and delivery. Pay attention to content, organization, the quality and appropriateness of visual aids, movement, body language, voice, and speaking style.

Chapter **6**

Stylistic Strategies

6.1 SUMMARY

This chapter introduces you to a range of stylistic features commonly found in technical documents that affect readability and rhetorical effectiveness. We address issues of style, focusing on the order of ideas within paragraphs and sentences and on clear, concise expression of thoughts. We also direct your attention to readers—to how people typically read and the problems your style may pose for them.

Our goal is to help you recognize various stylistic patterns that may exist in your writing so that you can revise for them most effectively. We offer you guidelines, not rules, because you must always judge stylistic choices within a rhetorical context; no advice applies equally in all situations. We also want to stress that no matter how carefully you read this chapter or how well you understand the concepts and strategies presented in it, you will improve your style only through practice. How much and how quickly your style improves depends on the time and effort you expend revising for style.

We have divided our discussion of style into four categories: *order and emphasis, connection, clarity*, and *conciseness*. These divisions overlap considerably; for instance, changing emphasis improves clarity, while improving clarity often improves conciseness, and so on. Nevertheless, these divisions serve a purpose, providing a way of organizing our examples and of highlighting general concerns.

1. Style for Order and Emphasis

We provide directions for analyzing paragraph structure and a sample analysis. We explain how locating the important, new information at the end of a sentence improves the flow of ideas and how inserting information between the subject and verb of a sentence distracts readers. We also explain why unnecessary passive constructions and empty openers—beginning with *there is* or with *it is* (when *it* has no noun to refer back to)—affect your ability to express your thoughts effectively. We point to the uses of passive constructions and explain how to identify them and how to reorder passive sentences to create active ones.

2. Style for Connection

A smooth flow of ideas requires well-ordered, clearly connected points. Repeating key words and phrases is one major means of creating coherence and a sense of flow in your writing. Parallel structure—presenting related ideas in the same grammatical form—is a form of repetition that is important in sentences as well as in lists. When parallel structure is violated, readers have difficulty comprehending what you have written. If you use *this* by itself as the subject of a sentence, readers will also have difficulty, so ensure that *this* and other vague subjects are followed by nouns or phrases that clarify meaning. You can also improve the flow of ideas and eliminate ambiguity by providing transitional words and phrases.

3. Style for Clarity

To help ensure you convey ideas clearly to your readers, we suggest an average sentence length of about 18 to 24 words for technical documents. You also want a fair degree of variety in both the structure and length of your sentences. You must also be aware of the large number of English words and phrases that are vague and largely devoid of meaning and eliminate this language from your documents. Further, when technical language is used to communicate with nontechnical readers, it becomes incomprehensible jargon. All documents written about technical subjects for readers unfamiliar with that technology should be revised to eliminate jargon. Runs of prepositional phrases and missing or misplaced commas can also affect the clarity of a sentence.

4. Style for Conciseness

Revising for conciseness is a matter of eliminating unnecessary words, but not of eliminating information, context, or useful repetition. We focus on a number of strategies for turning a style that depends on nouns into one that places more descriptive power in verbs. One of the characteristics of technical writing is the heavy use of nominalizations—using the noun form of a word that also has a verb form. Recognizing and converting these nouns into verbs can both shorten a document and make it easier to read. You should also learn to recognize *talkie verbs*—the most common verbs of conversation—and replace them with more powerful, descriptive verbs. Finally, we offer a list of common wordy phrases along with possible substitutions. But we also point out that you cannot use this list without attention to context; sometimes an apparently wordy phrase is the most appropriate choice.

A heuristic for analyzing writing styles, a checklist to use when revising for style, and exercises to help you improve specific aspects of your style are also provided at the end of this chapter.

6.2   STYLE FOR READERS

As a reader, you may be very familiar with the stylistic habits we identify in this chapter. You may even be surprised that we consider some of them problems because they are so common. Unfortunately, well-ordered, clear, concise writing seems to be the exception rather than the rule.

Think for a moment. How hard do you work when you read? Do you stumble over sentences? Do you suddenly find that you are not making sense of what you are reading? Do you need to reread passages in order to follow the author's train of thought? Does your attention wander? Not all such problems arise from the writer's style, but if well-ordered, clear, concise prose were more common, we would all be better readers.

Note that the stylistic features we identify in this chapter are not always problems; they can be appropriate in particular sentences in specific contexts. Problems arise when they are used *habitually* without thinking about whether they are appropriate in a particular context. For you to improve your style requires making conscious choices based on an understanding of how the words and grammatical structures you choose affect readers.

Also note that if you have difficulty actually seeing problems in your writing, this chapter may help you. Many people reread their own writing very carefully but fail to see typing mistakes or problems with punctuation, spelling, or sentence structure. Personal experience tells us that effort spent looking for specific patterns in your writing helps train you to see more of what is actually on the page. Taking the time to look carefully at what you have written by identifying a particular set of stylistic habits may not only help you see more of what is actually on the page, but also improve your ability to revise more effectively in terms of content, organization, persuasion, and tone.

Finally, please note that attending to your style is part of revising. If you attempt to break stylistic habits while drafting, you will most likely lose your train of thought. Write to get your thoughts down, and *then* revise. Of course, to revise effectively, you must know what to look for. As you read through the suggestions and examples that follow, we suggest you refer to some recent writing, taking note of the issues we raise and the habits we point out that are reflected in your writing.

6.3   STYLE FOR ORDER AND EMPHASIS

Organization is important not just at the document level, but also within paragraphs and even sentences. In this section, we look first at principles for ordering paragraphs and then at the important points of emphasis within sentences. These points of emphasis help you order information so that readers will be most likely to comprehend and remember what you have written.

6.3.1   Analyze Paragraph Structure

A clearly structured paragraph provides readers with a logical progression of ideas or with information relating clearly to a single topic. Unfortunately, this notion of a single topic can extend to an entire document and, therefore, is not always helpful in determining where one paragraph should end and another should begin. Paragraph boundaries are fairly arbitrary, determined by length and format as much as by shifts in topic. One useful rule of thumb is that paragraphs should be no longer than they are wide. Following this guideline, you will note that format helps determine paragraph length because documents formatted in columns should contain shorter paragraphs than those presented in full-page formats. But how can you determine where paragraph breaks are possible?

The best method we have found involves paying attention to levels of generality.[1] That is, a well-structured paragraph will almost always have one sentence with more general subject matter than all the others. Other, more specific sentences will elaborate on this general statement, provide examples, qualify the general point, or draw a conclusion. In turn, even more specific sentences may provide further detail.

---

[1] The notion of paragraphing according to levels of generality is derived from *Four Worlds of Writing*, a textbook written by Lauer, Montague, Lunsford, and Emig (2nd ed., New York: Harper & Row, 1985).

For example, the following list indicates the levels of generality for the paragraph you have just read:

Most General Statement

The best method we have found involves paying attention to levels of generality.

> **More Detailed Statements**
>
> That is, a well-structured paragraph will almost always have one sentence with more general subject matter than all the others.
>
> Other, more specific sentences will elaborate on this general statement, provide examples, qualify the general point, or draw a conclusion.
>
> In turn, even more specific sentences may provide further detail.

In this case, we use only two levels of generality. In theory, we could include many levels, but four is usually the limit.

Rather than moving beyond four levels of generality, you can cover topics that justify a high level of detail in more than one paragraph. Start a new paragraph at any sentence that is followed by two or more sentences offering more detailed information. The new opening sentence establishes the highest level of generality for that paragraph.

To illustrate how this notion of paragraph structure works, we begin by mapping the structure of the first paragraph in this section:

Level One: Most General

A clearly structured paragraph provides readers with a logical progression of ideas or with information relating clearly to a single topic.

> **Level Two: More Specific (qualifies what is meant by "a single topic")**
>
> Unfortunately, this notion of a single topic can extend to an entire document and, therefore, is not always helpful in determining where one paragraph should end and another should begin.

Level Two (more qualification of the topic)

Paragraph boundaries are fairly arbitrary, determined by length and format as much as by shifts in topic.

> **Level Three: Still More Specific (example of a format constraint)**
>
> One useful rule of thumb is that paragraphs should be no longer than they are wide.
>
> > **Level Four: Most Specific (further explanation and example of a rule of thumb)**
> >
> > Following this guideline, you will note that format helps determine paragraph length because documents formatted in columns should contain shorter paragraphs than those presented in full-page formats.

Level Two (return to more general statement about paragraph breaks)

But how do we determine where paragraph breaks are possible?

If the paragraph continued, the answer to the question posed at the end of this paragraph could also be counted as a second-level sentence, but we broke the paragraph at this point. The sentence beginning "*The best method we have found*" thus becomes the most general, level-one sentence in the following paragraph.

When structure at the paragraph level is viewed in terms of levels of generality, sentences are either somewhat more specific than the previous one or at the same level of generality as one or more of the previous sentences. Most importantly, one sentence—usually the first—is more general than all the others. Other sentences must clearly relate to the topic introduced in the most general sentence, or they must progress logically from one point to the next.

These principles of paragraph structure are restated and expanded on in the following directions for analyzing paragraph structures.

6.3.1.1 Directions for Paragraph Analysis

Begin by choosing two or three paragraphs from the middle of a document. Avoid introductory and concluding paragraphs because they often have a different structure. Pick paragraphs with four or more sentences (the longer, the better). Then follow the step-by-step instructions below:

1. Number the sentences in each paragraph.

2. Locate the most general statement in each paragraph and mark it as level 1. This statement should appear in the first or second sentence. If it appears in the second sentence, the first one should provide a transition that prepares readers for a shift in topic by restating a connection to remind readers of a point made earlier that relates to the topic of the upcoming paragraph.

3. Mark the remaining sentences as level 2, 3, 4, and so on, and note how sentences relate to one another in terms of subordinate relationships (i.e., more specific) or coordinate relationships (i.e., at the same level of generality). For example, all sentences at the second level of generality are coordinate to each other and subordinate to the first-level sentence.

4. To further clarify the relationships among sentences, briefly paraphrase the main point of each sentence you have mapped in the previous step.

5. For a final check that the order and relationship of ideas is logical and clear to readers, list the subjects of the main clauses of all the sentences. They should either refer to the same topic or suggest a logical progression from one topic to another.

The following sample analysis demonstrates this procedure.

6.3.1.2 Sample Paragraph Analysis

We begin by numbering the sentences in each paragraph of the passage:

(1) The computer currently in use by Mr. Boffo is an IBM clone and possesses several inherent limitations. (2) First, it is more than 15 years old and thus does not represent anywhere near state-of-the-art technology. (3) It uses an older-style cooling fan and as a consequence is noisier than more modern computers. (4) Second, the 8088 CPU (Central Processing Unit) used in the computer is severely limited in its processing speed. (5) The more modern 80486 and 80586 CPUs that have 32 bit I/O buses are much faster. (6) Third, the machine has a limited memory capacity—512 K. (7) Many current applications require a minimum of 640 K and some require several megabytes of RAM (Random Access Memory). (8) Given its memory limitations, the system will almost certainly fail when used with certain applications. (9) Fourth, the TSR (Terminate and Stay Resident) programs that Mr. Boffo has installed in order to overcome the limitations of DOS 3.3 have the effect of further reducing available memory and thus increasing the likelihood of a system failure.

(1) Consequently, we recommend that he consider upgrading the machine. (2) The simplest way to upgrade the machine is to install additional memory chips. (3) Installing a graphics card and a math coprocessor is also advisable because they would enable more efficient use of the word processing and statistical programs on which he relies. (4) However, purchasing a newer and faster computer along with a better operating system such as Windows 2000 would probably be more cost effective. (4) On the other hand, given Mr. Boffo's limited data entry speed, a faster system would likely be of little benefit. (6) Perhaps upgrading the operator is an option.

Mapping for Levels of Generality (with Paraphrasing)   Following the instructions for steps two and three, we identify the most general (level 1) sentence and map the levels of all other sentences using indentation to identify coordinate and subordinate relationships. As indicated in step four, we then paraphrase the main point of each sentence as a further check that each is logically related to the preceding one:

1. 1—limitations of computer
2.   2—old
3.   3—noisy fan
4.   2—limited processing speed
5.   3—new CPUs are faster
6.   2—limited memory
7.   3—too limited for some applications
8.   3—system will fail with those applications
9.   2—TSRs and DOS exacerbate these problems

1. 1—therefore, upgrade computer
2.   2—install more memory
3.   2—and add graphics card and coprocessor
4.   2—but a new computer and OS is a better solution
5.   2—however is more speed really needed?
6.   2—maybe the problem lies with the operator

Note that while the paragraphs have very different structures, both contain an opening sentence that is more general than the rest, is within the limit of four levels of generality, and involves a series of points that are logically related to one another.

Checking Subjects   As a final check that the paragraphs are logically ordered, we complete step five, listing the subjects of the main clauses. Note that in the first paragraph, grammatical subjects either refer to the original topic (Mr. B's computer) or to a closely related topic.

1. The computer
2. it [the computer]
3. It [the computer]
4. the 8088 CPU
5. The more modern 80486 and 80586 CPUs
6. the machine
7. Many current applications
8. the system
9. the TSR programs

Note that in the second paragraph, we include a couple of extra words to clarify the logical connections among sentences.

1. we [recommend]
2. The simplest way to upgrade the machine
3. Installing a graphics card and a math coprocessor
4. purchasing a newer and faster computer [more cost efficient]
5. A faster system [of little benefit]
6. upgrading the operator

We urge you to take the time to analyze a few paragraphs from a recent piece of your writing. Note that we are not suggesting you should analyze paragraph structure every time you revise. Instead, perform a sample analysis to determine how well you order ideas within paragraphs and then adapt the relevant steps of this analysis as strategies to help you overcome any problems you discover. Also note that you may find paragraph analysis useful any time you have difficulty ordering ideas.

6.3.2   Revise Sentences for Emphasis and Order

Many writers pay little, if any, attention to the order of ideas within sentences. And yet, what information comes where in a sentence plays an important role in how well writers communicate to readers. In this section, we present the appropriate order for ideas within a sentence as a matter of emphasizing and locating information the reader already knows in relation to the new information the writer provides.

6.3.2.1   Understanding Emphasis

For the purpose of this discussion, think of a basic sentence in terms of three points of emphasis: the beginning, the verb, and the ending. By placing information appropriately in each part of the sentence, you increase the chances of your readers understanding and retaining the information and ideas you present. In general, you want to begin with context and known information and then present important, new information at the end of the sentence. Figure 6.1 portrays the relative degree of emphasis readers usually accord to each part of a sentence.

The Beginning   The beginnings of sentences do not receive much emphasis but serve an important function by providing the reader with context and connection. As indicated in the previous section on paragraph structure, the beginnings of sentences guide readers by reinforcing a sense of order and connection. They connect what is to come with what came before and also provide the context needed to understand the new information at the end of the sentence.

For example, the subject identifies who or what the sentence is about. The addition of certain pronouns (*this* problem, *that* situation, *those* events) can reinforce the connection with previous sentences. Transitional words (*however, therefore, furthermore, next*) can further signal how a sentence relates to the previous one. Introductory phases (*In most cases, Until the 1980s, Under ideal conditions*) may also provide context and set limita-

Figure 6.1 Points of Emphasis in the Sentence

tions. Consequently, while the beginnings of sentences may not receive much emphasis, they are essential, providing the known information and the connections necessary for readers to make sense of what they read.

The Verb The verb achieves a certain degree of emphasis simply because it tells the reader *what is happening*. Without the verb, a reader has no idea what the subject of a sentence is doing or, for that matter, what the sentence means. The verb tells us how the beginning and ending of a sentence relate, allowing readers to interpret our meaning. Indeed, sentences can lack a subject (the beginning) or an object (the ending) and still make sense: *Construct bridges! Engineers construct.* However, if the verb is omitted, a sentence lacks meaning (i.e., *Engineers bridges*). If the verb is omitted, what do readers think the engineer is doing in relation to bridges? Inspecting them? Designing them? Planning them? Supervising the building of them? Simply admiring them?

The Ending In general, reading is least difficult and content most easily remembered when the most important or the new information is placed at the end of sentences. Therefore, a well-structured sentence should move readers toward the completion of a thought. As a somewhat exaggerated example, consider what happens when reading a mystery novel. Suspense is generated. We want to know *who done it?*—but not until the end of the book. Similarly, most sentences should end with the new information, with what readers do not know and are waiting to find out.

The end of a sentence also provides information needed to understand the next sentence. That is, the new information in one sentence may become the context for the following one. New information becomes known information, providing the context for more new information.

When our sentences are well ordered, with the point of most emphasis at the end, readers can easily keep the context and connection between ideas in mind and retain essential information in short-term memory. The longer information is retained in the reader's mind, the more emphasized it becomes (which, by the way, is also a good reason for repeating important information several times in a document).

The following strategies and examples suggest ways to ensure effective order and emphasis.

6.3.3 Place Important Information at the End

Our advice to place new or important information at the end of a sentence may seem just the opposite to what makes sense. Because we naturally focus on what is most important, we tend to put it first. However, when we do so, our intentions and the reader's expectations are likely to conflict. In our eagerness to communicate new information or to make an important point, we actually deemphasize this information. We break the flow by which new information becomes known information, which in turn provides the context for more new information.

Placing modifiers at the end of sentences is one way to break the flow of ideas. The following examples demonstrate how to restructure sentences ending with modifiers in order to locate important information at the end.

Original	Revised
The information provided in the report suggesting the crash was caused by mechanical failure is not persuasive *for the most part.*	For the most part, the information provided in the report suggesting the crash was caused by mechanical failure is not persuasive.

The modifying phrase, *for the most part*, belongs at the beginning of the sentence to modify the subject. When we move this phrase, then *not persuasive* is emphasized as the new, most important point in the sentence—which is what the writer intended.

Note a similar pattern in the following example:

Original	Revised
An explanation of atmospheric stability and a detailed evaluation of its application to this air quality evaluation *is contained in Appendix A.*	Appendix A contains an explanation of atmospheric stability and a detailed evaluation of its application to this air quality evaluation.

Again, *in Appendix A* is not the most important information in the original sentence. It merely tells the reader where to find information considered too detailed for the body of a report. Reversing the order of the sentence places the context—where the information can be found—at the beginning and relocates the new information in the position of most emphasis at the end of the sentence.

The revised sentence in the following example demonstrates that creating appropriate emphasis sometimes eliminates unnecessary repetition.

Original	Revised
The department had made a series of requests *in their letters of 15 February, 20 February, and 27 February 1999. The departmental requests may be listed as follows*:	In their letters of 15, 20, and 27 February 1999, the department made a series of requests:

Once more, the dates of the letters provide context—where something can be found. When we move this context to the beginning of the first sentence, we can simply eliminate the second sentence because ending with *a series of requests* now prepares readers for whatever follows the colon. We also eliminated unnecessary repetition by removing the repeated word *February*.

As the above example indicates, revising for emphasis can involve more than one sentence. When you revise one sentence for emphasis, you may see more clearly how other ideas within a paragraph relate and identify opportunities for further revisions. Consequently, after revising one sentence, read on to ensure no further changes are required.

6.3.4   Avoid Embedding Large Phrases

Another principle of order is that the subject and verb should be close to one another. Where possible, you should avoid complicating a sentence by inserting information between the subject and verb that you could place elsewhere. Embedding one idea within another can frustrate readers who expect the subject of a sentence to be followed by its verb.

Embedding may also mislead the reader because a large embedded phrase will appear more important than it really is. As the following examples indicate, you can often resolve the problem simply by moving the embedded information.

Original	Revised
Engineers, *because they inadvertently write in ways that, all too often, use technical jargon, lengthy noun strings, and the passive voice*, have problems communicating with non specialists.	Because they inadvertently write in ways that often use technical jargon, lengthy noun strings, and the passive voice, engineers have problems communicating with nonspecialists.

This practice, *while satisfying the client on whose project we are working*, leads to several of our other clients' needs not being met.	While satisfying the clients on whose project we are working, this practice leads to several of our other clients' needs not being met.

In these examples, we simply move the subject after the embedded clause or phrase to reunite the subject and verb. In the first example, a tendency to put the clause beginning with *because* in the middle of the sentence may result from the so-called rule to never begin a sentence with *because*. The writer could elect to move this clause to the end of the sentence, but that move may violate the principle of placing new information at the end. We assume that jargon, noun strings, and the passive voice were discussed in previous sentences, so that *problems communicating with nonspecialists* is the new information that is now appropriately placed at the end of the sentence.

Sometimes revising to eliminate unnecessary embedding involves breaking a sentence in two:

Original	**Revised**
The bipolar junction transistor, *a three terminal solid state device, which is often referred to simply as "the transistor*," is widely used in discrete circuits as well as in integrated circuits, *both analog and digital*, because the device's performance is remarkably predictable and quite insensitive to variations in device parameters.	The bipolar junction transistor is a three-terminal solid-state device that is often referred to simply as "the transistor." Because its performance is remarkably predictable and quite insensitive to variations in device parameters, this transistor is widely used in discrete and integrated circuits, both analog and digital.

In the revised sentence, we added a verb (*is*) after the subject, *bipolar junction transistor*, to create a separate sentence. We also moved the dependent clause (*because . . .*) to the beginning of the new second sentence, assuming that how widely the transistor is used is the important, new information. Note that to improve clarity, we changed the subject of the second sentence from *device* to *this transistor*.

If you would have retained *device* to avoid repeating *transistor*, remember the importance of repetition for ensuring your writing is coherent. Repetition is a particularly effective means of emphasizing the subject you are writing about. Of course, you can repeat a word too many times, but many writers worry too much about repeating words. When in doubt, repeat rather than substitute words.

We offer this final example to point out that embedding can also be caused by problems in your formatting.

Original	**Revised**
For inversion cases, the dispersion models *[Figures 3 and 4 placed here]* do not incorporate mixing height into E and F stability (e.g., inversion) ambient air quality predictions.	[Figures 3 and 4 placed here] For inversion cases, the dispersion models do not incorporate mixing height into E and F stability (e.g., inversion) ambient air quality predictions.

Placing even a small figure (or table) in the middle of a page so that it splits a sentence will distract readers enough that they lose their train of thought. In this case, the writer placed two full-page figures in the middle of a sentence, forcing readers to flip back and forth in order to follow the gist of the discussion. In most cases, this sort of formatting problem can be resolved by moving a figure so that it appears between sentences rather than in the middle of one.

6.3.5   Avoid Weak Sentence Openers

When revising for order, also check for weak sentence openers that fill two of the three points of emphasis in a sentence with words carrying no content (i.e., *It is . . . that* and *There is . . . that*). Revising a sentence with an empty opener is often as simple as eliminating a few unnecessary words. In the following examples, note that removing weak sentence openers not only shifts emphasis, but also increases clarity and conciseness.

Original	Revised
There are many aspects of the problem *that* have not yet been considered.	Many aspects of the problem have not yet been considered.
There are several problems *that* are caused by the slow response.	Several problems are caused by the slow response.

Or better yet

The slow response causes several problems.

Some writers believe they add emphasis to a sentence by using *it is* or *there are*. For example, in the last case above, the writer may have used the empty opener to emphasize *several examples*. However, the second revision that places this information at the end actually gives these words more emphasis and makes the sentence easier to read and remember.

Sometimes we can eliminate more than a few words:

Original	Revised
It is considered that implementation of a range of measures should lead to a reduction of the emissions.	Implementation of a range of measures should lead to a reduction of the emissions.

Or better yet

Implementing a range of measures should reduce emissions.

Note that our first revision removes the clause containing the empty opener (*It is considered that*) without changing the meaning of the sentence. For some writers, empty openers are a habitual way of beginning a sentence, functioning like the *um*s and *ah*s that punctuate some people's speech. The second revision eliminates more unnecessary words, reducing the verb phrase from *should lead to a reduction of* to *should reduce,* and thus producing a sentence with all the content but only half the words of the original.

The following examples also reflect habitual use of empty openers:

Original	Revised
It is clear *that* the kiln contributes to the problem.	*Clearly*, the kiln contributes to the problem.
It is probable *that* we should measure the wind characteristics at the site.	We should *probably* measure the wind characteristics at the site.

When empty openers are removed from the beginning of a sentence, we are often left with modifiers to deal with. In such cases, we can usually revise the sentence by turning an adjective (*clear* or *probable*) into an adverb (*clearly* or *probably*). While these adverbs most often introduce the sentence (*Clearly, the kiln contributes*), they sometimes modify the verb (*probably measure*).

Some empty openers reflect a reluctance to use first-person pronouns (*I, we*):

Original	**Revised**
It is our understanding *that* the clarifier does not form a direct part of the treatment system.	*We* understand that the clarifier does not form a direct part of the treatment system.

Given that *our understanding* includes a first-person pronoun (*our*), we see no reason for the writer to be reluctant to use the clearer, more concise expression, *we understand.*

Sometimes revising an empty opener out of a sentence involves combining sentences.

Original	**Revised**
There is a trust among peers and a growing trust with management. *This* encourages open communication and productivity.	The trust among peers and the growing trust with management encourage open communication and productivity.

We often see this combination of one sentence beginning with an empty opener and the next with *this* as the subject. Characteristically, what is left after the empty opener is removed from the original sentence can replace *this* as the subject of the following sentence.

Sometimes, one empty opener leads to another, and another, and another:

Original	**Revised**
It is true *that there* is now widespread acceptance *that there* is a solid market for our company's goods, but *there is* now increased competition for our products.	People generally agree our company has a solid market for its goods, but we also face more competition.

Four empty openers in one sentence are bound to cause problems for readers. Did you have trouble figuring out what the writer was saying in the last part of the original sentence? At first, we were confused by the phrase *competition for the company's products*, which can be read as more people wanting their products than ever before. However, on reading more of the document containing this sentence, we decided that the writer meant to say that the company faces more competition in the marketplace than used to be the case. We could not find a reference to explain who accepts that the company has a solid market for its goods, so we decided to add the generic noun, *people*. But the writer could have been thinking of *venture capitalists, senior management, stockbrokers, market analysts,* or *employees.*

The above example illustrates that revising to eliminate empty openers is not always straightforward. Sometimes you must play with a sentence. However, time spent revising for empty openers can be time well spent. As the above example indicates, empty openers can affect your ability to communicate thoughts effectively. Taking the time to identify and eliminate the majority of empty openers is one way to help ensure your readers are treated to well-ordered, coherent, clear, concise sentences.

6.3.6   Eliminate Unnecessary Passive Constructions

One of the *rules* of writing many of us learn along the way is to avoid using the passive voice. However, in technical and scientific writing, you cannot avoid passives. In fact, the passive voice can be a valuable means of reordering the elements of a sentence to create the appropriate emphasis. But while you must use passives, you should do so with care, for a specific purpose, rather than out of habit.

The next section is for anyone who is unsure of what passive constructions are or how to identify them.

6.3.6.1   Identifying Passive Constructions

In English, sentences can have one of two kinds of voice, active or passive, depending on how the subject relates to the verb. That is, a subject can do something or have something done to it. If the subject does something, it is the *agent* of an action, and the sentence is in the active voice. If the subject has something done to it, it is the *goal* of an action, and the sentence is in the passive voice.

	AGENT	ACTION	GOAL
ACTIVE:	**Some engineers**	**design**	**bridges**.
PASSIVE:	**Bridges**	**are designed by**	**some engineers**.
	GOAL	ACTION	AGENT

Note that in a passive construction, the verb phrase contains the verb *to be* followed by the active verb in its past participle form (usually the verb root with an *-ed* ending), and then the phrase usually ends with the preposition *by*: *is calculated by, was measured by, were delivered by, will be completed by, has been finished by.*

As you have likely noticed, the past participle (the passive form of the active verb) generally looks like the past tense form of the verb (*V-ed*). For some verbs, the past participle has its own form. For example, the past participle of *to write* is *written*, of *to speak* is *spoken*, and of *to begin* is *begun*.

If you are confused by grammatical terms, you may recognize the patterns more easily by studying examples.

Active	**Passive**
The boss *spoke to* her about being late.	She *was spoken to* by the boss about being late.
	Or
	She *was spoken to* about being late.
I *delayed* the project.	The project *was delayed* by me.
	Or
	The project *was delayed*.
They *finally completed* the job.	The job *was finally completed* by them.
	Or
	The job *was finally completed*.
The agency *has provided* the terms of reference for this study.	The terms of reference for this study *have been provided* by the agency.

Note that agents can be deleted in some passive sentences, as indicated by the alternative passive forms for the first three sentences.

6.3.6.2   Uses and Abuses of Passive Voice

Passive constructions serve a number of useful purposes. First, you can change the emphasis of a sentence by moving the goal to the subject of the sentence. For example, if the topic of a paragraph is a bridge rather than the engineers who built it, you can make the bridge the subject of the sentence by intentionally using a passive construction: *The Golden Gate bridge was designed to combine form and function to dramatic effect*. Because agents can be omitted, passive constructions are particularly useful when the agent is unknown (*The program was written in C++*) or when the agent is obvious or unnecessary (*The project was finally completed last night*).

Because the agent can be omitted, the passive voice also allows you to avoid appearing to blame someone. That is, you can say that a device *was damaged* without mentioning who damaged it. But you can also use passive constructions to avoid responsibility. That is, you can say that a decision *was made* without admitting that you made the decision. Unfortunately, if you omit agents without thinking, your readers may assume that you are avoiding responsibility when that is not your intention.

One of the reasons passive sentences are so common in technical and scientific writing is that they allow you to communicate an impression of objectivity. *After the experiment was completed, the data was analyzed* sounds more objective than *After I completed the experiment, I analyzed the data*. The notion that scientific objectivity should be reflected in scientific writing may well explain why some writers avoid using active voice and first-person pronouns.

You might also want to question the value of this objectivity in some situations. For example, compare *the procedure was changed* (a passive construction) with *we changed the procedure* (its active counterpart). The procedure did not change spontaneously; someone changed the procedure. Who did it may be useful information. Omitting the agent may leave readers wondering *who* made the change, distracting their attention by raising a question and thus disrupting the flow of information.

Another reason to pay attention to passive constructions is that they can be less clear than active ones. In fact, many writers unconsciously shift to passive sentences when they are unsure of what they are saying or are struggling to express their thoughts. In some cases, if you take a passive sentence and rewrite it as an active one, you will realize that the passive sentence did not express your thought correctly. If the active sentence is clearer for you as the writer, it will certainly be clearer for readers.

The uses and abuses of the passive voice are summarized in Table 6.1.

6.3.6.3   Examples of Unnecessary Passives

When revising your own writing, consider the context for the sentence and whether an active or passive sentence is most appropriate to achieve effective order and emphasis. For the following example, the passive sentence is appropriate only if the phrase *terms of reference* is the subject and *the agency* is the new information.

Original	Revised
The terms of reference for this study *have been provided* by the agency.	The agency has provided the terms of reference for this study.

Table 6.1   Uses and Abuses of the Passive Voice

Uses of Passive Voice	Abuses of Passive Voice
Changing the emphasis of a sentence	Creating a false sense of objectivity
Omitting unknown or unnecessary agents	Using habitually
Avoiding blaming someone	Avoiding responsibility
Communicating objectivity	Obscuring meaning

When the agent is named in a passive sentence (*by the agency*), revision is relatively simple: you reverse the beginning and ending of the sentence and use the active form of the verb. Revision is not quite so straightforward when the agent is missing:

Original	Revised
Therefore, the regional wind speed and direction patterns *are considered* in this air quality evaluation.	Therefore, this air quality evaluation considers the regional wind speed and direction patterns.
Kirchhoff's Voltage Law was confirmed by directly measuring the voltages in the circuit.	We confirmed Kirchhoff's Voltage Law by directly measuring the voltages in the circuit.

When the agent is omitted from a passive sentence, you can substitute another noun (*This evaluation*) or add the missing agent (*We*). One useful strategy is to add a dummy agent (*someone, something*) and then determine the appropriate agent or substitute subject:

Original	Intermediate	Revised
Kirchhoff's Voltage Law was confirmed.	*Someone* confirmed Kirchhoff's Voltage Law.	We confirmed Kirchhoff's Voltage Law.
Speed and direction are considered in this evaluation.	*Someone/Something* considers speed and direction in this evaluation.	This evaluation considers speed and direction.

Turning passive sentences into active ones sometimes requires other revision strategies:

Original	Revised
As part of the analysis of roadway improvements, the review of signal progression along Cambie Street *will be required*.	Analyzing roadway improvements requires reviewing signal progression along Cambie Street.

Making sense of the original sentence requires reading it in context. In context, *as part of* is unnecessary, so we can eliminate it and use a verbal phrase (*Analyzing roadway improvements*) as the subject of the sentence. Note that the active revision is both clearer and shorter.

If you use the passive voice without a good reason, you add unnecessary words and make your sentences less clear than if you used the active voice. Examine a recent piece of your writing—or run it through a grammar checker. Is the passive voice the default for your style? If so, you should learn to recognize passive constructions and revise to make your writing more active. But do not forget that passives can be useful and cannot always be eliminated. Instead they should be used for a purpose: to create appropriate emphasis, when the goal of an action is the desired subject, or when the agent is unknown.

6.4 STYLE FOR CONNECTION

Smooth and *flows well* are terms often used to describe an effective writing style. A smooth flow of ideas requires carefully ordered and clearly connected ideas. Of course, a poorly placed sentence disrupts the flow of thought. But even if your ideas are logically ordered, you must repeat key words and phrases to ensure readers follow your train of thought. You must also provide transitions that point back to what has already been written while preparing readers for what comes next.

The following revision strategies will help you lead readers smoothly from one idea or point to the next.

6.4.1   Repeat Key Words Appropriately

Repetition is a simple, effective way to create connections. You can repeat key words or phrases, substitute appropriate pronouns, or supply synonyms. The most common strategies involve repeating words and substituting pronouns. Synonyms must be used with care because readers may assume a new word applies to a new concept. Another reason to use synonyms cautiously is to avoid the tendency of many writers to repeat key words far too infrequently. Too little repetition of key words causes more problems than too much repetition. As a general rule, you can repeat the same word three times in a relatively short space without it seeming repetitious.

As an example of how often you can repeat a word, consider the paragraph above. We use *repeat* three times and words with the same root a total of eight times: *repetition, repeat, repeating, repeat, repetition, repetition, repeat,* and *repetitious.* Of course, you can overuse repetition. But when it comes to writing, anything you can do too much, you can also do too little.

Developing an effective style is like a balancing act: you do not want to lean too far in either direction. However, what counts as too much or too little repetition is not determined by a formula or magic number; rather, the appropriate balance depends on context and sentence structure.

The following example illustrates one type of excessive repetition to avoid.

Original	**Revised**
This critique is structured around the models identified in the MSAT Project Management course notes presented by Mr. G. Robinson of E.M. Sciences Ltd. A specific model identifying the organizational boundary and defining the internal and external components forms *the basis of this critique. The focus of this critique* is limited to the internal organization components that include the following subjects: Mission, Structure, Leadership, Rewards, Systems, and Relationships. The external components of the model —e.g., threats, opportunities, and demands—are beyond *the scope of this critique. This critique* focuses inside the organizational boundary.	*This critique* is based on a model identifying organizational boundaries and defining the internal and external components of the organization.[1] The external components of the model (e.g., threats, opportunities, and demands) are beyond the scope of *this critique,* which focuses solely on the following internal components: Mission, Structure, Leadership, Rewards, Systems, and Relationships.

[1]G. Robinson. 1995. *MSAT Project Management Course Notes.* E.M. Sciences Ltd.

Readers may find the repetition in the original awkward because it comes at the end of one sentence and then again at the beginning of the next. Repeating the closing words of one sentence at the beginning of the next—what we call *chaining*—should generally be avoided. Similarities among *the basis of, the focus of*, and *the scope of* further emphasize the repetitious nature of the original passage. When bombarded with so much poorly placed repetition, readers lose track and become frustrated. Note that to eliminate chaining, the revised version uses a footnote and substitutes *which* for a repetitive subject. At the same time, *the basis of* becomes *is based on* and *the focus of* becomes *focuses.*

The next example indicates just how much of a habit chaining becomes for some writers:

Original	Revised
The length of the transistor that can function as a long channel device is strongly related to its *junction depth*. The *junction depth*, of the boron source/drain implant, can be reduced by *lowering the implant energy*. However, *lowering the implant energy* increases *the channeling probability for boron ions. The channeling probability for boron ions* . . .	The length of the transistor that can function as a long channel device is strongly related to its junction depth. Lowering the implant energy (and thereby increasing the channeling probability for boron ions) reduces the junction depth of the boron source/drain implant. The channeling probability for boron ions . . .

Even these few sentences demonstrate that this style is monotonous and difficult to follow. Imagine the effect of several pages written in this style. Beyond a certain point, over-connecting ideas through repetition has the same general effect as a lack of connections. Readers get lost, lose interest, and tune out.

To resolve this problem in the above example, we used several strategies to restructure and combine sentences. For instance, we shifted from the passive to the active voice, reversing the order of the second sentence to eliminate two of the repetitious chains. We also embedded the third sentence in the second one to eliminate yet another repetition. (Earlier in this chapter we suggested that embedding is not a good idea, but remember that we also mentioned that we offer guidelines, not rules. A problem in one situation may be a solution in another.)

Note that eliminating unnecessary repetition is often simply a matter of replacing the subject of the second sentence with *which* to combine sentences:

Original	Revised
The length of the transistor is strongly related to its *junction depth*. The *junction depth* can be reduced by lowering the implant energy.	The length of the transistor is strongly related to its junction depth, which can be reduced by lowering the implant energy.

Of course, when joining sentences by replacing a subject with *which* or *who*, you must also pay attention to sentence length. When the sentences are relatively short, this revision strategy works well. But for longer sentences, you should consider other options. You might, for instance, join sentences using *which* to eliminate the chaining, but then break the sentence into two new ones at some other point to avoid creating one that is too long or too complex.

6.4.2   Use Parallel Sentence Structure

Parallel structures are a particularly powerful way to communicate information because they create structural repetitions that emphasize and connect ideas. In grammatical terms, two structures are parallel if they have the same grammatical form (both verbs or both nouns, for example). These parallel items or ideas are often coordinated as items in a list.

Identifying lists that are not presented in parallel structures is much more difficult for writers than for the readers of their work. Writers know how their ideas relate and tend not to notice when points are not parallel. Without parallel structures, readers may fail to recognize how ideas connect and may be confused by what they read.

Do you see or hear a problem in the following original example?

Original	**Revised**
In order to improve our facilities, we must do the following:	In order to improve our facilities, we must do the following:
• *repair* our existing PCs	• *repair* our existing PCs
• *purchase* 11 more hard drives	• *purchase* 11 more hard drives
• *security* is insufficient	• *increase* system security
• *expand* our operating hours	• *expand* our operating hours

As you probably noted, the items in the original list are expressed inconsistently. While the first, second, and fourth items begin with verbs, the third begins with a noun. The list becomes parallel when the third item is revised so that it also begins with a verb.

Ideas within sentences that are joined by coordinating conjunctions (*and, but, or, nor, for, yet,* or *so*) must also be parallel. For example, *gadgets* and *gizmos* are joined by coordinating conjunctions in each of the following phrases. These structures are appropriately parallel because the items coordinated are the same part of speech (i.e., *gadgets* and *gizmos* are both nouns):

both gadgets *and* gizmos

not gadgets, *but* gizmos

neither gadgets *nor* gizmos

either gadgets *or* gizmos

Table 6.2 lists the most commonly used ways of indicating that two or more ideas are coordinated. When these coordinators appear in a sentence, the pairs or lists of items they connect must have the same grammatical form.

If the structures in a list or on either side of the coordinators do not share the same grammatical structure, readers will have difficulty making sense of what is written. To find problems with parallelism, read what you have written out loud. Your ears are much more practiced at finding this sort of problems than your eyes. To help your eyes recognize problems with parallel structure, in the following example we have emphasized relevant coordinating connectors and the first few words of the structures that should be parallel.

Original	**Revised**
The system survey report will describe, in general, *the current* system activities, *the problems* that the current system is experiencing, AND *to present* options and make recommendations to you and other decision making authorities.	The system survey report will describe, in general, the current system activities and the problems that the current system is experiencing. It will also present options and make recommendations to you and other decision-making authorities.

Table 6.2   Examples of Coordinators in Parallel Structure

And	Rather Than
Or	Neither . . . Nor
But	Either . . . Or
As Well As	Not Only . . . But Also

Note that this example appears to present a list of three items: *the current system, the problems*, and *to present options*. In this case, however, the writer expected us to make a different connection. The writer intended us to read that the report *describes* activities and problems and *presents* options and *makes* recommendations. Creating two sentences clarifies this intended connection with minimal revision.

The following example also suffers from faulty parallelism:

Original	Revised
This discussion prompted me to contact PTE for further information and the support package that PTE can offer for PC/FOCUS.	This discussion prompted me to contact PTE *for* further information and *for* the PC/FOCUS support package.

Did you have difficulty making sense of the original sentence on first reading? Read the original sentence once more and then the revised version.

To make the sentence clearer on a first reading, we repeated the preposition: *for* information and *for* the support package. Repeating prepositions (*for, in, of, by*, etc.) at the beginning of coordinated phrases both clarifies and emphasizes connections.

Our final example includes another dropped list:

Original	Revised
The purpose of this paper is to critique our current project management structure. This critique will discuss the following topics:	The purpose of this paper is to critique our current project management structure. This critique will discuss the following topics.
• My vision of the company as an optimal organization.	• My vision of the company as an optimal organization.
• The strengths and weaknesses of current practices.	• The strengths and weaknesses of current practices.
• The effects of the current structure on projects, the employees, and the organization as a whole.	• The effects of the current structure on projects, the employees, and the organization as a whole.
• Finally, I will recommend ways to improve our structure and bring us closer to our mission.	Finally, I recommend ways to improve our structure and bring us closer to our mission.

As you should have noticed, the only change we made in revising was to make the final point a separate sentence, rather than part of the list. The first three points list the three topics for discussion in point form. The original last point is not only a complete sentence, but also the content is not coordinated with the other points. It points to recommendations drawn from a discussion of the three topics rather than to another topic. (Note that this sentence may have been added to the list by mistake when the writer forgot to turn off the bulleted list feature.)

6.4.3 Avoid Vague *This* Subjects

Many writers develop the habit of using the single word *this* (or *that, these, those*, or even *it*) as an all-purpose subject to point back to *what I just said* or to *everything I've been talking about*. The problem with using *this* on its own as a subject is that writers assume the reference is obvious. But far too often, the reader concludes something different or must reread a passage to figure out what the writer had in mind.

To help ensure that your subjects establish necessary connections, develop a habit of always following *this* with a noun or descriptive phrase: this *problem*, this *situation*, this *state of affairs*, this *antiquated approach*, and so on. As you read through the following sentences and their revisions, note how vague subjects make you reread or lead you to a different conclusion from that suggested in the revision. The first example demonstrates what can happen when *this* is habitually used as a subject.

Original	Revised
The use of lower wind speeds in the dispersion model tends to predict higher ambient concentrations. *This* is an additional reason to use Northland meteorological data in the dispersion model.	The use of lower wind speeds in the dispersion model tends to predict higher ambient concentrations. *This conservative approach* is an additional reason to use Northland meteorological data in the dispersion model.

How did we decide that *this conservative approach* was the subject of the second sentence? Actually, the writers had to tell us. *This* does not refer back to something earlier in the paragraph, or even to something earlier on the page, but to a concept mentioned 10 pages earlier in the report.

Clearly, readers could not be expected to make a connection with references to a conservative approach that was last mentioned that many pages earlier. And yet, because the writers knew the point they were trying to make, they failed to notice the vague subject. Adding a noun or phrase after *this* ensures that key words are repeated often enough to keep readers on track.

The next example demonstrates a more common problem relating to vague subjects. Read the original; then try adding a noun or phrase to clarify the subject; then check the revision.

Original	Revised
The incremental contribution of the emissions to the ambient hydrocarbon contributions would be only 31% of the average annual hydrocarbon concentrations measured by the monitor at the site. *This* would only occur for a relatively short period of time under the scenario of the "worst case" meteorological conditions.	The incremental contribution of the emissions to the ambient hydrocarbon contributions would be only 31% of the average annual hydrocarbon concentrations measured by the monitor at the site. *This extra contribution* would only occur for a relatively short period of time under the scenario of the "worst case" meteorological conditions.

You may have noticed that the revised subject (*This extra contribution*) helps clarify the point of the previous sentence. Avoiding vague *this* subjects can thus reduce the chance that readers must reread the previous sentence in an attempt to make sense of what you have written. Ensuring precise subjects is particularly important when readers have little technical expertise or little knowledge of a subject. In either case, they need frequent repetition and restatement of key points to understand and retain information.

Without prior knowledge, readers may be unable to decide how to interpret a vague subject. Note the number of possible interpretations for the second sentence in the following original example:

Original	Revised
RNB is a large, well-managed technology company with dedicated people possessing generous amounts of skill and experience. *This* has been externally validated by our ISO 9000 certifications and SEI level 2 rating.	RNB is a large, well-managed technology company with dedicated people possessing generous amounts of skill and experience. *This good management? dedication? skill? experience?* has been externally validated by our ISO 9000 certifications and SEI level 2 rating.

Or

These attributes? strengths? have been externally validated by our ISO 9000 certifications and SEI level 2 rating.

Note that the most likely references are plural. Vague *this* subjects can be doubly difficult to interpret when a singular *this* is meant to refer to a plural concept.

As demonstrated in the following example, the subject should sometimes be something different from *this* (or *these*) plus a noun or phrase. Also note that vague subjects often appear in several sentences in a row, further reducing the chances of clear communication.

Original	Revised
The next recommendation is to clarify the rewards structure. *This* is required to maintain motivation on projects where many extra hours are required and no overtime policy exists. *This* can be critical for commitment from employees.	The next recommendation is to clarify the rewards structure *in order* to maintain motivation on projects where many extra hours are required and no overtime policy exists. *Providing rewards* can be critical for commitment from employees.

Readers need clear connections and repetition of key points. Revising your work to remove the vague subjects can significantly improve how well your ideas flow for readers.

6.4.4 Use Transitions

You can often improve coherence by providing transitional words or phrases to clarify how ideas relate to one another. Many of these transitional devices are simply the conjunctions used to join clauses together in order to create compound or complex sentences. Others are add-ons that direct readers' attention to the relationship between ideas.

Table 6.3 provides a partial list of transitions, indicating some of their more common uses. Use these transitional words and phrases wherever you feel the reader will benefit from a signpost indicating how ideas relate. Note, however, that any stylistic device you can use too little of, you can also use too frequently. Sometimes the connec-

Table 6.3 Transitions and Their Uses

Uses	Transitions
To add to a previous point.	and, or, nor, furthermore, indeed, also, moreover, in fact, first, second, in addition
To illustrate or expand on a point.	for instance, for example, for one thing, similarly, likewise
To summarize or emphasize a point.	therefore, thus, so, and so, hence, consequently, on the whole, all in all, in other words, in short, in conclusion
To qualify or illustrate a point.	frequently, occasionally, in general, specifically, in particular, usually
To shift to a different point of view or signal a contradiction.	but, however, yet, on the contrary, not at all, surely, no
To make a concession.	although, though, whereas
To connect an explanation to a statement.	because, as, since, for
To qualify and restrict a more general idea.	if, provided, in case, unless, lest, when

tions between ideas are clear from the context. Not every sentence needs a transition. But transitions are particularly important when writing for an audience composed of nonspecialists so that they can follow the logic and organization of new and sometimes difficult information.

Transitions are powerful aids to understanding because they help readers identify the relationships between ideas. They are so powerful that readers may rely on them to connect ideas even where no such connection exists. Read the following paragraph:

> *Isaac Newton wrote* Principia *to explain his theories of motion and gravitation. However, tomatoes grow best in full sunlight. So doctors are required to take the Hippocratic oath. Consequently, we must learn to more clearly explain what we mean.*

As you read the passage for the first time, did you, at least momentarily, seek logical connections because of the transitions? When we read the items in a list without the conjunctions, the lack of relationship between the ideas is immediately evident:

- Isaac Newton wrote *Principia* to explain his theories of motion and gravitation.
- Tomatoes grow best in full sunlight.
- Doctors are required to take the Hippocratic oath.
- We must learn to more clearly explain what we mean.

If transitions can create even a fleeting appearance of connection among this strange collection of sentences, imagine how useful they can be in clarifying connections among logically related points.

Sometimes simply adding *for example* or *in other words* at the beginning of a sentence will improve the flow of ideas. Without such signposts, readers may assume that what they are reading is a new point, become confused, and need to reread a passage to make sense of it. Every time readers stop and go back, they lose some of the information held in short-term memory and thus have more difficulty comprehending and remembering what they are reading.

Table 6.4 Partial List of Prepositions

Spatial:	**Temporal:**	**Compound:**
about, above, against, around, at, behind, below, beneath, beside, between, beyond, by, down, from, in, inside, into, on, off, out, outside, over, through, to, toward, up, upon, with	after, before, during, since, till, until **Others**: except, for, like, of	according to, along with, due to, except for, in addition, in front of, in order to, in spite of, on account of, instead of, with regard to

Tables 6.4 and 6.5 list the prepositions and conjunctions most commonly used to signal how ideas are related. Exercise some caution when using *since* and *as* to substitute for *because*. Used inappropriately, *since* and *as* can confuse readers, as demonstrated by the following examples:

Original	**Revised**
As the electrode was connected, a short circuit occurred on the board.	*Because* the electrode was connected, a short circuit occurred on the board.

Or?

When the electrode was connected, a short circuit occurred on the board

Original	**Revised**
Since the electrode was connected, a short circuit occurred on the board.	*Because* the electrode was connected, a short circuit occurred on the board.

Or?

After the electrode was connected, a short circuit occurred on the board.

A simple way to resolve this ambiguity is to develop a habit of using *because* when you are writing about causes or effects:

Original	**Revised**
The majority of this paper's recommendations point to the need to engage in this type of project *as* they have a larger management component.	The majority of this paper's recommendations point to the need to engage in this type of project *because* they have a larger management component.

Table 6.5 Partial List of Conjunctions

Coordinating:	**Adverbials:**	**Subordinating:**
and, but, for, nor, or, so, yet **Correlative**: both/and, either/or, neither/nor, not only/but also **Relative**: that, which	also, consequently, finally, firstly, further, furthermore, hence, hereafter, however, indeed, initially, likewise, moreover, nevertheless, previously, secondly, subsequently, thereafter, thereby, therefore, therein, thus, while	after, albeit, although, as, as a result of, as far as, as if, as soon as, as well as, because, before, even though, if . . . (then), inasmuch as, insofar as, once, only, since, so as, so far as, though, unless, when, whenever, where, whereas, whereby, whereof

As well as clarifying the relationship between ideas by specifying their logical connections, you can also use conjunctions to increase sentence variety (or to lengthen your sentences if you tend to write excessively short ones). In the following examples, note how subordinating conjunctions (*because* and *even though*) are used to clarify the relationships between ideas (and to combine short sentences):

Original	Revised
Injuries of this type have become a severe problem for the pharmacy. There is no equipment available on the market for the IV admixture.	Injuries of this type have become a severe problem for the pharmacy *because* no equipment is available on the market for the IV admixture.
There are real risks to whistle blowing. Sometimes we must blow the whistle.	*Even though* whistle blowing has real risks, sometimes we must blow the whistle.

The order of clauses in the revised sentences depends on context: important, new information should come at the end. Therefore, whether the first example is best ordered with the dependent (*because*) clause at the end or the beginning depends on what information should be emphasized: the lack of equipment or the problem for the pharmacy.

The following example demonstrates that some writers can improve their style dramatically by adding transitional words and phrases:

Original	Revised
There is not much peer-to-peer interaction. *This* has resulted in strong relationships between the R&D manager and individuals on the team and weak relationships among team members. People in R&D are not used to interacting as peers. The focus has been on domain expertise, which has resulted in pockets of expertise that lead to bottlenecks because expertise is not shared. Individuals start to feel ownership for individual components of the system. *This* creates technical dedication, but creates situations where individuals are taken as the authority that discouraged the technical exchange of information among the group. Because of the technical focus in R&D, *it is* easy to ignore the concerns and needs of the customer. On many occasions, with customer-focused projects, *there has been* direct contact between R&D and the customer. In order to maintain a customer focus, *it is* important for R&D teams to maintain a good relationship with the customer.	A lack of peer interaction has resulted in strong relationships between the R&D manager and individuals on the team *but* weak relationships among team members. *Although* the focus on domain expertise encourages people to feel ownership for individual components of the system, it *also* creates situations where individuals are viewed as authorities. *Consequently*, technical dedication is encouraged *while* exchange of technical information is discouraged. *In other words*, we have pockets of expertise and bottlenecks in the flow of information because expertise is not shared. Technical focus in R&D can *also* lead to problems in relationships with customers whose concerns and needs are often ignored. *Note, however*, that on many occasions our customer-focused projects have provided direct contact between R&D and customers. *If* we pay more attention to customer relations, we shift the focus in R&D from domain expertise to customers *and thus* encourage much needed peer interaction.

As the revised version indicates, adding transitions often involves more than adding a few words. In this case, it involves fairly major revision. The original contains a number of empty openers and vague subjects that are eliminated in the revised version. Once we removed the empty opener (*There is*) from the first sentence and the vague subject (*This*) from the second, we combined these sentences, using what was left of the first sentence as the subject of the following one. We also changed *not much* to *a lack* and substituted *but* for *and*, eliminating the need for the original third sentence. We then reorganized the next three sentences to create a more logical flow from issues relating to individuals to overall effects.

The connections among ideas were also clarified by a series of transitions: *although, also, consequently, while,* and *in other words*. Next, we eliminated *because* in the seventh sentence of the original, adding the phrase *can also lead to problems in relationships with customers* to link the problem with customers to the phrase *weak relationships among team members* in the revised first sentence. We then added *Note, however*, to the next sentence to emphasize the shift from problems to solutions. The *if* clause in the final sentence reinforces this shift and the addition of *and thus, encourage much needed peer interaction* adds useful repetition to remind readers of the subject of this paragraph: a lack of peer interaction.

Sometimes revising for readers requires a lot of work.

To ensure ideas are clearly connected, keep the strategies offered above in mind: repeat key words and, use parallel structures; avoid vague subjects; and use transitions. Using connections to create a smooth, flowing style can greatly improve clarity. A number of other strategies for ensuring clear communication are offered in the following section.

6.5 STYLE FOR CLARITY

Writing clearly and unambiguously can be crucial for engineers. For example, we know of a case where a subcontractor misinterpreted ambiguous instructions and dumped land-fill in the wrong location. The resulting loss of time, money, and goodwill (and potential lawsuits) could have been avoided if the instructions had been clearly written. You must communicate as clearly and unambiguously as possible. Engineers who cannot communicate clearly do not generally progress as far or as quickly as those who can. The farther you progress, the more important this ability becomes because you will spend increasingly more time writing for nonexpert readers. The more complex your ideas and the greater your knowledge compared to your readers', the more important clarity becomes.

In this section, we point to a number of revising strategies that can help ensure you convey information and ideas clearly. Note, however, that the revising strategies already discussed for improving order and connection (and others we will discuss for improving conciseness) can also improve clarity. To some degree, all effective revisions improve clarity.

6.5.1 Strive for Relatively Short Sentences

To ensure your ideas are expressed clearly, aim for an average sentence length of 18 to 24 words—as a very rough rule of thumb—and very few sentences longer than four lines of text. Keep in mind that we are suggesting *averages*; some sentences will necessarily be longer and some should be significantly shorter. Be particularly careful, however, to avoid runs of short sentences.

Some readers unconsciously make assumptions about the competence or intelligence of writers based on the length of their sentences (among other things). Note how easily you can deduce the approximate age of the writer of the following letter:

> *I watched your show called Discovery. It was about spiders. Don't kill the spiders. I know you have to catch them for medicine. But only catch some spiders. I just don't think it's fair. Also, you have to be very, very careful around tarantulas and black widows. Another reason why is because if you hurt them they'll hurt you. I like spiders. But to tell you the truth, not very much.*

The writer's age is probably quite close to her average sentence length—about 7. If your sentences average less than 15 words, you may persuade readers that you are young and inexperienced. Or your readers could assume that you are oversimplifying things because you think their ability to understand is limited. Either way, too many short sentences can alienate readers.

On the other hand, if your sentences are too long, you risk confusing and frustrating readers. Writing is like a balancing act: you do not want to lean too far in either direction. However, what counts as too long or too short depends on a number of things, including the age, education, and prior knowledge of your readers.

The *range* of your sentence lengths is at least as important as their average length. If all your sentences are of roughly the same length, you risk putting your reader to sleep. You take the same risk if all your sentences are similarly structured. Variety helps maintain readers' interest.

In the following example, note how all five sentences in the original are relatively short, begin with a subject (noun), have similar simple structures, and lack transitions.

Original	**Revised**
Market conditions are very unstable. Business reacts quickly to the price of pulp and paper. Capital spending by pulp and paper mills is known to be cut off immediately that pulp prices drop. Engineering projects can be stopped with little notice. The technology is relatively stable.	Market conditions are very unstable *as indicated by* how quickly business reacts to the price of pulp and paper. *Because* mills sometimes cut off capital spending immediately when pulp prices drop, engineering projects can be stopped with little notice. The technology is relatively stable, however.
(Average Length = 9 words.)	(Average Length = 15 words.)

In the revised version, we combined two pairs of sentences (using *as indicated by* and *because*) and added a transition (*however*) to the last sentence to indicate a shift to a different point of view. We also rewrote the third sentence in the active voice. While the average sentence length in the revised version may still appear low, two of three sentences are within the 18 to 24-average word-length range. The shorter, six-word sentence is now well placed at a natural point of emphasis where a shift from unstable market conditions to stable technology is indicated. Keep in mind that short sentences are generally more emphasized than long ones. In the first version, too many sentences are short to allow this kind of emphasis.

Note the following example:

Original	**Revised**
Leadership in an individual is the degree to which several qualities are exhibited. These qualities include innovation, organization and motivation. Our managers and associates exhibit varying degrees of leadership but commonly are high in innovation.	Leadership in an individual is the degree to which several qualities *such as* innovation, organization and motivation are exhibited. Our managers and associates exhibit varying degrees of leadership but commonly are high in innovation.

Or

Indivduals exhibit leadership through *such qualities as* innovation, organization and motivation. Our managers and associates exhibit varying degrees of leadership but commonly are high in innovation.

Both revisions combine sentences. In the first revision, we combine the first two sentences of the original, leaving the third sentence in its original form. Then, in the second revision, we eliminate an unnecessary passive construction from the new first sentence. If the topic were leadership, we would retain the passive construction. But in this case, the topic is managers, and the second revision with the active voice is the best choice.

Note that while the sentence length is similar for the original passage and the second revision, the revised version packs much more content into the first sentence. Often the problem with runs of short sentences is that each one presents very little information in relation to the number of words used. Also note that because we are dealing only with fragments of text, average lengths are useful only as general reference points and for comparison.

We suggest that you take several pages you have written recently (preferably from one document) and determine your average sentence length and overall range. Are your sentences too long or too short? Do you vary the length or have runs of sentences of approximately the same length?

Paying attention to sentence length is one way to improve the clarity of our writing. Others involve taking a close look at the kinds of words you use and identifying habits in your use of language. The next few strategies concentrate on language use.

6.5.2 Avoid Overly General Language

Technical readers expect you to express yourself as precisely as possible. To some degree, technical and professional writing depends more or less upon your ability to use somewhat precise language. As we hope you noticed, the previous sentence demonstrates an inappropriate style relying heavily on overly general language: *to some degree, more or less*, and *somewhat*. Readers of technical and scientific documents become frustrated by this kind of language because it leaves them with questions. To *what* degree? *How much* is "more or less?" *How* can anything be "somewhat precise"?

Expressing numerical values accurately (e.g., $x = .065 \pm .005$) is generally more straightforward than expressing thoughts precisely in words. Nevertheless, many engineering students spend much time ensuring the numerical expressions they use are correct, but little, if any, time identifying and revising overly general language. If you have a habit of using general language, you must become aware of this tendency and learn how to revise.

Table 6.6 lists some of the more common words and phrases to watch for when revising for general language. The English language provides a wide range of words and phrases that can be used in ways that are vague or even meaningless. Note how general language functions in the following example and how the revised versions are clearer and more concise.

Original	**Revised**
Approximately 60% of our work is of a repeat *nature* with only minor variations. This repeat work is specifically of the inspection and quality control *type* and does not require higher education.	Approximately 60% of our work repeats with only minor variations. This repetitive inspection and quality control does not require higher education.

Or

Approximately 60% of our work involves repetitious inspection and quality control that does not require higher education.

Table 6.6   List of General Words

General Word	Example Sentence
Aspect	One *aspect* of the theory is correct.
Basically	*Basically*, the results were accurate.
Big/Little	There is a *big* difference between our results.
Clear/Unclear	The reason for the error is *clear*.
Good/Bad	The accuracy of the measurements was quite *good*.
Important	This is an *important* result.
Interesting	This is an *interesting* idea.
Kind of/Sort of	The results are *kind of* confusing.
Large/Small	The results show a *small* deviation from theory.
More or less	The experiment went *more or less* as expected.
Number of/Lots of	There are a *number of* problems with this approach.
Partly	My results were *partly* right.
Perfectly	The results are *perfectly* consistent with the theory.
Situation	The *situation* was unexpected.
Somewhat	I think the theory is *somewhat* misleading.
Thing	I learned two *things* in this experiment.
This	*This* was an inaccurate result.
To a certain extent	*To a certain extent*, I was surprised by the results.
To some degree	*To some degree*, the equipment malfunctioned.
Type	This *type* of experiment is interesting.
Very much/Very little	*Very much* was learned from this experiment.
Way	The *way* the values were obtained is incorrect.

In the first revision, we eliminate *nature* and *type* and change *repeat* from a noun to a verb. In the second revision, we combine the sentences, assuming that the expression, *repeats with only minor variations*, is unnecessary and repetitious detail.

Much general language divides into two categories: excess or vague modification (*more or less, to some degree,* etc.) and unnecessary nouns (*nature, type, environment,* etc.). Some writers habitually use one or both kinds. Examine a recent piece of writing to determine how much you rely on such imprecise modifiers and unnecessary nouns.

6.5.3   Keep Jargon to a Minimum

Jargon is another type of language you must be aware of using. But first, you must distinguish between *jargon* and *specialized technical language*, a distinction found not in particular words, but in the contexts in which these words are used. A number of terms and acronyms that are useful *specialized language for an expert* in a field are incomprehensible, irritating *jargon for a nonexpert reader*.

Used in an appropriate context, technical language helps you communicate precisely with other specialists who possess a level of expertise similar to your own. Thus, when Stephen Hawking uses the acronym *GUT* in an article written for physicists, he employs specialized language familiar to his readers. But if he failed to define the same term in a magazine article for the general public, he would be using jargon that might leave some readers wondering about the relationship between the digestive tract and the origins of the universe.[2]

In other words, useful technical terms used *reflexively* (i.e., without thought) for the wrong readers become jargon. If you have any doubts about your reader's level of expertise, you should err on the side of caution, minimizing your use of technical language. If specialized terms are unavoidable, define them in the text, in a footnote, or in a glossary.

Most often, we use jargon simply because we fail to consider how well our readers know the subject we are writing about. Some writers, however, use jargon purposefully to make a simple idea seem more complex or to make a self-evident statement sound more impressive. This use of jargon is often described as *gobbledygook*. Reflect on its effect as you read the following passage:

> *In effect, it was hypothesized that certain physical data categories including housing types and densities, land use characteristics, and ecological location, constitute a scaleable content area. This could be called a continuum of residential desirabilities. Likewise, it was hypothesized that several social strata categories, describing the same census tracts, and referring generally to the social stratification system of the city, would also be scaleable. This scale could be called a continuum of socioeconomic status. It was also hypothesized that there would be a high positive correlation between the scale type of each continuum.[3]*

The author of this passage may have unconsciously used the specialized language of his or her field, but what we read is jargon. From our nonspecialist perspective, the language seems meant to impress readers rather than to communicate. A jargon-free translation reads:

> *Wealthy people live in nice homes in desirable neighborhoods while poor people live in substandard housing in run-down neighborhoods.*

Because writers of technical documents are most concerned with communicating information, they rarely exhibit this extreme use of jargon. We do, however, find gobbledygook in college assignments when students put impressing teachers above communicating content. (Teachers are rarely impressed.) The risk for these writers is developing habits that carry forward into workplace writing.

Jargon can also be used as a form of deception. This potentially sinister use of jargon is known as *doublespeak*. We invented the following passage using examples we heard on news channels during the Persian Gulf War. Note how the passive voice contributes to the overall effect:

> *To service the target, the theater was entered and a package of ordnance was delivered. Subsequently, a BDA was undertaken in order to assess the softening of enemy assets. It was observed that the ordnance was incontinent thereby leading to some collateral damage.*

Translated into jargon-free language, the passage reads:

> *We dropped bombs on the enemy and then later looked to see if we killed them, but we missed and blew up civilians instead.*

---

[2]In case you are wondering, *GUT* stands for *Grand Unified Theory*, which is a theory that unifies three of the four fundamental forces of nature: the electromagnetic force, the weak nuclear force, and the strong nuclear force. The fourth force, gravity, is theoretically unified with the other forces in *Superforce Theory*.

[3]Excerpted from N. Green, "Scale Analysis of Urban Structures: A Study of Birmingham Alabama," *American Sociological Review*, 21:1, 1956, 9.

While doublespeak is not a common concern in technical writing, you can expect to encounter it in other situations. In engineering contexts, inappropriate use of technical language is the form of jargon that is all too common.

6.5.4   Use Acronyms Carefully

While acronyms help you write about technical subjects in a concise manner and are necessary in technical reports, they are often overused, especially in high technology fields. Two technical experts may happily chat away or send each other e-mail about *a pseudorandom generator on the LSB of the CB and CR* or about *rerouting OAM messages using UDP/IP in a CDPD Radio.* But for nonexperts, this language might as well be alphabet soup.

To avoid turning your document into gobbledygook, keep the following three strategies in mind:

1. Introduce acronyms, providing the words they replace the first time you use them.

2. Limit your use of acronyms, restricting the number in individual sentences.

3. Repeat the words acronyms replace periodically and use few if any acronyms in introductions, conclusions, abstracts, and executive summaries.

If an acronym is well known by your readers (e.g., *IBM, FBI*), then it can be used without concern. Also, the greater the technical expertise of your readers, the higher their tolerance for acronyms.

Ensure that you provide the words an acronym replaces the *first* time you use it. While this point may seem too obvious to need stating, we often find acronyms used several times before they are explained, as the following example from the first page of a letter report demonstrates:

Original	Revised
This letter provides a summary of the results of dispersion modeling of *ETBM* emissions from the site, a description of the approaches taken to estimate the *ETBM* emissions, to conduct the dispersion modeling and to prepare the contour plots of ambient *ethyl tertiary butyl methelyne (ETBM)* concentrations.	This letter provides a summary of the results of dispersion modeling of *ethyl tertiary butyl methelyne (ETBM)* emissions from the site, a description of the approaches taken to estimate the *ETBM* emissions, to conduct the dispersion modeling, and to prepare the contour plots of *ETBM* concentrations.

Or better yet

This letter *summarizes* the results of dispersion modeling the ethyl tertiary butyl methelyne (ETBM) emissions from the site and *describes* the approaches for estimating ETBM emissions, conducting dispersion modeling, and preparing contour plots of ambient ETBM concentrations.

In many cases—as in the original version—writers revise their work but fail to notice that the explanation of the acronym is no longer connected with the first use. Note that the second revision also improves the clarity and conciseness of the passage by using more descriptive verbs (*summarizes* instead of *provides a summary of* and *describes* instead of *a description of*). This revising strategy is discussed in the section on revising for conciseness.

If at all possible when addressing nonexpert readers, avoid using more than one acronym in a sentence. In any case, do not use more than two (unless presenting a list). If a third acronym in a sentence seems unavoidable, use the full term for one item. Note how acronyms interfere with understanding the following sentence from the introductory section of a report prepared for a nonexpert audience:

Original	Revised
Any requirements agreed upon by *DOD* and *FAA* after the baseline of the *SS* and *IRDs* are placed into concept papers called *ORDDs*.	Any requirements agreed upon by *DOD* and *FAA* after the baseline of the *Segment Specifications* and *Interface Requirement Documents* are placed into concept papers called *??* (*ORDDs*).

We retained two of the acronyms, *DOD* (*Department of Defense*) and *FAA* (*Federal Aviation Administration*), because they had been clearly defined a few sentences earlier and were terms the audience could be expected to easily understand and remember. To avoid overloading the revised sentence with acronyms, we then used the full terms for *Segment Specifications* and *Interface Requirements Documents*.

We would have liked to use the full term for *ORDD* in our revision, but we (and probably most other readers) could not determine what the acronym meant (aside from knowing generally that it stands for some type of concept paper). Although the report from which this sentence was taken had a glossary, *ORDD* was not defined in it. This omission was particularly unfortunate because the term was used again several pages later, by which point readers had probably forgotten even the general sense of what it meant. After looking in the glossary to no avail, all the hapless reader could do would be to flip through the report looking for a definition of the term.

To avoid frustrating your readers in this way, use the full term of the acronym when beginning a new chapter or major section or when the term has not been used for several pages. Also use the full term for acronyms that are used infrequently.

Finally, in the introduction and conclusion of a report, write out the full term for all but the most common acronyms. If your document has an executive summary or abstract, it should contain few, if any, acronyms. Executive summaries and abstracts are typically standalone documents that should be written in plain language.

6.5.5   Avoid Lengthy Noun Strings

Another form of jargon found in technical and scientific documents involves stringing adjectives and nouns together (i.e., *the event list buffer overflow error* or *the predicted ambient total suspended particulate concentrations*). These *noun strings* are a useful aspect of technical language, providing a group of experts with short expressions for complex ideas. Noun strings become a problem, however, for readers who are not familiar with the concepts being discussed. When nouns are used to modify other nouns, readers have difficulty determining the logical relationships among the words in the string; consequently, readers may interpret an expression in ways the writer did not intend. Note, for example, the ambiguity in the following example:

Original	**Revised**
The parent company is *a worldwide data network provider service company.*	The parent company is a service company that provides data networks worldwide.

Or?

The parent company provides and
services worldwide data networks.

Either revision is plausible and only a reader familiar with the company could interpret the noun string with confidence. As this example demonstrates, the less previous knowledge or context the reader has, the more important revising for noun strings becomes.

The examples in this section contain the sort of noun strings typically found in technical documents. The writers of these examples habitually used noun strings as part of their everyday technical language and failed to realize that their readers (who were not experts in the field) would have difficulty making sense of the documents. Their failure to notice this problem is understandable. When we know what we are saying and are concentrating on the content of a document, we express concepts in the language used to talk about them.

All three of the following examples were excerpted from reports intended for readers who lack technical expertise, and the writers are using expressions that are part of the technical language of their working groups. Other people who work closely with them have no trouble understanding the original. But this technical language may be ambiguous, confusing, or incomprehensible for readers outside the group. The less the reader knows about the topic under discussion, the greater the problem.

Original	**Revised**
The wind pattern was further illustrated following *a recovery boiler electrostatic precipitator fire* at the site in 1988.	The wind pattern was further illustrated following a fire in the electrostatic precipitator of a recovery boiler at the site in 1988.
The dispersion model calculated twenty-four hour averaged concentrations assume continuous operations over that twenty-four-hour period.	The averaged twenty-four hour concentrations calculated using the dispersion model assume continuous operation over that twenty-four-hour period.
Because *the predicted ambient total suspended particulate concentrations* met *Department of Environment, Lands, and Parks' Level B ambient air quality objectives*, a further evaluation was not required.	Because the predicted total concentrations of ambient suspended particulates met the level B ambient air quality objectives mandated by the Department of Environment, Lands, and Parks, a further evaluation was not required.

Note that the original versions contain fewer words than the revised versions, but you can read the revised text more quickly and comprehend more of what you read more easily. You should be less concerned with how many words you use than with how quickly a reader can read and comprehend what you have written.

Keep in mind that noun strings and the other stylistic issues discussed in this chapter should be attended to while revising. If you try to change your language to accommodate readers while you draft, you will lose track of your own thoughts and find the task of writing frustrating. That we see so many noun strings in documents written by technical experts for nonexpert audiences suggests many writers either do not revise sufficiently or simply fail to recognize the problem.

Whenever readers have different fields of expertise or are unfamiliar with the specific technology or project under discussion, revising for noun strings is a potentially effective way to improve clarity. Revising involves finding ways to indicate the logical connections among the words in a string. You can establish these connections in four ways. First, you can add prepositions and, where appropriate, change the order:

Original	Revised
a recovery boiler electrostatic precipitator fire	a fire *in* the electrostatic precipitator *of* a recovery boiler
the predicted ambient total suspended particulate concentrations	the predicted total concentrations *of* ambient suspended particulates

Second, you can add verbs or turn one of the words into a verb, reordering as required:

Original	Revised
the Department of Environment, Lands, and Parks' Level B ambient air quality objectives	the level B ambient air quality objective *mandated by* the Department of Environment, Lands, and Parks
the dispersion model calculated twenty-four-hour averaged concentrations	the averaged twenty-four-hour concentrations *calculated using* the dispersion model

Third, you can hyphenate nouns that function as adjectives. For example, consider a string such as *event list buffer*. In this case *event list* functions as a single adjective in that the buffer is not an event buffer or a list buffer but an *event list* buffer. In this case, you can use hyphenation: *event-list buffer*. Note, however, that if you wrote simply *event list* without the following noun (i.e., *The committee compiled an event list.*), you could not hyphenate because only *event* functions as an adjective explaining what kind of list is being compiled.

Hyphenation makes noun strings easier to read by indicating relationships and reducing the number of words in a string:

Original	Revised
Team members must understand that *behavior oriented activities* are just as important as *task oriented activities.*	Team members must understand that *behavior-oriented* activities are just as important as *task-oriented* activities.
We must improve the *project oriented product development environment.*	We must improve the *project-oriented* product development environment.

Or better yet

We must improve the *project-oriented* environment *for* product development.

In the second example, we could have hyphenated *product development*. But double hyphenation is difficult to read (*project-oriented product-development environment*), so we chose to break up the string with a preposition (*for*). Like so many other stylistic features, hyphenation can be overused as well as underused.

Fourth, you can sometimes avoid awkward noun strings by removing unnecessary words, which is the approach used in the following example:

Original	**Revised**
Hence, a method must be devised to alleviate *the slow response performance problem*.	Hence, a method must be devised to alleviate *the slow response*.

Or better yet

Hence, a method must be devised
to speed up the response.

In this case, we revised in two steps. First, we noted that *performance* is redundant because a *slow response* defines the nature of the performance. Further, in the context in which the original sentence was written, a *slow response* is clearly a *problem*, making this final word redundant as well. We could have stopped revising at this point, but when we read our first revision, we noted that *alleviate the slow response* was more directly expressed as *speed up the response*. In this case, revising for the noun string was just the first step in communicating more clearly.

When you are writing about technical topics for nonexpert audiences, revise for noun strings using these four methods: add prepositions, add verbs, hyphenate, and eliminate unnecessary words.

6.5.6   Avoid Runs of Prepositional Phrases

Just as you may be unaware of your reliance on noun strings, you may not notice long runs of prepositional phrases in your writing. While prepositions are useful for breaking up lengthy noun strings, they create problems for readers when overused. As you read the original version below, note where you begin to lose track.

Original	**Revised**
More detailed analysis, using Cepstrum technique, is applied based *on* computation *of* the power spectrum *of* the logarithm *of* the power spectrum *of* the vibration data obtained *from* the accelerometer *on* the truck frame *in* different positions.	A more detailed analysis *uses* Cepstrum technique *to compute* the power spectrum as a logarithm of the various vibration spectra *measured* by the accelerometer at different positions on the truck's frame.

When faced with more than three consecutive prepositional phrases, readers are likely to become confused. Given *eight* prepositional phrases in a row, even readers with some knowledge of the experiment will not easily comprehend how the phrases connect. The problem in this example is further exaggerated by a string of four prepositional phrases beginning with *of*. In the case of this particular preposition, two consecutive *of* phrases are awkward and three are confusing.

Keep in mind that because you know how your ideas connect, you are unlikely to notice a problem with prepositional phrases unless you edit carefully for this particular feature or read your work out loud. A grammar checker set to find three or more prepositional phrases in a row is one way to find runs of prepositional phrases (also note that you can often set your grammar checker to identify lengthy noun strings). This check can help you determine whether you use this pattern habitually. If you have this habit, you should revise specifically to eliminate runs of prepositional phrases, avoiding more than three prepositional phrases in a row and eliminating consecutive *of* phrases.

Our use of words such as *avoid* and *eliminate* may suggest that we think you should *never* create runs of more than three prepositional and *never* use two *of* phrases in a row. We acknowledge that on occasion, a revision limiting consecutive prepositional phrases *could* be more awkward than the original string. But such exceptions are rare. If you want to break a habit, you must be vigilant. Therefore, if you habitually use strings of prepositional phrases, you should assume that all of them could be reduced to fit the guidelines: limit runs of prepositional phrases to three and avoid sequential *of* phrases.

If you find most of your revisions less appealing than the original string, then you may be choosing inappropriate revision strategies *or* you may be using sentence structures that seem odd only because you are not familiar with them. Because new sentence structures, like new fashions, sometimes take some getting used to, you may want to ask an experienced writer for his or her opinion of your revised sentences.

While the example at the beginning of this section demonstrates the potential problem with consecutive prepositional phrases, it is not particularly useful in demonstrating revision techniques given the length of the string and the extensive revision required. The following example offers a more straightforward demonstration of how to use verbs to reduce the number of prepositional phrases.

Original	**Revised**
Another component needed *for* the completion *of* the project is the reassignment *of* project team members *to* the next project and the rewarding *of* team members *for* their role *in* the project.	Another component needed *to* *complete* the project *involves* *reassigning* project team members to the next project and *rewarding* them for their role in the project.

Note how nouns in the original become verbs in the revision. Identifying nouns with verb roots and converting them to verbal forms eliminates four prepositions:

Original	**Revised**
for the completion *of* the reassignment *of* the rewarding *of*	to complete reassigning rewarding

No further revision is required.

In the following example, we also change a noun (*a lack*) into a verb form (*lacking or lack*):

Original	**Revised**
One example *of* the effects *of* such a lack *of* effective strategies, project orientation and interdepartmental cooperation is our low morale.	For example, one result of *lacking* effective strategies, project orientation, and interdepartmental cooperation is our low morale.

Or better yet

For example, because we *lack* effective strategies, project orientation, and interdepartmental cooperation, morale is low.

In the first revision, we remove the first *of* phrase by turning *one example of* into *for example*. We then changed *effects* to *one result* to maintain the sense of *one* example and to avoid the awkward repetition of *effects . . . effective*. We could stop here, having solved the problem of consecutive *of* phrases, but the verb (*is*) remains near the end of the sentence. By creating a dependent clause (*because we lack . . .*), we add a verb near the beginning of the sentence and emphasize the new information (*morale is low*) at the end.

As the above example indicates, revising is often best accomplished in more than one step. First fix obvious problems and then scrutinize the revised sentence for other problems. Once you make one change, further revisions are often much easier to identify.

Sometimes, you can eliminate a string of prepositional phrases by breaking a sentence and starting a new one.

Original	Revised
The emission requests cannot be accommodated *at* this time because the emission inventory is neither sufficiently advanced nor intended to allow *for* temporal variation or for identification *of* a rationale *for* each occurrence *of* a high emission rate and/or ambient concentrations.	At this time, the emission requests cannot be accommodated because the emission inventory is not sufficiently advanced and does not allow for temporal variations. Moreover, this inventory does not permit identifying the cause of each high emission rate and/or ambient concentration.

While prepositional phrases are not a problem in the first sentence of the revision, we have moved *at this time* to the beginning of the sentence in order to demonstrate another technique for breaking up prepositional strings. That is, you can move a prepositional phrase to the beginning of a sentence so that it functions as an introduction.

To review, when confronted with long or awkward strings of prepositional phrases, search for nominalizations and replace them with verb forms, break the sentence in two, or move a phrase to the beginning to serve as an introduction. But as the first example with eight consecutive prepositional phrases demonstrates, sometimes more major revision is required. To a large extent, learning how to revise effectively and efficiently is a matter of practice. The more you revise, the easier revising becomes.

6.5.7   Pay Attention to Punctuation

Something as simple as a misplaced or missing comma can affect the clarity of a sentence. For instance, because a comma is missing in the following sentence, some readers will assume the italicized portion is one idea, will reach a point in the sentence (generally the verb) where they realize that something is wrong, and will then be forced to reread the sentence.

Original	Revised
A precision full-wave rectifier is constructed using precision diodes and a square-wave generator is assembled using a schmitt trigger.	A precision full-wave rectifier is constructed using precision diodes, and a square-wave generator is assembled using a schmitt trigger.

In the above example, the conjunction *and* separates two sentences. But without a comma before *and*, readers may assume that *precision diodes and a square-wave generator* are coordinated (i.e., joined by the *and*). Whenever you join two complete sentences with a coordinating conjunction (*and, but, or, nor, yet, for*, or *so*), place a comma before the conjunction to signal where one sentence ends and the next begins.

The following example demonstrates another punctuation problem that affects clarity by encouraging readers to misinterpret the grammatical relationships between words. Again, note what happens when you read the words in italics as one idea:

Original	Revised
By participating more in scheduling release dates can be planned so that projects dovetail.	By participating more in scheduling, release dates can be planned so that projects dovetail.

In this case, a comma is needed to signal the end of an introductory phrase and to clarify that the subject of the sentence is *release dates*. Because writers rarely recognize clarity problems of this sort while writing or revising their work, you should make a habit of placing a comma at the end of an introductory phrase.

You should also place a comma before an *and* (or any other coordinating conjunction) that signals the last item in a list, as in the following example:

> ABC Corp. is offering three summer co-op positions: market analysis, testing and debugging and customer support.

We can be sure that one position involves market analysis, but does the second position involve testing or testing and debugging? We cannot be sure from the example. To clarify the groupings, a comma is needed to signal which *and* comes before the last item in the list.

> ABC Corp. is offering three summer co-op positions: market analysis, testing, *and* debugging and customer support.

Or

> ABC Corp. is offering three summer co-op positions: market analysis, testing and debugging, *and* customer support.

Because lists with compound items are fairly common in technical documents, you will encounter many situations in which you could confuse readers by omitting the comma before the final item in a list.

The problem—and confusion—many writers encounter when punctuating lists may result from conflicting advice. For example, take the case of a group of three researchers who were editing a proposal. Each had learned a different "rule" for punctuating lists:

- Put a comma before *and* at the end of a list.
- Do not put a comma before *and* at the end of list.
- Either put a comma before *and* at the end of a list or not, but be consistent.

In the end, they agreed to put the comma before the *and* at the end of a list because this habit is most likely to ensure clarity.

For those of you who have further difficulties with punctuation, we have included a punctuation guide later in this chapter (pages 234–240).

6.6   STYLE FOR CONCISENESS

You write concisely when you convey information or express your thoughts in as few words as possible. However, conciseness is a relative concept. Depending on your readers' needs, you may require considerably more or fewer words to communicate with them. In general, the more familiar readers are with a topic, a concept, or the history of a problem, the less background and explanation they need. Conversely, the less they know, the more background information and explanation they require.

How readers will use the information you provide also helps determine how much detail to include. For example, senior managers often want only the big picture in a condensed form they can read quickly. Therefore, executive summaries are most often pared to the bone and contain a point-form list of recommendations. On the other hand, engineers responsible for installing complex systems need specific, detailed information. These readers are often most interested in the minute technical detail provided in appendices.

While purpose and audience are keys to determining how much information to include, the style of your sentences affects how many words you require to make a particular point. In this section, we focus on conciseness as a stylistic issue at the sentence level. Note, however, that you cannot view sentences in isolation. The nature of the topic and the needs of readers remain important considerations, sentence by sentence.

Conciseness is a relative concept. If your sentences can be too wordy, they can also be too concise. Although we have used the words *too wordy*, we caution you not to equate conciseness with some vague notion of having used too many words. Writers who are advised that their writing is too wordy sometimes take this advice literally and pare down the number of words in a sentence to the point where they omit the context and eliminate the repetition readers need to make sense of the words on the page. Revising for conciseness is a matter of eliminating unnecessary words, but not of eliminating information, context, and useful repetition.

At the sentence level, revising to create a more concise style involves learning what to look for. Strategies discussed in the sections on revising for order and clarity also improve conciseness, such as revising for passive constructions and empty openers. In this section on conciseness, we focus on three common stylistic habits that tend to occur together: a noun-based style, reliance on talkie verbs, and habitual use of wordy phrases. Revising for these three features can make your writing more concise as well as easier to read and comprehend.

6.6.1   Write in the Verbal Style

In the previous section on clarity, we pointed out that eliminating strings of prepositional phrases often involves turning nouns into verbs or verbal forms. Some of the examples we presented involved changing *the completion of* to *complete, the reassignment of* to *reassigning*, and *the rewarding of* to *rewarding*. In each case, we created a more verbal style by eliminating unnecessary *nominalizations*. A nominalization is the noun form of a word that also has a verb form. For example, *completion* and *reassignment* are nouns corresponding to the verbs *complete* and *reassign*.

You can often (but not always) identify nominalizations by their endings. Table 6.7 lists some of the more common suffixes for nominalizations. (The suffixes are in bold.) Although the list of endings helps identify many nominalizations, as Table 6.8 indicates, others are formed without these suffixes.

Overuse of nominalizations is widespread in technical writing, affecting both clarity and conciseness. But note that the noun forms of verbs are also very useful for communicating abstract concepts. While you cannot eliminate nominalizations, you should limit their use.

Table 6.7   Common Suffixes for Nominalizations

Verb Form	Noun Form
require	require**ment**
impose	imposi**tion**
discuss	discus**sion**
resemble	resembl**ance**
remove	remov**al**
conform	conform**ity**

The following intentionally exaggerated example compares the differences between texts written in the noun-based and verb-based styles.[4] The noun forms in the noun-based example and corresponding verb forms in the verb-based example are in bold.

Noun-based Style

Everything today has *the* **requirement** *of the* **conformity** *of* people to some standard. There is *the* **requirement** *to* be similar in their **speech** and in their **beliefs**. If you have **different** ideas, people have *the* **thought** you are odd. Our **fear** stops *the* **expression** *of* our **thoughts** about our surroundings. We must have *a* **resemblance** *to* our neighbors or friends. Because there is so little we can do to be individuals, it is my **thought** that we would commit *an* **error** by *the* **imposition** *of a* dress standard on students. It would be *a* **loss** *to* their **expression** *of* individuality. (100 words)

Verb-based Style

Everything today **requires** people **conform** to some standard. Similarity is **required** in how they **speak** and what they **believe**. If your ideas **differ**, people **think** you are odd. We **are afraid** to **express** what we **think** about our surroundings. We must **resemble** our neighbors or friends. Because there is so little we can do to be individuals, I **think** we would **err** by **imposing** a dress standard on students. They would **lose** a way **to express** individuality. (77 words)

Note that the revised version cuts close to 25 percent of the words without eliminating content. Also note that some of the nominalizations in the above example are preceded by

Table 6.8   Some Nominalizations without Suffixes

Verb Form	Noun Form
reward	reward
think	thought
analyze	analysis
speak	speech

[4]Adapted from R. L. Hake & J. M. Williams, "Style and Its Consequences," *College English*, 43, 1981, 448–449. Used with permission.

an article (*a, an, the*) and followed by a preposition (typically, *of.*) These words are italicized in the above example to draw attention to this pattern:

Original	Revised
the requirement *of*	require
the conformity *of*	conform
the requirement *to*	require
the expression *of*	express
the resemblance *to*	resemble
the imposition *of*	impose
a loss *to*	lose
[their] expression *of*	express

Because nominalizations frequently follow this pattern, you can identify a majority of them in your own writing by searching for it.

You may also have noted that not all moves to a more verbal style affect nouns. For example, the adjective *different* in the original becomes the verb *differ* in the revised version. Developing a verbal style is not just a matter of revising for nominalizations, but of revising to move from a reliance on nouns and adjectives toward a reliance on verbs, verbals, and adverbs. Think of this move in terms of a continuum with nouns at one end and verbs at the other, as presented in Figure 6.2.

In the following examples, note how the revisions become more verbal by changing nouns and adjectives into verbs, verbals, and adverbs. In the first example, also note the article-noun-preposition pattern discussed above.

Original	Revised
The removal of reduced sulfur compounds from in-mill processes and *the prevention of* anaerobic formation of reduced sulfur compounds will improve both air and effluent quality. (26 words)	*Removing* reduced sulfur compounds from in-mill processes and *preventing* anaerobic formation of reduced sulfur compounds will improve both air and effluent quality. (22 words)

In this case, we simply replaced nominalizations with verbal forms. A saving of four words may not seem worth the effort, but note that the revised version is somewhat clearer and easier to read. Because frequent small changes of this nature add up, what appears as an insignificant change in a single sentence can make a significant difference when repeated throughout a document.

Also consider conciseness from the reader's perspective. Because readers can process information more quickly and easily when it is presented in a verbal style, they will perceive

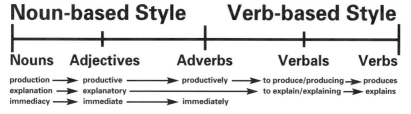

Figure 6.2 Nominal/Verbal Continuum

the document as shorter or more concise because they can read it relatively quickly. In terms of revising strategies, clarity and conciseness are often two sides of the same coin.

The next example demonstrates another common feature of a noun-based style:

Original	Revised
Our request is that *on your return,* you *conduct a review of* the data and *provide an immediate report.* (19 words)	We request that when you return, you review the data and report immediately. (13 words)

In the original, possessive pronouns (*our, your*) precede a nominalization. When these possessive pronouns are replaced by subjects (*we, you*), *our request is* becomes *we request* and *on your return* becomes *when you return.* Also note how the verbs *conduct* and *provide* are unnecessary once the nominalization is identified:

Original	Revised
conduct a review of	review
provide an immediate report	report immediately

Adding less descriptive verbs to stand in for the verb that is being used as a noun is another characteristic of a noun-based style. Note this pattern in the following example:

Original	Revised
Many engineering personnel have *made comments regarding a lack of knowledge* about what new information is available in the library as literature is filed into the library without *any sort of notification.* (32 words)	Many engineers have commented about not being notified when new literature is filed in the library. (16 words)

In this case, we can omit *made* and convert the noun *comments* into the verb *commented.* Note that we also converted *without any sort of notification* to *not being notified* and eliminated the redundant phrase, *regarding a lack of knowledge.* Simply by revising for nominalizations, we have cut the number of words in half without affecting the meaning.

As the following example demonstrates, developing a more verbal style can also help eliminate unnecessary technical language from documents written for readers who lack technical expertise or who are unfamiliar with technical terms.

Original	Revised
Hence, there is diurnal and seasonal *variation in* the air mass stability and dispersion characteristics. (17 words)	Hence, the air mass stability and dispersion characteristics vary daily and seasonally. (12 words)

Note the empty opener (*there is*) in the original version. To make the style more verbal, we eliminated the empty opener, turned *variation* into a verb and located a new subject (*air mass stability and dispersion*). We then changed adjectives (*diurnal* and *seasonal*) into adverbs (*daily* and *seasonally*) to complete the transition from a noun-based to a verb-based style. Note that while *dispersion* is also a nominalization of the verb to *disperse,* we have not attempted to revise it because *dispersing an air mass* does not carry the same meaning as *air mass dispersion.*

Nominalizations exist because they serve a purpose. Note in the following example how the original is more appropriate than the revised version:

Original	**Revised**
The development of quieter submarines created a demand for research by the Navy into active sonar.	*Developing* quieter submarines created a demand for research by the Navy into active sonar.

In this case, we want the abstraction (*development*) rather than the action (*developing*) because we are not talking about developing quieter submarines but rather about the *process* of developing them.

As with passives, you cannot eliminate nominalizations from your writing, particularly from your technical documents. But you should revise for unnecessary ones, especially when writing for readers who are unfamiliar with your topic or who lack technical expertise. Even expert readers will appreciate the increased clarity and conciseness of a verb-based style.

6.6.2   Use Descriptive Verbs

Developing a verb-based style is particularly important because verbs largely control the power and clarity of written communications. Like the engine of a car, verbs are the power plants of your sentences. Verbs move sentences. Like a car with a worn-out engine, a sentence with a worn-out verb does not function properly. It stalls. And if you stall too often, readers become irritated and start honking their horns. Excessive or inappropriate use of the verb *to be* and *talkie verbs*[5], which we explain below, are often encountered in writing. Happily, they are relatively easy to revise for, as long as you can extract more descriptive alternatives from your memory.

If you make a habit of using *to be*, you want to break it. This verb simply asserts that something exists. If you rely on it heavily, revising to increase your use of descriptive verbs will help eliminate vague, imprecise expressions while increasing clarity, precision, and conciseness. For example, consider the following sentence: The technician *is* in the server room. All the reader knows is the location of the technician. (The technician *exists* in the server room.) The reader does not know what the technician is doing there.

Maybe the technician *hides* in the server room or *snoozes* there. Or perhaps the technician *solves* a problem or *fixes* equipment. Although readers might interpret the verb correctly from the previous context, they all too often misinterpret the intended meaning of the verb *to be*. Using a more precise verb increases the likelihood that a reader will understand what you are attempting to communicate. Breaking the *to be* habit also eliminates unnecessary words and places more emphasis on verbs with descriptive power:

Original	**Revised**
The meteorological evaluation *is intended to provide* the basis for the dispersion modeling. (13 words)	The meteorological evaluation *provides* the basis for the dispersion modeling. (10 words)
The following analysis *is aimed at highlighting* the pertinent aspects that affect dispersion. (13 words)	The following analysis *highlights* the pertinent aspects that affect dispersion. (10 words)

---

[5]We would like to thank Don Wilson (Selkirk College, Castlegar, BC) for devising the term *talkie verb* and for showing us how often these verbs interfere with clear, concise communication.

Original	Revised
Additionally, the most environmentally significant compound will not occur at concentrations that *are indicative of* environmental or health problems. (20 words)	Additionally, the most environmentally significant compound will not occur at concentrations that *pose* environmental or health concerns. (18 words)

Note that the original examples contain verb phrases containing several words (each beginning with some form of *to be*) while the revised versions contain a single verb in their place. In the first two examples, revising is simply a matter of identifying the appropriate verb already present in the phrase. As the third example demonstrates, sometimes you need a substitute verb.

You may have noted that we saved only a few words in each of these examples. Keep in mind that conciseness is not just a matter of the number of words on the page, but also of how quickly a reader can comprehend those words. Packing more descriptive power into the verb speeds comprehension.

Did you also note how changes in the first two sentences affect the tone of the sentences, making them potentially more persuasive? Readers may be less than enthusiastic about reading a report that *aims* or *attempts* or *intends* to do something. They want to know what the report succeeds in doing. In other words, the original versions of the first two sentences may undermine what the report actually achieves. Our change makes the sentence more assertive and to the point.

Despite the benefits of using descriptive verbs, most writers' tendency to overuse *to be* may be explained by our heavy reliance on it when we speak. In fact, when we speak, we rely heavily on a whole family of *talkie verbs* (*do, make, give, go*, etc.). Because we use them so frequently, they are the verbs most likely to come to mind while we are writing. Unfortunately, they are so imprecise that they can be interpreted in too many ways. Thirty of these worn-out talkie verbs are listed in Table 6.9. Revise to limit them in your writing.

Many of these verbs have accumulated a vast array of meanings over the centuries. Listing the various meanings of the verb *make*, for example, requires a full 13 pages in the *Oxford English Dictionary*; listing the meanings of the verb *go* requires a similar number of pages. This potential ambiguity is not particularly evident in speech because listeners can ask clarifying questions and speakers can monitor the audience to guard against misunderstandings. In writing, you can provide no such clarification and must keep your use of talkie verbs to a minimum.

To make our point, try to determine what *made* means in the following sentence: Yesterday Josey *made* three hundred dollars. Perhaps Josey is a forger and *counterfeited* three hundred dollars. Or he may have worked a little overtime and *earned* some extra

Table 6.9 List of Talkie Verbs

Be	Drive	Go	Look	Seem
Bring	Face	Grow	Make	Show
Carry	Feel	Have	Place	Take
Come	Find	Hold	Put	Think
Deal	Get	Keep	Say	Try
Do	Give	Know	See	Turn

cash. Josey might even work for the Mint and have *printed* (or *manufactured*) the money. The point is that without sufficient context, the reader has little idea what *made* really means. So readers supply their own meanings—meanings that may differ from what the author intended. In this case, Josey sold some shares at a profit.

If you have trouble thinking of alternatives for talkie verbs, a good thesaurus will help. However, when using a thesaurus, only choose substitutions you are fully familiar with. Words listed as synonyms are not always good replacements because they may be suited to other contexts or have a connotation or meaning not suited to your purpose.

In the following example, note how the conversational tone of the original version shifts in the revised version to a level of formality more appropriate to technical documents.

Original	**Revised**
Because of the team's ignorance, no one could *know at the beginning* what the product *was going to look like* or how long *it was going to take* them to create it. (32 words)	Because of the team's ignorance, no one *initially understood* how the product *would appear* or how long they *would require* to create it. (23 words)

In this example, we replaced talkie verbs (*know, look, take, go*) with common everyday verbs that are not so overused (*understand, appear, require*).

The next example also illustrates the technique of replacing talkie verbs with less-imprecise alternatives. Note how the descriptive power of verbs increases in the revised version:

Original	**Revised**
When a company *grows* at such a rapid rate, there may not *be* enough structure or direction for the new employees to *make* decisions on these matters. They also may not *know* who to *go* to for advice. (38 words)	When a company *expands* rapidly, it may *lack* the structure or direction new employees *require to handle* these matters. They may also *be uncertain* whom to *ask* for advice. (29 words)

Note that we removed an empty opener (*there may not be*) and then reintroduced the verb *to be* (*be uncertain*). We draw attention to this point to again remind you that we are offering guidelines, not rules. We are not suggesting that you should entirely avoid the verb *to be*, but that you should use it and other talkie verbs with care. After considering alternatives, you generally do best by trusting your judgment and common sense rather than by relying on rules.

The next two examples demonstrate a common coupling of talkie verbs and nominalizations. Note that the revisions are both more concise and easier to process mentally.

Original	**Revised**
Her report *makes a recommendation* that we *do a study* of the problem. (13 words)	Her report *recommends* that we *study* the problem. (8 words)
The committee *had made the following series of recommendations*: (9 words)	The committee *recommended* the following: (5 words)

When talkie verbs and nominalizations appear together, revising is often a fairly straightforward process of eliminating the talkie verb and turning the nominalization into a verb.

The verb *to be* and other talkie verbs also appear frequently in unnecessarily long verb phrases.

Original	Revised
The term "domain expertise" *is used to describe* some of the responses we *can make*. (15 words)	The term "domain expertise" *describes* some of our *potential* responses. (10 words)

In this example, we simply eliminated an unnecessary passive (*is used*) and then replaced *we can make* with *our potential* to remove the remaining talkie verb. Earlier we suggested that nominalizations (i.e., *responses*) often signal an opportunity to shift to a more verb-based style. Note, however, that you must pay attention to what else is going on in a sentence to determine the best revision strategy. In this case, we chose to avoid the talkie verb and not to worry that *response* could be revised as *respond*. Keep in mind that all the structures we identify are problems only when misused or overused. They all serve useful purposes. You do not want to eliminate any verbs from your writing, but you do want to break the habit of relying heavily on *to be* and talkie verbs.

We offer one more example as a reminder that talkie verbs are habit forming:

Original	Revised
They *go* ahead and *make* the purchase because they *know* they are *doing* the right thing. This *gives* people ownership as well as the responsibility to *make* decisions that directly impact their project. (33 words)	They purchase the *equipment* because *management supports* them. This *approach* enables ownership as well as responsibility for decisions directly impacting their project. (22 words)

Imagine the effect of a long document written in the style of the original. Also note that in this case, the revision is not only shorter but provides information missing from the original (as indicated by the words in italics in the revised version).

As part of your overall revision strategy, identify talkie verbs, note whether you use *to be* habitually, and replace as many of these imprecise verbs as possible. Revising to increase the descriptive power of your verbs has the added benefit of increasing the number of verbs in your working vocabulary. That is, revising for weak verbs increases your on-hand cache of strong verbs, improving the chances that the first verb you think of will be something other than *to be* or a talkie verb.

6.6.3   Avoid Wordy Phrases

The advice given in the above heading is useless in and of itself. The problem for many writers lies in recognizing the source of wordiness and developing effective revision strategies. Many writers have created major problems in their writing by responding inappropriately to the criticism that they are too wordy. *Wordiness* is too subjective a notion to guide revision. Instead, you must revise for specific features such as nominalizations, talkie verbs, passive constructions, and empty openers.

You should also become familiar with common wordy phrases. Some of the more common expressions and possible substitutions are listed in Table 6.10 Revise as many of these wordy phrases as possible from your writing.

Table 6.10   Wordy Expressions and Concise Substitutes

Wordy Expressions	Concise Substitutions
are indicative of	indicate
at a rapid rate	rapidly
at that point in time	then
at the present time	now, today
aware of the fact that	know
bring the matter to a conclusion	conclude
do not find any difference	find no difference
due to the fact that	given that, because
form a consensus of opinion	agree
has the ability to	can
have to	must
in a great number of cases	often
in close proximity	near
in regard to	about
in spite of the fact that	even though
in the event that	if
in this day and age	today
informs as to the fact that there is	tells of
is able to	can
is unable to	cannot
it is anticipated that	we anticipate
it is believed/considered by many	many believe/consider/etc.
it is clear that	clearly
it is evident that	obviously
it is necessary that someone . . .	someone must . . .
it should be assumed that	assume that
it will be noted that	note that
make a decision	decide
on a personal basis	personally
seem to be	seem
take the place of	substitute
the majority of	most
the preponderance of	most
the reason why is that	because
there is also the likelihood that	likely
until such time as	until

Note how many of these wordy phrases are examples of problems discussed earlier in this chapter. Others are simply phrases to remember. For example, the following revision uses a substitute provided in Table 6.10.

Original	Revised
This meteorological data file was chosen for use in this dispersion modeling evaluation *due to the fact that* the mill is also located on the north side of the valley. (33 words)	This meteorological data file was chosen for use in this dispersion modeling evaluation *because* the mill is also located on the north side of the valley. (29 words)

While substitution is often the best option for a wordy phrase, in some contexts, you can simply omit them. Consider, for example, the phrase *be able to*. In a particular context, different verb phrases may carry the same meaning (i.e., *She is able to draft, She can draft*, and *She drafts*). In such cases, omitting the wordy phrase is your best option.

On the other hand, the phrases listed in Table 6.10 are not necessarily wordy in a particular context. For example, consider the following example:

Original	Revised
The operator should *be able to* set the speed and number of complete cycles in both the X and Y directions.	The operator should set the speed and number of complete cycles in both the X and Y directions.

Note how the meaning changes in the revised version. In this case, the original *be able to* is the appropriate choice so that readers do not assume that the operator *must* set the speed and number when the point is that they *should* be able to do so *when necessary*.

The above example offers a good point on which to conclude our discussion of revising for style. It provides an opportunity to repeat an important notion: we are not providing you with rules for revising, but with *guidelines*. Context, intended meaning, readers' prior knowledge, readers' expectations, and your purpose for writing should largely guide the choices you make. The advice and examples offered throughout this chapter reflect our observations as to the habits writers tend to develop and the ways experience has taught us to revise for them.

Because punctuation is also an aspect of style and can be a help or hindrance to readers' understanding of what you write, we include a punctuation guide as the next section of this chapter. For many writers, revising for punctuation is just as important as revising for style.

6.7 A FUNCTIONAL PUNCTUATION GUIDE

Throughout this text, we have repeatedly made the point that, rather than following rigid rules, successful writers choose among options based on considerations of audience and purpose. However, when you shift your focus from these higher-level rhetorical concerns to the level of mechanics (grammar, spelling, and punctuation), you move much closer to the realm of rules. That is, rather than your choices being more or less effective, they can be right or wrong. For example, if you join two complete sentences with a comma or end a question with a period, you commit a punctuation error.

However, learning to punctuate effectively is only partly a matter of following rules; it is also an art—a matter of choosing among options for rhetorical or stylistic effect. For example, the decision to replace a pair of commas with either parentheses or dashes can affect clarity, emphasis, and tone. The choice of punctuation may also be an expression of personal preference. For instance, some people love dashes while others despise them.

Punctuation is not, as many believe, a means of indicating where to take a breath while reading, but a function of sentence structure (which is also referred to as syntax). To understand the rules of punctuation, you must be able to identify the building blocks of sentences (clauses and phrases) and other structural units. Punctuation functions to alert readers to syntactic boundaries: to signal the end of an introductory phrase and the beginning of the main clause; to separate the items in a list; to mark the intrusion of nonessential information between a subject and its verb, and so on. In other words, punctuation marks tell a reader which words to group together and which ones to separate.

Given that the following discussion of punctuation rules and strategies takes this functional perspective, we have not taken the usual approach of discussing each punctuation mark individually. Instead, we discuss various methods of punctuating clauses, phrases, and lists. We also provide brief definitions for and examples of the grammatical terms we use.

Note that in all the examples, the structure under discussion is presented in boldface. Those of you who are allergic to grammar should read the explanations but not worry too much about making sense of them. You can clarify punctuation patterns by studying the examples.

6.7.1   Sentence Level Clauses

To punctuate correctly, you must distinguish between main and dependent clauses. A *clause* is a group of words with a subject and a main verb. All complete English sentences contain at least one *main* (or *independent*) clause. While it may be preceded in the sentence by an introductory word, phrase, or another clause, a main clause begins with a subject. Two main clauses may be joined by a coordinating conjunction (*and, or, for, nor, yet, but, so*) as in the final example below.

Tomorrow, **the team meets at noon**.

I know that the team meets at noon.

The team will meet at noon because the room is unavailable in the morning.

The team will meet at noon, but **I am busy until 12:30**.

Sentences may also include *dependent* (*subordinate*) clauses that begin with a subordinating conjunction (*because, although, if, as, since*, etc.) or a relative pronoun (*that, which*, etc.):

The team will meet at noon **because the room is unavailable in the morning**.

The meeting is scheduled **so that the team can use the conference room**.

6.7.1.1   Rules for Punctuating Clauses

When you join two main clauses with a coordinating conjunction, place a comma before the conjunction:

The team will meet at noon, **but** I won't arrive until 12:30.

If you wish to join two closely related main clauses (or more complex, complete sentences) without a conjunction, use a semicolon:

The conference room is only available from noon until two**;** the team needs the conference room**;** therefore, the team will meet at noon.

When a dependent clause follows a main clause, you normally do not need any punctuation:

The team will meet at noon **because** the conference room is only available then.

When a dependent clause comes before a main clause, place a comma before the subject of the main clause:

Because the conference room is only available from noon to two**,** the team will meet at noon.

6.7.2 Conjunctive Adverbs

Conjunctive adverbs are often confused with subordinating conjunctions. But whereas subordinating conjunctions must come at the beginning of a clause, conjunctive adverbs can be used at the beginning, inside, or at the end of a clause:

> **However**, the design isn't finished yet.
>
> The design, **however**, isn't finished yet.
>
> The design isn't finished yet, **however**.

If you are not sure whether you are dealing with a conjunction or an adverb, try moving the word to the other side of the subject or to the end of the sentence. If the word makes sense only at the beginning of the clause, it is a conjunction (see the rules in the previous section). If the word makes sense in more than one place in the clause, then it is likely a conjunctive adverb and the following rules apply. The most commonly used conjunctive adverbs include *however, therefore, thus, hence, nevertheless*, and *consequently*.

6.7.2.1 Rules for Punctuating Conjunctive Adverbs

If the conjunctive adverb begins the sentence, place a comma after it:

> **Therefore,** this problem deserves further study.

If the conjunctive adverb comes in the middle of a clause, place commas on either side of it:

> This problem, **consequently,** deserves further study.

If the conjunctive adverb ends a clause, place a comma before it:

> This problem deserves further study, **however**.

If the conjunctive adverb comes between two main clauses or complete sentences, place a semicolon or period before it and a comma after it:

> Our solution has proven unsuccessful**; therefore,** this problem deserves further study.
>
> Our solution has proven unsuccessful**. Therefore,** this problem deserves further study.

6.7.3 Phrases

A *phrase* is a group of words that belong together, but lack a verb, a subject, or both. Other than verb or noun phrases, which include either a verb or a noun somewhere in their structure, phrases generally take their names from the first grammatical element:

- *Prepositional phrases* begin with a preposition: **in** the software, **on** the desktop, **of** the designer, **with** the customer.
- *Verbal phrases* begin with an infinitive (*To V*), a present participle (*V-ing*), or sometimes a past participle (*V-ed*): **to run** the program, **going** at top speed, **conceived** by the designer.
- *Adverbial phrases* begin with adverbs: **before** distribution, **after** development, **during** the design phase.

6.7.3.1 Rules for Punctuating Phrases

When a phrase precedes the subject of a main clause, place a comma after the phrase:

In the morning, I catch up on paperwork, but **in the evening,** I visit the site.

When a phrase follows the subject or verb, you generally do not punctuate:

I catch up on paperwork **in the morning**, and then visit the site **in the afternoon**.

However, when a present participle (*V-ing*) phrase comes after a complete clause, a comma usually precedes it:

He left early, **forgetting his 5:30 appointment**.

The temperature fell below zero in the early afternoon, **delaying the procedure for yet another day**.

6.7.4 Relative Clauses Modifying Nouns

A *relative clause* begins with a relative pronoun, such as *whom, who, which*, or *that*. These clauses modify nouns and are therefore distinct from dependent clauses, which are major sentence elements. Relative clauses are embedded within another clause and can be one of two types.

Restrictive relative clauses provide information about the subject that is essential to the meaning of the clause or that defines or distinguishes the subject:

Team leaders **who lack patience** intimidate new members. (Not all team leaders intimidate new members.)

The engineer **who lives in L.A.** has the farthest to travel. (Experts are being flown in from across the country; the one who lives in L.A. has the farthest to travel.)

Nonrestrictive relative clauses provide additional information about the subject, but neither affect the meaning of the main clause nor provide information necessary to define or distinguish the subject under discussion:

The replacement part, **which was promised for two weeks**, was finally shipped yesterday.

The young man on the skateboard, **who looks about 12 years old**, is a programming wizard.

6.7.4.1 Rules for Punctuating Relative Clauses

Because a restrictive clause is an integral part of the subject or object, do not punctuate it:

Employees **who wear t-shirts to work** make a poor impression on foreign visitors.

Because a nonrestrictive clause provides additional information that could be omitted from the clause, place it within a pair of commas:

My preferred solution, **which we discussed yesterday**, has been chosen by the team.

Note the use of a comma before *which* in this example. If you have difficulty deciding whether to use *which* (with a comma before it) or *that* (without a comma), keep in mind that you use *that* with a restrictive relative clause but *which* with a nonrestrictive clause.

>   The team has chosen my preferred solution, **which we discussed yesterday**.

Also note that you can radically change the meaning of a sentence by using the wrong punctuation with a relative clause:

>   Environmentalists, **who have no respect for industry,** annoy her.

The nonrestrictive clause is appropriate if all environmentalists annoy her because she believes they all fail to respect industry.

>   Environmentalists who have no respect for industry annoy her.

The restrictive clause is appropriate if *only* those environmentalists who have no respect for industry annoy her.

6.7.5   Lists

Three punctuation marks are commonly used to punctuate lists: colons, commas, and semicolons.

6.7.5.1   Rules for Punctuating Lists

Use a colon following a complete sentence that prepares the reader for a list. Note, however, that a colon is necessary only after a grammatically complete sentence. Many writers misuse this punctuation mark by placing it before any list:

Incorrect	Correct
The primary locations are: **London, Rome, and New York**.	The primary locations are **London, Rome, and New York**.
	Or
	We have chosen three primary locations: **London, Rome, and New York.**
You can only get there by: **determination, hard work, and good luck.**	You can only get there by **determination, hard work, and good luck.**

Place a comma before the conjunction signaling the last item in the list. If you make a habit of including the final comma—as we have done in the previous examples—you not only clarify the boundaries between items, but also cut down on the chances of ambiguity. To illustrate this point, we offer the following example:

>   The Montana, Maine, and Ohio offices all require new equipment: **work stations, laptops and a printer and a server**, respectively.

A writer in the habit of omitting the final comma might overlook the fact that the reader cannot tell whether the Maine office requires just laptops or laptops *and* a printer. On the other hand, writers who habitually include the final comma avoid this problem:

>   The Montana, Maine, and Ohio offices all require new equipment: work stations, laptops**,** and a printer and a server, respectively.

Now we know that Maine receives only laptops and that Ohio requires a printer and a server.

Note that you can also punctuate a list with semicolons after a colon when the items in that list are either relatively long or already include commas:

> They have three very different approaches to the problem: **scrap the design and start over; hire two more people to debug the existing system; eliminate the features that don't yet work and begin production**.

6.7.6   Explanations Following Complete Sentences

Use a colon at the end of a grammatically complete sentence to separate it from the answer to an implied question, a subsequent explanation, or a quotation. What comes after the colon can be anything from a single word to one or more complete sentences.

> There was only one thing left to do: **celebrate!**

> Their performance was exceptional: **despite supply problems and unseasonably wet weather, they completed the project on time and under budget**.

> In response to that kind of question, she always made the same reply: **"Use your common sense. Don't expect me to do your thinking for you."**

Note than in each of the above examples, the initial sentence leaves the reader with a question that is answered or an expectation that is met after the colon.

6.7.7   Inserted Explanations and Interrupting Comments

Commas, parentheses, and dashes can all be used to separate an inserted explanation or comment from the rest of a sentence. Note that if you skip the information contained between pairs of commas or dashes or enclosed in parentheses, the sentence should still be grammatically complete.

When the interruption is relatively minor, use a comma:

> Jill Smith, **President of MBI,** will visit the site next Tuesday morning.

> This procedure**, which is clearly superior to the existing one,** will be unpopular with certain departments, such as Purchasing and Shipping.

Parentheses can also be used to enclose optional information that clarifies, comments on, or provides alternative terminology for whatever precedes it:

> Some organisms live in an anaerobic **(airless)** environment.

Parentheses can also be used to enclose numbers (or letters) inserted before the items in a list:

> The procedure involves three basic steps: **(1) strip, (2) dip, (3) dry.**

Dashes provide a greater sense of separation and/or emphasis than that afforded by commas or parentheses:

> However valid or useful this approach—**and it certainly has an extensive application within the field of civil engineering**—it has limited application within the present context.

> All three models—**XL, XM, and XN**—are in stock.

Dashes are also used to signal an aside (extraneous information that briefly interrupts the flow of thought):

Dashes are effective—**if not overused**.

Everyone must now wear identification cards—**a consequence of the recent rash of thefts**.

We've reached the end of our punctuation guide. Because we have focused on punctuating according to the structural units within sentences, we have not covered all aspects of punctuation. We have, however, covered the majority of the punctuation issues that affect the clarity and readability of technical documents.

6.8   ANALYZING WRITING STYLES

One of the challenges most professionals face is the need to master a wide range of writing tasks. Becoming proficient writing for frequently encountered rhetorical situations is one thing, but being faced with the task of writing a new type of document in a new context is quite another. For example, imagine that you were asked to revise a research report as an article for a professional journal. Making the appropriate changes would be very frustrating and time-consuming if you were unaware of the potential stylistic differences between the two situations.

The following set of questions will help you analyze samples of unfamiliar types of documents as a first step toward writing in that genre. (A similar set of questions for analyzing format is included near the end of Chapter 7.)

Paragraph Length and Structure

- What is the average length of paragraphs in the documents? Do paragraph lengths vary significantly?
- How are the paragraphs structured? Do they usually begin with an opening generalization? If not, where is this general statement located?
- Do paragraphs often start with a sentence that connects with ideas in the preceding paragraph?

Order and Emphasis

- Is either the active or passive voice used extensively or are both commonly employed?
- In what situations does the passive voice predominate?

Coherence

- What is the average length of sentences in the documents? Do the sentences vary significantly in terms of length?
- How are the sentences structured? Are they generally simple sentences or are they complex, relaying heavily on dependent clauses, embedded phrases, or the like?
- Is parallel structure frequently or rarely used? Are lists common?
- Does the style depend on many transitions to create coherence?
- What kinds of transitions are used? Where are they placed?
- Are transitions relatively infrequent and is some other feature such as headings used to direct the reader?

Clarity

- Is the style characterized by many or by few noun strings?
- Are runs of prepositional phrases rare or common?
- Is everyday English used or is the language specialized?
- How formal is the writing? Does first person (*I* or *we*), second person (*you*), or third person (*they, the team*, etc.) predominate? Are contractions used or avoided?

Conciseness

- Is the style noun based or verb based?
- Are talkie verbs common or rare?
- Is brevity or elaboration favored? (Are points made briefly or are they developed in great detail?)
- Are statements direct or are points qualified?
- Are adjectives and adverbs used regularly or sparingly to modify nouns and verbs?
- Is the word-to-content ratio low or high?

Answering these questions will provide you a fair degree of insight into the style of the document(s) you have analyzed. These insights should also suggest questions to add to the list.

Once you have analyzed the style of the type of document you want to write, avoid the temptation to simply mimic the style. The crucial step is to determine why certain choices have been made. What do you know about the rhetorical situation for this document and what do an author's stylistic choices suggest about it?

Your ultimate goal is to understand why particular stylistic features predominate. You can then decide what the most important characteristics are and determine how closely you should model your document after the one(s) you have studied. Which features are characteristic of the style and which reflect a writer's personal preferences? To answer this question, you will need to analyze more than one document.

Suggested exercises based on this heuristic for analyzing style are included at the end of this chapter.

6.9   STYLE CHECKLIST

We provide the following style checklist to help you determine and remember what to consider when revising your documents for style. We suggest that you consider the points these questions raise just *before* you revise a document. We further suggest that you make a list of the bad habits in your style—those issues you recognize from the examples in this chapter and those that others (or your grammar checker) have pointed out. Review the strategies that will help you revise appropriately and highlight the features you have not yet mastered. You can ignore those features in the following lists that are characteristic of your writing and require no special attention.

A reminder about grammar checkers is perhaps in order: Because your grammar checker cannot account for context, it will point to many potential problems that are actually the best choice in a given situation. We suggest turning off all but a few features that you know are habits in your writing and judging the suggestions your grammar checker makes in the context of a particular sentence and a specific rhetorical situation.

6.9.1   Revising for Order and Emphasis

Do you employ the general to more specific structure for most of
your paragraphs?   _____

Do you place the important or new information at the end of most
of your sentences?   _____

Do you avoid breaking up sentences with embedded clauses?   _____

Do you avoid weak sentence openers?   _____

Do you use passive constructions judiciously?   _____

6.9.2   Revising for Connection

Do you repeat key terms to keep your readers on track?   _____

Do you avoid excessively repeating the end of one sentence as the
beginning of the next?   _____

Are your lists parallel and have you used parallel structures on either
side of coordinators (*and, or, rather than*, etc.)?   _____

Do you follow *this* with a noun or descriptive phrase?   _____

Do you use sufficient transitions to clarify how ideas relate to
one another?   _____

Do you generally use *because* rather than *since* or *as* when discussing causal
relationships?   _____

Do you avoid excessively short sentences and use conjunctions to combine
sentences and connect ideas?   _____

6.9.3   Revising for Clarity

Is your average sentence length relatively short (18–24 words)?   _____

Are your sentences varied in length and structure?   _____

Do you use precise rather than general language?   _____

Do you adjust your use of technical language according to the expertise
of your readers? That is, are you using appropriate specialized language
while avoiding jargon?   _____

Do you define all acronyms the first time you use them?   _____

Do you revise your work to eliminate any lengthy noun strings your readers
do not use regularly?   _____

Do you avoid using three or more prepositional phrases or consecutive
of phrases?   _____

Do you habitually use commas after introductory phrases, before *and* or other coordinating conjunctions used to join sentences, and to signal the last item in a list?   _____

6.9.4   Revising for Conciseness

Do you avoid unnecessary nominalizations?   _____

Do you use talkie verbs sparingly?   _____

Do you avoid using wordy phrases?   _____

6.10   EXERCISES

When revising your documents, we recommend that you use the style checklist provided earlier in this chapter, concentrating on those features that will help you overcome unfortunate habits in your style. We also suggest that you review the functional punctuation guide provided on pages 234–240 of this chapter and note those situations in which you characteristically commit punctuation errors. If you know what to look for, you can edit more efficiently and effectively.

The following exercises should also prove useful for improving specific aspects of your style and adapting your style to new rhetorical situations. For a number of these exercises, you could use the same piece of writing. If you do, begin each exercise with the original draft and create separate revisions for each exercise to avoid concentrating on too many issues at once.

1. Analyzing Paragraph Structure
Using the directions for paragraph analysis provided on pages 190–194 of this chapter, analyze the paragraph structure of a two-to-three-page document you have written—or choose a portion of a longer document. If you are unsure what to do, refer to the sample analysis on pages 192–194. Keep in mind that introductory and concluding paragraphs may not follow the suggested general-to-specific pattern. After you have finished your analysis, revise the document you analyzed to improve paragraph structure.

2. Analyzing Sentence Length
Chose a two- or three-page document you have written—or two or three pages of a longer document—and number the sentences in each paragraph. Then list the paragraphs and sentence numbers providing the word count for each sentence as demonstrated below. Also note the average length of the sentences in each paragraph and comment on the variety in sentence lengths and structures and on the content of particularly short or long sentences.

Paragraph 1

1. 15 words	Average Length: 16
2. 20 words	Comments: The length is reasonable and variety is good.
3. 23 words	The short concluding sentence is appropriate because it
4. 7 words	emphasizes an important point.

Paragraph 2

1. 19 words	Average Length: 26
2. 28 words	Comments: This paragraph would be improved if the second
3. 32 words	sentence were rewritten in the verb-based style to reduce the word-to-content ratio and lower the average sentence length. Because the third sentence contains a list, its length is acceptable.

When you have finished this chart, revise the document to improve sentence length and structure. Alternatively, write a paragraph describing general changes you should make to improve this aspect of your style.

3. Revising for Order and Emphasis
Choose two pages of something you have recently written and review the section on revising for order and emphasis pages (194–202). Check each sentence for the following stylistic problems:

- Sentences with the new information at the beginning
- Clauses or phrases of more than a few words placed between the subject and verb

- Weak sentence openers such as *there is* or *it is*
- Unnecessary use of the passive voice

Revise to eliminate these problems using the strategies discussed in the section on order and emphasis.

4. Revising for Connection

Choose two pages of something you have recently written and review the section on revising for connection (pages 202–212). Check the entire passage to ensure you have repeated key words and phrases often enough to keep readers on track. Also check each sentence for the following stylistic problems:

- Too much repetition of words and phrases at the end of one sentence and beginning of the next
- Lists or sentences lacking parallel structure (read your work out loud to find this problem)
- Vague subjects, such as *this* followed directly by a verb

Improve the flow of ideas by revising to create stronger connections.

5. Revising for Clarity

Chose two pages of text from the middle of a technical document you have written. If none is available, then choose a technical article in a magazine that is written for a reasonably wide range of readers. Review the section on revising for clarity (pages 212–224) and then identify the following:

- General language
- Acronyms and specialized language (determine whether any are inappropriate for the intended audience)
- Noun strings (determine whether any are inappropriate for the intended audience)

- Runs of prepositional phrases or consecutive *of* phrases

If you analyzed a piece of your own writing, revise to improve clarity. If you analyzed a piece of someone else's work, provide the author with comments on how to improve the clarity of his or her writing.

6. Revising for a more Verbal Style

Chose two pages of text from a document you have written and review the section on revising for conciseness (pages 224–234). Check each sentence carefully for the following:

- Nominalizations
- Talkie verbs
- Wordy phrases

Determine whether each nominalization is appropriate. Then revise for a more verbal style by rephrasing sentences to replace nominalizations with verbs, to substitute more descriptive verbs in place of talkie verbs, and to eliminate wordy phrases.

7. Analyzing a Writing Style

Identify a kind of document you might want to write but are relatively unfamiliar with, and then obtain one or more samples (of published works, if appropriate). Using the set of questions for analyzing writing styles provided on pages 240–241 of this chapter, analyze complete documents if they are short or choose specific sections of longer documents. If you analyze more than one document, also note any features that vary significantly from one author to another. Write a page or two describing the features of this style. If you reviewed more than one document, also discuss variations among authors, commenting on whether you think these differences reflect specific rhetorical contexts or personal preferences.

Chapter 7

Format Strategies

7.1 SUMMARY

This chapter introduces you to some of the conventions you must follow in order to format your documents appropriately. Our goal is to help you produce documents that are professional in appearance and easy for readers to understand. We also draw attention to the relation between format and rhetorical situation, to how the World Wide Web and Hypertext transform notions of form, and to the ethics and mechanics of referencing other people's words and ideas.

1. Introduction to Rhetorical Form

The nature of form is all too easily misunderstood. This section asks you to reconsider your understanding of form by thinking of it as a set of strategies for informing and persuading your readers. In many formal writing situations, your mastery of form can be as important as your mastery of the content you wish to communicate.

a. The Relationship between Form and Content

In a metaphorical sense, form is often simply thought of as a container that the writer fills with content. We argue that this metaphor for the relationship between form and content is misleading. A better metaphor for the nature of the relationship between form and content is that of a growing plant. Forms change through time, and they evolve in ways that support the evolving goals and methods of a discipline.

b. Rhetorical Form and Community

The forms that you choose to use are a reflection of the disciplinary community to which you belong. The degree to which you have mastered the forms preferred within your specific discipline indicates to an experienced reader whether you should be accepted as a member of that discipline. Consequently, the forms you use can serve an important persuasive function.

c. Rhetorical Form as Dynamic

Rhetorical forms change through time and in response to the need to communicate different kinds of information to different kinds of audiences. Consequently, you need to remain alert to these changes. Just as you must stay current with the subject matter of your field, so too you must learn new forms.

d. Form and the World Wide Web

Because of our growing dependence on the Internet and World Wide Web, we also discuss how computer-mediated communication is transforming how we think about documents, and specifically, the need to rethink form when creating on-line documents.

e. Rhetorical Form as Static

While many forms evolve in response to changing audiences and purposes, you must also recognize that many forms are widely accepted by the members of your discipline as conventions that have withstood the tests of time. Employing these widely accepted conventions will help you meet your readers' expectations and avoid confusing them.

2. Mastering the Technology

You must learn a wide range of technologies in order to produce professional-looking documents. In this section, we focus on the advantages to working with style sheets and templates. If you master the use of style sheets and templates, you will efficiently produce consistent and professional-looking documents.

3. Principles of Format

You must master an array of different principles related to format in order to produce high-quality documents. We draw your attention to the areas of the page without text and to the importance of considering print quality, paragraphing, and margins. For readers, the global organization of a document and how that organization is communicated to them are critical elements of a document. We outline four key strategies for formatting documents to help readers see the overall organization: section headings, pagination, headers and footers, and tables of contents. We also outline basic issues to consider when working with figures, tables, lists, and equations.

4. Referencing Information Sources

Most documents that we write depend on information derived from a range of sources. Ethically, we must acknowledge the sources for this information. We outline strategies for using acknowledgments, footnotes, and references in order to appropriately attribute the sources of the information you cite in your work or synthesize with your own line of reasoning. We also include a guide to referencing that provides a set of conventions, a discussion of various styles of referencing, and advice for adapting to a required style.

The format heuristic in this chapter is a companion to the style heuristic in the previous chapter. Use this list of questions to analyze the format of sample documents in preparation for writing in a new rhetorical situation. We also include a checklist to use before you format a document and exercises for mastering various elements of format.

7.2 INTRODUCTION TO RHETORICAL FORM

Reflect on the following situation for a moment: two recently graduated engineers are applying for employment with an established corporation. The first is quite well qualified for the position. She has relevant work experience and a good academic record. However, she wears a dirty T-shirt, cutoff blue jeans, and torn running shoes to the interview. Further, she has not bathed for several days and has long unkempt hair. The second engineer is only moderately qualified for the position. His experience is limited and his academic record is mediocre. However, he dresses appropriately for the interview and is generally well groomed.

If these were the only two applicants, which of the two would most likely obtain employment with the company? The well-groomed applicant may well be successful. Why? Simply because appearances matter. By dressing appropriately, the second engineer demonstrates both a professional attitude and respect for the interviewers.

Obviously, the above example is extreme. No one who is seriously seeking employment would dress in such an inappropriate fashion. Nonetheless, people frequently submit sloppy résumés and covering letters to prospective employers. These individuals are also less likely to obtain employment than those who submit professional-looking applications.

Sadly, some documents submitted by both engineers and engineering students are also sloppy; the content is good, but little attention is paid to the form of the document. We have seen reports lacking page numbers, tables of contents, lists of figures, and reference lists. We have seen other reports that were nearly illegible because of a poor choice of font or inferior reproduction. In other cases, authors fail to label figures, place figures in inappropriate locations, or submit crude pencil-drawn figures. These documents reflect an underlying attitude: *I am being evaluated on the content, so why should I bother worrying about how the report looks*? Anyone with this attitude needs to understand that form and content are closely related.

7.2.1   Relationship between Form and Content

A "who cares?" attitude toward format relies on the assumption that form and content are easily separable. They are not. At one time, form was viewed simply as an empty container to be filled with content. According to this container metaphor, form was secondary to content and required little if any attention. But, does this assumption hold true? Why for instance, do we not drink our morning coffee or tea out of a two-gallon bucket? Obviously, the bucket is too large and cumbersome to drink from easily; the large surface area allows hot liquids to cool too quickly, and so on.

As every engineer knows, form and content must work together. What is sometimes forgotten is that the relationship of form and content applies to documents as well as to physical phenomena. Without some type of form, be it well or poorly structured, no content can be communicated. Even what is sometimes described as *empty form* (a document that looks appropriate in form but lacks substance) communicates information, if only to suggest that the writer has very little to say or does not care much about the topic. Even the word "in-*form*-ation" implies that ideas must be structured in some fashion or other.

Both form and content communicate information, but they do so in different ways. To better understand the relation between form and content, consider the act of speaking: your words communicate information in an explicit, linear fashion while your body language communicates information in a more subtle, multilayered fashion. To be effective, the two types of communication must complement and reinforce each other. Clearly, different kinds of content require different forms. For example, love letters are not suited to a business letter format.

A more accurate metaphor for the relation between form and content is that of form as a growing plant.[1] Just as plants grow in different shapes and sizes to support different kinds of fruits and seeds, so too the forms of documents come in different shapes and

---

[1] We would like to thank Professor Richard Coe (Simon Fraser University, Burnaby, BC) for devising this metaphor and for demonstrating how it applies within various disciplines.

sizes to accommodate different kinds of information and to accomplish different purposes. Throughout history, form has grown organically in a way that supports the content. The two are inseparable. Over many years, various forms for communicating information have evolved to address various functions within the discipline of engineering.

Moreover, rhetorical forms are not easily interchangeable. For example, the genre of the essay used extensively in the humanities is rarely used in the discipline of engineering. Conversely, the genre of the lab report that is so frequently relied on in engineering and the sciences is rarely used in the humanities. Further, if you were to submit a lab report written using the essay form, you could expect to receive a significantly lower grade than if you used the lab report form. While no one is likely to confuse these two genres, many students fail to pay sufficient attention to those features of format appropriate to the documents they are writing.

All else being equal, students submitting well-formatted documents using the appropriate rhetorical forms receive grades that are 10 to 15 percent higher than those submitting poorly formatted documents using inappropriate rhetorical forms. Assuming both documents demonstrate a mastery of the required engineering concepts, this 10 to 15 percent could be the difference between an A and a B or B+.

If we look beyond the realm of the classroom, into the world of work, we can speculate that the difference between submitting documents of an appropriate or inappropriate form to your supervisors may well be the difference between being promoted and receiving a raise or being passed over. If for no other reason, you should master form in order to elevate your grades and eventually to increase your earnings.

7.2.2 Rhetorical Form and Community

Why does form have this impact on grades and promotions? In part, the answer lies with the individual who reads the document. From their past experiences within a discipline, most readers have acquired a set of expectations about how information will be organized and presented in a document. For both writers and readers, the form of a document helps structure the search for information and helps determine how easily that information will be comprehended. Moreover, appropriate form also communicates your *abilities* and *attitudes* to the reader:

- Appropriate form communicates attention to details. (You are *able* to deal with all the subtle details and implications of a given topic.)
- Appropriate form communicates attention to organization. (You are *able* to think in the logically organized ways expected of engineers.)
- Appropriate form communicates interest. (You have a sufficiently positive *attitude* toward the topic that you want to present it in the best possible fashion.)
- Appropriate form communicates respect. (You have a sufficiently positive *attitude* toward your reader that you are willing to present the information in the most easily comprehended fashion.)
- Appropriate form communicates dedication. (You have a sufficiently professional *attitude* that you are willing to devote the time to producing first rate documents.)

Form, therefore, communicates whether you are a member of the professional community of engineers, whether you understand engineers' preferred ways of organizing and presenting information and, thus, whether you engage in *engineers' preferred ways of thinking*. As a student seeking admission to this professional community, your attention (or lack of attention) to form may well communicate to members of this community the degree to which you deserve admittance. In fact, we can argue that the thesis or the project report you submit to complete your degree requirements tests your mastery of both the forms and concepts of engineering.

Form matters—not only by influencing the success of your career as an engineer, but also by influencing your success in gaining admission to the community of engineers. You are thus well advised to master the various rhetorical forms required in engineering as soon as possible.

By using the appropriate forms, you communicate that you are a professional who shares certain beliefs, values, ideas, and methods with your audience. You demonstrate your authority. Moreover, you communicate to the audience that you are attentive, organized, interested, and dedicated. On a more subtle level, proper format displays respect for readers. You help them identify with you as an individual who can be trusted. In other words, the form you choose for a document serves a persuasive as well as an informative purpose.

The details of form make an important first impression on the reader. In fact, under some circumstances, the appearance of a document may make a more important impression on the reader than the information contained within the document. For example, busy senior engineers or administrators may lack the time to read an entire report, particularly if the report is lengthy or extremely complex. They may only read the table of contents, the executive summary (or the abstract), the introduction, recommendations, and the conclusion. Then they may scan the document, examining some of the figures and tables.

If the document appears professional and the readers see no problems (including, by the way, spelling or grammatical errors) in the sections they do read, you can reasonably expect they will accept the report. On the other hand, if the document looks unprofessional, it may be returned for major revisions—even if the content is acceptable.

7.2.3   Rhetorical Form as Dynamic

The metaphor of form and content as a growing plant is useful both for indicating the organic relationship between form and content and for drawing attention to the fact that rhetorical forms are dynamic, that they evolve over time in response to changing rhetorical situations. In fact, we are currently witnessing major changes in rhetorical form as we shift from paper documents that are linear and static to electronic documents that are based on dynamically changing information and *hypertext links* (connections between two separate locations in a document or between documents).

You need only peruse the *World Wide Web* or read some ongoing conversations in *Usenet Newsgroups* to observe the rapidly changing nature of rhetorical form. The use of links on the Web is changing the way we read. Instead of reading a document in a linear fashion, from beginning to end, we now follow links to create our own pathways through a document or to access related information in other documents. Because links are frequently added and deleted, sometimes on a daily basis, electronic documents are much more dynamic than paper documents. When following the links of electronic documents, what you read tomorrow will almost certainly be different from what you read yesterday.

On *Usenet*, documents such as FAQs (Frequently Asked Questions) change in response to questions and feedback from readers. Unlike paper documents, the ability to communicate directly with the authors of such documents (and others who read the newsgroups) leads to a more interactive way of reading. Rather than simply reading the document and assuming that it might be difficult or time consuming to contact the authors, e-mail now provides easy and speedy contact. Of course, this immediacy of contact does have a downside in that it now becomes all too easy to *flame* (to insult using e-mail) someone whose ideas or viewpoints you disagree with.

Despite the rapid pace of change, form has not become any less important. Indeed, we could argue that it has become even more important. You might note, for example, the increasingly widespread need to learn Hypertext Markup Language (HTML). You can use this system of tags to structure, link, and cross reference parts of an electronic document on the World Wide Web so that people using different operating systems can view them. But producing on-line documents involves much more than *marking up* a document. New media are challenging traditional notions of document form.

7.2.4   Form and the World Wide Web

This chapter focuses largely on *format* as a major issue in producing professional print-based documents, but when we turn our attention to documents produced for the Web, we cannot sustain this focus on format. For on-line documents, the reader, not the writer, largely determines format. Web pages are interpreted by software on users' computers so that three users with three different browsers (i.e., Netscape Navigator, Microsoft Explorer, and Mosaic) will each see a different version of a web page. How different these versions are is further determined by the degree to which each user has set his or her own preferences in terms of fonts and other format features.

Rather than controlling format elements such as font types and sizes, line spacing, margins, or color, HTML provides tags for structural elements such as headings, paragraphs, and lists. You can gain some control of format and increase the chances that what readers view resembles what you intended by using style sheets to specify layout. As with print-based documents, style sheets also facilitate global changes to on-line documents. But when preparing documents for the Web, format issues should be secondary to structural ones.

When producing documents for the Web, your first concern should be determining how to *structure* information so that readers can find what they are looking for quickly and easily. Structuring information for the Web requires finding a suitable organizational strategy that readers can use to navigate the site, such as structuring information like chapters in a book, like the branches of a tree, or like the rooms in a building. When structuring information for on-line viewing, you must also keep in mind that readers will characteristically be in a hurry and will allow very little time (probably less than a minute) for your page to load and for them to locate the information they want.

Information must therefore be structured so that readers can quickly determine the content of your site and easily find their way to the information they seek. You must provide an overview that allows readers to see the overall structure and to survey the contents of your site quickly. A table of contents often serves this purpose and can also provide a means of navigating the site.

An on-line document should be further structured in terms of what a reader will view on a standard-sized computer screen. Information should be conveniently chunked for viewing on the screen. Every screen should also have a similar look and feel, accomplished by ever-present features such as a table of contents or menu bar (which is often presented in a left-hand margin that remains constant while the rest of the screen scrolls or changes), by using standard headers and footers, by a unifying theme for graphics, by consistency in the way links are presented, and so on.

One of the challenges when designing a web page is adapting to the potential technical and physical limitations of readers. Because those visiting a web page may have limited technical capacity, anything requiring extra download time or requiring software or hardware that may not be supported on a typical reader's machine should be omitted or offered as a link. By providing some indication of the size of a file or of the software required to view a file, readers can decide whether they should follow a particular link.

Readers may also be using a wide range of technologies that may or may not involve using a mouse. Consequently, identifying links by asking readers to *Click here* is best avoided. Creating links by underlining text that describes the content of a link is a more appropriate approach. Likewise, readers may be disabled and using Braille readers or speech synthesizers. These readers (and those using outdated technology) will appreciate text descriptions of graphics.

Many novice web page designers become so captivated by the ability to link to other pages and to incorporate audio, graphics, video, and animation, that they lose track of the importance of having form and content support one another. When you first consider adding a link or augmenting your site with graphics, sound, or animation, ensure that you are adding value to your message and not just adding bells and whistles for their own sake. When deciding what to include on a web page, keep in mind your readers' needs, your purpose, and the nature of the information you are providing.

In this section, we have touched on major issues relating to the form of web documents. Perhaps the best way to learn about web form is to surf the Net and to pay attention to the structure and features of those web sites you find useful and easy to navigate and of those sites you find disappointing or frustrating. Develop a habit of asking yourself what makes the good sites effective and what would make the poor sites better.

7.2.5   Rhetorical Form as Static

The more things change, the more they stay the same certainly holds true for technical documentation. The Web provides opportunities to experiment and to create your own solutions to the many challenges of on-line writing. But despite the popularity of the Web, much technical writing remains print based, and you must master forms that are dictated by long-standing conventions.

Forms that have not changed for some time, which have become *form*-ally accepted in a field or discipline, are referred to as conventions. Each discipline has its own set of conventions (indeed, even within a discipline such as engineering, we find several overlapping sets of conventions). Standard formatting conventions, such as how to reference information sources, how to display equations, and how to label tables and figures, have developed over many decades in engineering and serve important purposes in terms of helping readers easily process information.

However, given the many subtle differences between the conventions you may be expected to use and because some of these conventions are changing even as we write this text, we cannot hope to provide you with the *best* way to format your documents. In fact, that *best* or *correct* way to format a document is an illusory goal. No matter what formatting convention is proposed, someone will find an alternative approach that makes more sense in a specific context.

The recent change from placing the glossary in an appendix at the end of a document to placing it near the beginning of the document is a good example of how formatting conventions sensibly evolve. At some point, a writer observed that placing the glossary at the end of a document fails to alert the reader to its existence soon enough. This problem was traditionally overcome by using boldface for the first technical term in a report and by adding a footnote that told the reader about the existence of the glossary. But placing the glossary in the prefatory pages of the report is more likely to ensure the reader sees it. This approach also eliminates the necessity of footnoting a technical term, which itself improves the appearance of documents that use footers (a repeated line of text at the bottom of a page throughout a document).

Rather than attempting to provide you with the *correct* approach, this chapter provides you with several generally useful principles about formatting documents. As you consider the format of your documents, try to do so from the perspective of the reader. To make things as easy as possible for the reader, ask yourself specific questions about your format. For example, if you were the reader, would you really want to read a document produced with an 8-point font size? Probably not. Extended reading of material using such a small font would quickly lead to eyestrain. Usually, 11 or 12 point is a more appropriate choice.

You must also develop an eye for the appearance of your document. In particular, examine your document for the consistency of such things as heading levels, page numbering, and margins. Inconsistency in the levels of headings, for example, can prove an irritating distraction to the reader. Whatever methods you use to differentiate among heading levels (font size, capitalization, bold face, or indenting), ensure that you are consistent throughout the document. Initially, you will develop this eye for the appearance of your document by examining the samples of well-formatted writing produced by others.

To improve the format of your own documents, we strongly recommend that you print drafts so that you can most effectively edit for consistency. Eventually, *WYSIWYG* (*What You See Is What You Get*) technology will be perfected and you will be able to do most of your format editing on-screen. But until then, a paper copy remains essential. This text, by the way, was produced in precisely this fashion. Formatting corrections were made on paper copies.

7.3 MASTERING THE TECHNOLOGY

The task of formatting your documents can be greatly simplified if you learn how to correctly operate the word processing and graphics software as well as the printers and photocopiers that you use. We encourage you to enroll in one of the many word processing workshops offered by most universities. Also, if you do not already have one, obtain a good manual for your word processing package.

Whatever word processing package you use, we strongly recommend that you spend some time learning how to use *styles* and *templates*. A *style* is any formatting that you use repeatedly throughout a document and a *template* is a collection of styles that you apply to documents of the same type, such as a *letter template*. The advantages of using styles and templates rather than directly formatting your documents are threefold:

1. Styles ensure that your documents will have a consistent appearance in terms of headings, captions, paragraph spacing, fonts, etc. This characteristic is particularly important when writing long documents or when writing collaboratively.

2. Styles make changing the format of a document relatively easy. This characteristic is especially useful when a document undergoes several iterations or editions.

3. Styles save time by enabling the use of speed formatting keys, by eliminating the need to constantly reconsider what format was applied previously, and by allowing tables of contents to be generated automatically. This characteristic is particularly helpful when working on long, complex documents.

As an example of how a style works, several years ago, when we wrote a draft of this text as a handbook for our students, the items in the bulleted list immediately above this paragraph had a style with the name, *List 1*, associated with them. This particular style had a set of characteristics that we defined prior to beginning work on the handbook: 11-point, *Book Antiqua* font, flush-left margin, indented 1 inch, hanging indent ¼ inch, 13-point line spacing, and widow/orphan control. Indeed, that handbook had 31 distinct styles associated with it, covering everything from the appearance of the basic paragraph format to the appearance of the table of contents. Figure 7.1 provides a screen capture of the Style Dialogue Box for *Microsoft Word 2000*.

Because the handbook was about 200 pages in length and was co-authored, consistency was particularly difficult to maintain and would have been nearly impossible if we had simply formatted the document directly. By assigning styles to the various paragraph formats throughout the handbook, we ensured consistent formatting. Moreover, by defining a series of styles and then saving those styles as a *template* called *handbook.dot*, we ensured that all subsequent handbooks would follow the same format.

As well as ensuring consistency, the use of styles enabled us to make global formatting changes from edition to edition as we improved the format of the handbooks. For example, because we wanted our students to make marginal notes, we decided to increase the size of the margins by decreasing the width of the paragraphs. This change was easily accomplished by modifying the format of the style called *Body*. Simply making this one modification resulted in a consistent change to the format of most paragraphs in the handbook. If we had not had styles associated with the document, we would have had to change each paragraph individually—a time-consuming and haphazard process.

To save even more time, we assigned speed keys to the various styles so that we could apply them to any paragraph simply by pressing a few keys. For example, we applied the *Body* style simply by typing *Alt-B*. Moreover, by using various built-in heading levels, our word processor could automatically generate and update the table of contents, which represents a significant time saving compared to manually compiling the table of contents. This approach is also more accurate than manually looking up each page number before entering it into the table of contents.

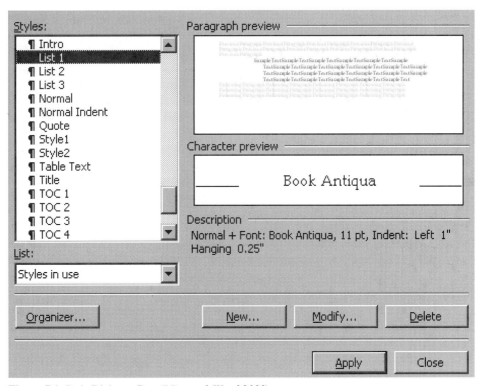

Figure 7.1 Style Dialogue Box (*Microsoft Word 2000*)

However, this last point, that of using built-in heading levels, reveals one potential risk of employing styles: They do not address the needs of any particular rhetorical situation. *Do not* simply accept the default styles that your word processor proposes for headings (or for that matter, for anything else). Modify them so that they meet your overall goals for the appearance of the document.

For example, the default heading styles used UPPER CASE and underlining for several of their levels. We intentionally wanted to avoid capitalization because text written in all uppercase is more difficult to read than text in mixed upper- and lowercase. We also wanted to avoid underlining because we dislike the way that the line cuts off the descenders on letters like *g, j, p, q,* and *y.*

Although the details of how to set up styles vary from program to program, and thus will not be discussed here, these techniques are worth mastering because they can greatly enhance your proficiency and consistency. Once you have created the necessary style sheets or templates and defined speed formatting keys for several different document types, applying these styles to new documents becomes a simple matter.

In other words, you will need to expend some time and effort to set up your basic formats, but all subsequent documents can be produced quite easily. We strongly recommend that you read the manual for your word processor, paying particular attention to the specifics about styles, templates, speed keys, auto numbering, and so forth. The long-term time savings will prove substantial and are well worth the effort.

7.4   GENERAL PRINCIPLES OF FORMAT

As you format your work, you must attend to a wide range of issues. In particular, you should pay special attention to white space, organization, graphical elements, and referencing conventions.

7.4.1   White Space

Generally, when most writers think about format, they only consider the text of the document and not the white space surrounding that text. However, to generate the best possible format for any particular document, you must pay careful attention to the areas *without text* as well as to those with text. The white space in a document serves several important functions. Perhaps the most obvious function of white space is to allow instructors (and sometimes peer reviewers and editors) space for comments. For this reason, many instructors and editors insist that you provide adequate margins (1 to 1 ¼ inches) and double space your documents.

A more important general function of white space is that it enables readers to clearly see what information belongs together. White space can be used to separate various elements of a report, such as paragraphs, sections, and figures. For example, you might allow for ½ inch of white space between the end of one section and the next heading, but only allow for ¼ inch of white space between that heading and the following section. This sort of spacing helps the reader identify that the heading belongs with the following text. Figure 7.2 contrasts a text with poorly spaced headings (on the left) with one that properly spaces headings (on the right).

Careful spacing is especially important with figures and tables. If the figure label is located an equal distance between the figure and the text, at first glance a reader cannot tell whether the label is, in fact, a label or perhaps a new heading. The label should be located closer to the figure than to the text. The same is true for the labels you use with tables. Figure 7.3 contrasts a text with properly spaced figure labels (on the right) with a text that has improperly spaced figure labels (on the left).

Unclear Relationship between
Headings and Text

Clear Relationship between
Headings and Text

Figure 7.2 Spatial Relationship between Headings and Text

 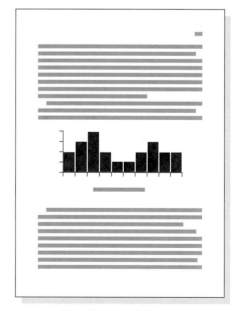

Unclear Relationship between
Figure Label and Text

Clear Relationship between
Figure Label and Text

Figure 7.3 Spatial Relationship between Figure Labels and Text

Further, should you ever need to place two figures on a single page, ensure you space them so the reader will have no difficulty identifying the first label as belonging with the first illustration. If the label is an equal distance between the two figures, the reader will momentarily be confused about which figure the label identifies. Obviously, the reader will eventually see the second label and solve the problem.

But please remember, we are suggesting that you should use your format to make things easy for readers. Even a momentary confusion causes readers to lose their focus on the information you are communicating. When a document is formatted properly, formatting features should not draw readers' attention. Effective format is more or less invisible.

White space also subtly indicates to the reader the degree of complexity of the information contained within a document. When readers first glance at documents with dense text and little white space, they often assume that the information being communicated is complex or difficult to understand. Excessively long paragraphs communicate the same impression.

Reflect on, for example, the differences between books written for children versus books written for adults. Children's books have wider margins, wider spaces between lines, larger type, shorter paragraphs, shorter sentences, and more illustrations—overall more white space—than books written for adults. As children become older, and thus more proficient at processing information, the textual density increases and the white space decreases in relation to their age.

The key point is that most of us have learned to associate textual density with difficulty. Depending on your audience, you might also want to vary the textual density of your documents. When writing for nonspecialists, it is probably appropriate to increase the amount of white space in your documents by shortening your paragraphs, including more figures, and using indented lists more frequently. These format features may well increase your readers' comprehension (or at least their willingness to read the material). At the same time, you should be careful not to oversimplify a document; otherwise, you may unwittingly irritate your audience by appearing to talk down to them.

Further, you should exercise some caution when shortening your paragraphs. One-sentence paragraphs are usually not a good idea (although when used *occasionally*, one-sentence paragraphs can prove an effective form of emphasis). An average paragraph length of four to five sentences may be appropriate for the nonspecialist audience while an average paragraph length of six to seven sentences may be more appropriate for the specialist audience. As well, for any audience, *variety* in your sentence lengths and paragraph lengths will help maintain interest.

Figure 7.4 provides a graphic example of how the amount of white space in a text influences the perception of complexity. If you were faced with reading texts with these differing amounts of white space, which would you prefer?

7.4.1.1   Print Quality

The quality of the print you use in your documents impacts significantly on the willingness of readers to read them closely. The more effort readers devote to deciphering the print, the less attention they pay to comprehending the information. You have probably had the unpleasant experience of trying to read a photocopy that was so faded you had to strain to decipher the individual words. When you must devote too much effort to translating individual letters and words, you cannot pay sufficient attention to the ideas and may give up in frustration. The quality of the print depends on a number of factors: font size and type, line spacing, and justification, as well as the capability of your printer and the quality of your reproductions. Each of these factors impacts your readers.

Single Spaced without Graphics   Double Spaced with Graphics

Figure 7.4 Texts with Contrasting Amounts of White Space

Many thousands of different fonts are available, but not all are acceptable for use in technical documents. Table 7.1 provides a sample of some common fonts. To some degree, the choice of a font is a matter of personal preference. But readability and rhetorical purpose must also be considered. Obviously, using a font such as *Linotext* or *Caslon Open Face* in a technical document would be inappropriate. Both fonts are far too ornate and are best reserved for special uses such as signs or invitations.

In general, you would be well advised to use *serif* fonts rather than *sans-serif* fonts (literally, fonts *without serifs*) for most of the text in print-based technical documents, particularly lengthy ones. (A sans-serif font such as Arial is considered more appropriate for on-line documents, however.) *Serifs* are simply the small lines at the tops and bottoms of letters. Figure 7.5 enlarges the difference between a *serif* and *sans-serif* font.

The reason for avoiding *sans-serif* fonts for the body of a document is that the lack of serifs in these fonts makes recognizing the patterns of words more difficult for readers. This difficulty becomes especially apparent when reading lengthy documents. Eyestrain sometimes results. By cutting off the lower half of the words, Figure 7.6 exaggerates how *sans-serif* fonts interfere with character recognition.

Table 7.1 Selected Examples of Fonts

Text (Serif)	Headings (Sans Serif)	Special Purpose
Times New Roman	Arial	Courier
Cheltenham	Arial Narrow	Lucida Fax
Century Schoolbook	Avant Garde	Linotext
Bookman Old Style		Caslon Open Face

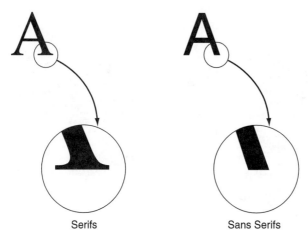

Serifs   Sans Serifs

Figure 7.5 Detail of Serifs

We suggest that you reserve *sans-serif* fonts for headings. But even for this use, some caution is advisable: a narrow *sans-serif* font, such as *Arial Narrow*, is more difficult to read than its uncompressed counterpart, *Arial*. Nevertheless, you may occasionally want to use a narrow font when space is limited, as in the cells of a spreadsheet, for example.

When we originally produced this text as a handbook for our students, we chose a font called *Book Antiqua* for most of the text because it photocopied better than the font called *Times New Roman* and thus made the text much easier for students to read. For similar reasons, we generally use a font called *Lucida Fax* when sending a facsimile transmission (fax). To demonstrate the differences, Figure 7.7 presents several lines of text using three different fonts. The font size in all cases is 10 point.

As you may have noted in Figure 7.7, *Lucida Fax* is the largest and least compressed of the three fonts. Consequently, it works well when faxing documents to a machine that may not clearly reproduce smaller, more tightly compressed fonts. *Bookman* is intermediate in size, while *Times New Roman* is the smallest. In fact, because *Times New Roman* is so small, we recommend that you do not use less than 12 point for text. In general, we suggest using a font size of 11 or 12 points, although you might use smaller sizes for footnotes (i.e., 8 or 9 point) and larger sizes for headings (i.e., 14, 18, and 22 point).

You would also be unwise to use italics or boldface for an entire document: reserve these font styles for special purposes such as indicating emphasis or creating headings. Similarly, you should avoid using all uppercase whenever possible. Overuse of uppercase is less of an issue for reports than for e-mail and oral presentations.[2] In e-mail, all uppercase screams at the reader; in presentations, all uppercase is simply more difficult to read. Reading text formatted in all uppercase is inherently slower than reading text formatted in

Illumination Illumination

Figure 7.6 Word Pattern Recognition for Sans Serif versus Serif Fonts

---

[2]In fact, we know of one local company that insists that all overheads prepared for presentations use only uppercase. Although the engineers who work for the company recognize that all uppercase is more difficult to read, they must nevertheless follow the practice.

Lucida Fax is an easy to read font that is sometimes used when sending FAXes. Lucida Fax is an easy to read font that is sometimes used when sending FAXes. Lucida Fax is an easy to read font that is sometimes used when sending FAXes.

Bookman is an easy to read font that is sometimes used for photocopied documents. Bookman is an easy to read font that is sometimes used for photocopied documents. Bookman is an easy to read font that is sometimes used for photocopied documents.

Times New Roman is a general purpose font that can sometimes be difficult to read. Times New Roman is a general purpose font that can sometimes be difficult to read. Times New Roman is a general purpose font that can sometimes be difficult to read.

Figure 7.7 Comparison of the Readability of Selected Fonts

mixed upper- and lowercase. Figure 7.8 demonstrates the difficulty that using all uppercase causes readers.

Finally, another factor beside readability may influence your choice of font: how modern the font looks. Although *Courier* was once popular (mainly due to printer limitations), it is rarely used anymore because it has a decidedly old-fashioned appearance. Today, *Courier* is used exclusively for special purposes such as listings of program code. Similarly, some people now feel that *Times New Roman* and *Arial* have been used for so many years that they also appear dated—something to remember when responding to the specific rhetorical situations of your technical documents.

7.4.1.2   Paragraphing

In general, the average paragraph length in technical documents lies between three and seven sentences depending on the expertise of your audience, the width of the paragraphs, and whether you single or double space the document. In addition, we advise that you avoid overusing one-sentence paragraphs because too many will lead to a lack of connection among ideas. The occasional one-sentence paragraph, of course, can be an effective way to add emphasis to a point. Also note that paragraph length should be influenced by your audience and purpose. In some situations, longer or shorter paragraphs may be appropriate.

TEXT FORMATTED USING ONLY UPPERCASE LETTERS IS MUCH MORE DIFFICULT TO READ THAN TEXT FORMATTED USING A MIXTURE OF UPPER- AND LOWERCASE LETTERS. THIS DIFFICULTY OCCURS BECAUSE WE DO NOT READ VERY EFFICIENTLY LETTER BY LETTER. WE READ MORE EFFICIENTLY WHEN WE CAN RECOGNIZE THE PATTERNS OF WORDS AND THUS CAN READ WORD BY WORD. BECAUSE UPPERCASE LETTERS ARE ALL EVEN IN HEIGHT, WE CANNOT USE PATTERN RECOGNITION, AND WE ARE FORCED TO READ LETTER BY LETTER. THAT SLOWS US DOWN.

Figure 7.8 The Problem with Using Only Uppercase

As well as considering the length of your paragraphs, you should also consider spacing between lines, dealing with orphaned lines, and indicating the beginning of a new paragraph. In general, we recommend that you double space documents written for instructors or for anyone else who will review a document. The double spacing permits reviewers to write comments between the lines, thus facilitating later editing of the document.

Final versions of reports, and documents such as letters and résumés, are usually single spaced. Yet even here you may want to consider the effect of spacing. Using a technique called *leading* (pronounced *ledding*[3]) increases the readability of most text. With leading, you use a line spacing that is two points greater than the font size. For example, if your document uses a 12-point font, you might set the line spacing to 14 points. (Printers would describe this as *12 on 14 spacing*.) The larger space between lines improves readability.

Because 2-point leading is the default setting for many word processors, most documents you produce will already employ this feature. Nevertheless, being aware of this practice can be useful should you ever need to squeeze one or two extra lines into a short document. Reducing the leading from 2 points to 1 or 1.5 points may allow you to fit those extra few lines onto the page. We do not recommend, however, that you totally eliminate leading (*12 on 12 spacing*) because the *descenders* of some letters (*g, j, q, p,* and *y*) will run into the *ascenders* of other letters (*b, d, f, h, k, l,* and *t*) from the line below.

To ensure your format looks professional, avoid *widows*. A widow is a single line from the beginning or end of a paragraph that is isolated from the rest when the paragraph splits at a page break. Although widowed lines may not interfere with a reader's comprehension, they are visually distracting, and most word processors allow you to set up your paragraphs so that widowed lines are not permitted.

A final point to consider in relation to paragraphing relates to how you communicate a paragraph break to the reader. In most published work, the start of a new paragraph is indicated by a small indent. For documents produced using a word processor, however, two conventions compete: using the small indent or leaving a blank line between paragraphs. Figure 7.9 shows the difference between the two conventions.

Although you can follow either convention, the indent convention is more typically found in documents written for the humanities while the blank line convention is more typically found in documents written for the applied sciences. Remember that the form you adopt for your documents subtly indicates your membership in a professional community. In other words, using a blank line rather than an indent may serve a persuasive function when working as an engineer.

In any event, you should avoid using both conventions at the same time (if for no other reason than because doing so is redundant). Whichever convention you choose, ensure you are consistent throughout the document. Switching back and forth between the two conventions in a single document could be rather disconcerting for your readers.

Finally, for those rhetorical situations in which the indent convention is appropriate, ensure that the indent is no more than ¼ inch long. Some people use a ½ inch indent (the default setting for tabs[4]). This larger space is a holdover from the days of typewriters and early printers that used *monospaced* fonts (i.e., each character took up the same width in a

---

[3]The term *leading* originates from the practice of placing strips of *lead* between rows of characters during the early days of printing.

[4]We suspect that the default setting for tabs has a lot to do with the continued use of ½ inch indents. It is intrinsically much easier to simply use the tab rather than to set up a paragraph style that defines the paragraph indent as ¼ inch. We suggest that you learn to use styles so that you have much finer control over the appearance of your documents.

Indent Convention   Blank Line Convention

Figure 7.9 Alternative Paragraph Conventions

line of text—*Courier* is an example of a monospaced font). The larger space looked fine with these fonts. However, most modern word processors and printers use *proportionally spaced* fonts (i.e., the spacing is adjusted depending on the width of the character so that an *i* takes up less space than an *m*). More than a ¼ inch indent is excessive with proportionally spaced fonts.

7.4.1.3   Margins

In general, you should provide margins of at least 1 inch on all sides of your documents. However, in documents that are submitted to reviewers or instructors, and on which you expect comments to be written, consider increasing your margins to 1¼ inches. In some circumstances, you may even want to increase one of the margins to 2 or 2½ inches to accommodate binding or note taking. We followed this practice in the handbook version of this textbook, partly to create a more visually appealing format, but mainly to give the students room to write comments during our lectures. This wider margin is sometimes known as the *Scholar's Margin*. In addition to providing room for written comments, the scholar's margin also enables some people to read the document more quickly because they can read each line with a single glance rather than with two or three. Figure 7.10 contrasts the standard margin with the scholar's margin.

Paragraph Justification   Our final point about white space involves considering whether to *justify* your text or to leave the right margin *ragged*. Figure 7.11 compares full justification with a ragged right margin. Readers find it easier to maintain their place in a document with a ragged right margin; consequently, where comprehension is a key concern, avoid justifying your text. However, sometimes the professional appearance of justified text serves a persuasive function, and an even margin is the better choice. As with so many issues, your choice depends on your rhetorical situation.

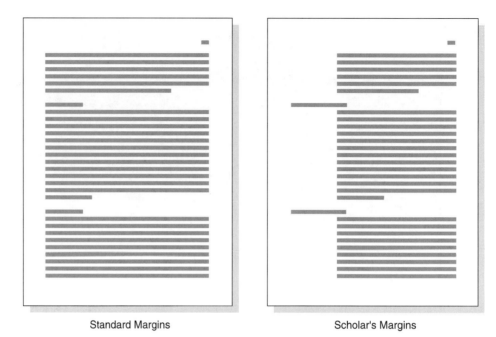

Standard Margins   Scholar's Margins

Figure 7.10 Comparison of Standard Margins and Scholar's Margins

In any case, avoid justifying text that is formatted in narrow columns. Justifying narrow columns of text frequently results in unsightly gaps between the individual words (sometimes called *rivers of white*). Unfortunately, *not* justifying narrow columns often results in an overly jagged right margin. To avoid this problem, we suggest using the hyphenation option included in most word processing programs.

Full Justification   Ragged Right Margin

Figure 7.11 Full Justification versus Ragged Right Margin

Hyphenation   Should you choose to use hyphenation to avoid an overly ragged right margin, do not rely on the *automatic* setting for hyphenation. Use the *manual* setting that allows you to review proposed changes. Or, at the very least, carefully proofread any text that has been automatically hyphenated. The following points indicate some of the issues to consider while hyphenating documents:

- Do not hyphenate headings or subheadings.
- Avoid hyphenating three or more lines in a row because excessive hyphenation tends to slow down readers (much like justification) and creates unsightly parallel lines at the right-hand margin.
- Do not hyphenate proper nouns (i.e., names of people and places) or compound words that already have a hyphen in them.
- Avoid hyphenating the last word in a paragraph.
- Ensure you have broken the word at the correct location depending on its grammatical context (i.e., *pro-ject* is a verb while *proj-ect* is a noun).

7.4.2   Organization

The organization of your documents is critical because most readers of technical documents have certain expectations about how the document should be structured. As you organize technical documents, pay attention to the traditional sections for your type of document, to the headings used to communicate the document organization to the reader, to the pagination of the document, to the use of headers and footers, and to the table of contents and lists of figures and tables.

7.4.2.1   Sections

The following list of sections presents a standard organization for reports that suits a wide range of technical writing situations. Note, however, that you may discover good reasons to use a different format—good reasons created by the nature of your audience, purpose, or subject matter. For example, when you are writing a short report for an audience that is quite familiar with your subject matter, you might consider writing a *letter report*. This type of report begins with the standard salutation (*Dear Ms. Smith*:) and ends with the standard closing (*Sincerely*,). The letter report omits the prefatory pages and appended pages, and provides only the body of the report (which is usually, but not always, divided into subsections).

Although you may not need all of them, the sections of a report are conventionally tenfold (in the following order):

1. Title Page
2. Copyright Page or Revision History Page
3. Executive Summary or Abstract
4. Acknowledgments
5. Table of Contents
6. List of Figures and Tables
7. Body of the Document
8. Glossary (now often placed before the Body)
9. References
10. Technical Appendices

We outline the content and layout of these pages in more detail in Chapter 8.

7.4.2.2 Headings

The formats for headings and subheadings vary from writer to writer and from company to company. Once on the job, you may find it necessary to adopt the particular format favored by your employer, but in the meantime, you can develop your own system—within certain limitations.

In your own reports, you can use any combination of numbering, indentation, capitalization, boldfacing, and font size, as long as your format clearly distinguishes among levels and is consistent throughout. Note that underlining is decreasing in popularity as a method for indicating heading levels because it cuts off the descenders on certain lowercase letters (*g, j, q, p,* and *y*). Figure 7.12 indicates the problem.

Capitalization is also becoming less popular because reading text that is formatted in all uppercase is more difficult than reading text that is formatted in mixed upper- and lowercase.

You should also consider the following points as you create your own headings:

- Headings should be as descriptive as possible (i.e., rather than just using a heading such as *Tungsten Carbide*, which contains little information about the content of the section, you should add descriptive words such as *Failure Modes in Tungsten Carbide Impellers*).
- The format for headings and subheadings must clearly indicate whether a particular heading introduces a major section, subsection, sub-subsection, or paragraph.
  - Major section headings—or in the case of long reports, chapter headings—should begin on a new page.
  - Except for chapter headings, each heading should be followed by text.
  - To create a subsection, you should be able to divide a larger section into at least two parts (i.e., generally avoid a single subsection).
- Omit all end punctuation except a question mark—with the exception of paragraph headings, which should be followed by periods.
- Avoid using acronyms or abbreviations in headings wherever possible.
- Either capitalize only the first letter of the first word in a heading or capitalize the first letter of all *major* words in a heading (i.e., do not capitalize articles, prepositions, and conjunctions unless they are the first word in the heading). Consistently follow whichever convention you choose.

Also carefully proofread to ensure that you have not isolated a heading or subheading at the bottom of a page.

The best formats are as simple as possible while still clearly distinguishing among the different heading levels. You should not, for example, find it necessary to bold, to underline, and to capitalize a heading, or even to bold and to underline it. Figure 7.13 illustrates one possible format for reports with four levels of headings. If you set up your headings as specific styles within your word processor, you will easily maintain consistency in your heading levels. Furthermore, you will be able to easily produce the table of contents.

Ways of presenting data graphically

Figure 7.12 The Problem with Underlining Headings

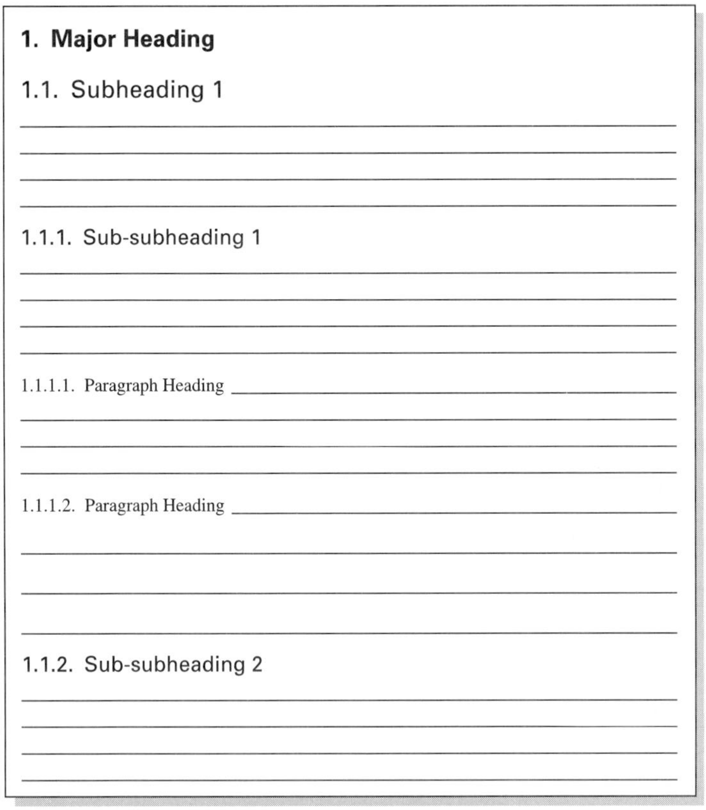

---

1. Major Heading

1.1. Subheading 1

_____
_____
_____
_____

1.1.1. Sub-subheading 1

_____
_____
_____
_____

1.1.1.1. Paragraph Heading _____
_____
_____
_____

1.1.1.2. Paragraph Heading _____

_____
_____
_____

1.1.2. Sub-subheading 2

_____
_____
_____
_____

Figure 7.13 Accommodating Four Heading Levels

7.4.2.3   Pagination

Page numbers are essential in lengthy reports and manuals because they help readers find specific topics mentioned in the table of contents. Yet even in short reports that do not require a table of contents, you should always number your pages. Keep in mind that documents are often photocopied for distribution. Should the person doing the copying mix up the pages (which is easy to do), a document without page numbers may well be reordered incorrectly. Few things are more disconcerting for a reader than receiving a report that lacks page numbers and is not in the correct order.

You should also consider the following conventions as you number your pages. Traditionally, the page numbers for the prefatory pages of most reports (i.e., the executive summary or abstract, the table of contents, the lists of figures and tables, and the acknowledgments) are located at the bottom center of the page. Use lowercase _Roman numerals_ for these pages (i.e., _ii, iii, iv_, etc.) but do not number the title page, even though it counts as page _i_.

Page numbers for the body of a report are generally located in the upper right-hand corner of the page (or on the upper outside edge of the page if you are printing on both sides of the page—that is, alternating from top right to top left corners). Occasionally, page numbers are placed in the lower corners of documents, but we discourage this practice because most readers expect to find the page numbers in the upper corners and look there first. In the body of a report, always use Arabic numbers (i.e., _1, 2, 3, 4_, etc.), and note that the page number on the first page of a report as well as on the first page of each chapter in a manual or handbook may be suppressed.

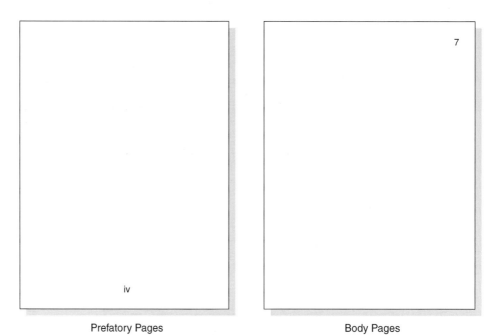

Prefatory Pages   Body Pages

Figure 7.14 Conventions for Page Numbering

Figure 7.14 indicates the conventional placement of page numbers in the prefatory and the body sections of a document. Note, however, that the convention of using lowercase Roman numerals located at the bottom center for the prefatory pages seems to be changing. In industry, reports and project documents are often numbered using only Arabic numerals, starting with the title page (which may or may not have a page number on it).

In this case, the change in convention has probably occurred because numbering pages in that fashion is easier than setting up different formats for the prefatory pages and the body of the document. In other words, the change may have a lot to do with the writers' mastery (or lack thereof) of their word processing software.

7.4.2.4   Headers and Footers

Headers and footers can be an elegant way to add a professional appearance to your documents. Perhaps more importantly, headers and footers are used to communicate important information to the reader. In longer documents, chapter or section headers can help readers easily find a particular section in a manual or a handbook by identifying the chapter or section title and number. In other cases, headers are used to provide the company name and/or logo. In the case of a résumé, headers often include the name of the applicant. Generally, page numbers also appear in headers. Footers, on the other hand, are often used to provide information about the date the document was prepared and its revision number. In addition, footers are sometimes used to indicate that a report contains confidential or proprietary company information.

A final point to consider about headers and footers is to ensure that they do not crowd the text on the page (this crowding can be a particular problem if you are using both footers and footnotes). Headers and footers are generally placed in the upper and lower margins about ½ inch from the edges of the paper. Ensure that you leave about ½ inch of empty space between the header and footer and the text on the page. If you are also using footnotes, you should generally leave ¾ to 1 inch of blank space between the footer and the footnote.

7.4.2.5   Tables of Contents

A table of contents is usually included in a document that exceeds seven or eight pages in length. The table of contents should list at least two levels of headings down the left-hand margin of the page and the page number down the right-hand margin. The format should clearly indicate the level of each heading. Note that you do not include the title page or table of contents among the headings listed, but you do include the headings and page numbers for the executive summary or abstract, lists of figures and tables, and acknowledgments. If you have more than one appendix, you should provide descriptive headings for each one. Figure 7.15 provides a sample table of contents.

If you are using the automatic table of contents feature of your word processor, you should find it relatively easy to produce the table of contents. However, you should not simply accept the default style provided with your word processor. Doing so will often result in some of the headings and page numbers in the table of contents being different sizes and styles (i.e., some may be 12-point bold while others may be 10-point italic, etc.). The effect looks decidedly odd. We recommend that you set up the table of contents in a relatively consistent and plain format and then manually bold main headings if you wish.

We also suggest that you use dot leaders[5] to connect the title with the page number, especially if the table of contents is lengthy. The use of dot leaders helps ensure that readers identify the correct page with the section heading when they are looking up information.

Table of Contents

Abstract	ii
List of Figures and Tables	iv
Introduction and Background	1
Kirchhoff's Laws and Norton's Network Theorems	2
Part 1: Verifying Kirchhoff's Current and Voltage Laws	5
Experimental Procedures	5
Results & Discussion	6
Part 2: Determining Norton Equivalent Circuits Directly	7
Experimental Procedures	7
Results & Discussion	8
Part 3: Determining Norton Equivalent Circuits by Measuring Current vs. Voltage Characteristics	9
Experimental Procedures	10
Results & Discussion	12
Conclusions	14
Glossary	16
References	17
Appendix	18

Figure 7.15 Sample Table of Contents

[5]A *dot leader* is a tab that is underscored with dots. Do not simply type in a row of periods to connect the heading with the number. Aside from being a time-consuming task, you will never get the page numbers to line up correctly. We mention this feature because we have seen so many of our students waste a lot of time typing in periods due to their unfamiliarity with their word processor.

Figure 7.16 Sample List of Figures and Tables

7.4.2.6 Lists of Figures and Tables

Although figures (including graphs) and tables should be listed separately, you can include both lists on the same page if they are relatively short (as illustrated in Figure 7.16). Exclude from these lists any figures or tables included in appendices. Note that in the body of your document, you refer to graphics located in appendices by appendix and page number (i.e., *See Appendix A, p. 18*).

Also note that if you have only one or two figures in a relatively short document. (i.e., an eight-page report), you can omit the list of figures and tables (but if you provide a list of figures, you should list even one table, or list one figure if you need a list of tables). We also suggest that you use the automatic feature of your word processor for producing the list of figures and list of tables because it simplifies the task.

7.4.3 Graphical Elements

Entire textbooks have been written about how to use graphics effectively in the field of engineering. To provide the level of detail necessary to help you master the production of graphics is beyond the scope of this text; nevertheless, we draw your attention to a few basic points that you should consider when working with graphics.

7.4.3.1 Figures and Tables

Look at the alternative formats presented in Figures 7.17 and 7.18 for reporting the change in the temperature of a liquid over a period of time.[6] Although the information presented in both formats is identical, most people find the table easier to understand than the text, in part because the table organizes the data visually for the reader. In

---

[6]Adapted from Gopen and Swan, "The Science of Scientific Writing," *American Scientist*, 78, (1990), 550–558. Used with permission.

t(time) = 0′, T(temperature) = 25°; t = 3′, T = 27°; t = 6′, T = 29°; t = 9′, T = 31°;
t = 12′, T = 32°; t = 15′, T = 32°;

Figure 7.17 Data Presented in Text Form

other words, wherever possible, you should present data as a table or a graph rather than as text.

However, simply presenting information in tabular form does not ensure that the information will be easily understood. You must carefully consider the organization of the information that you intend to present in figures and tables. Study, for example, Figure 7.19. Although the same information is presented as in table 7.18, most people find the table in Figure 7.19 more difficult to understand. Because we read from left to right, and because the table in Figure 7.18 places the context (time) on the left and the new information (temperature) on the right, we process information presented in this format more easily. Figure 7.19 is more difficult to understand because it violates the expectation that the context (or independent variable) will precede the new information (or dependent variable).

Because readers generally pay very close attention to figures and tables, you must also devote sufficient attention to their layout and labeling. By convention, the label is placed *below* a figure but *above* a table, as demonstrated in Figure 7.20. However, some companies have recently adopted the practice of placing the labels above for *both* figures and tables. Although this change may cause some confusion for readers who expect the more traditional convention, it makes sense from a cognitive perspective. As we mentioned earlier, readers process information most easily when they are provided with a general context before specific details. By naming the figure prior to its presentation, this newer practice may assist readers. Nevertheless, you should follow the established convention in most situations.

In addition to considering the organization of your figures and tables, you should adhere to the following guidelines for tables and figures:

- Leave approximately ½ inch of white space between graphics and text.
- Always mention figures and tables in the text *before* you present them.
- Avoid placing a figure or table in the middle of a sentence.
- Number and provide informative titles for all figures and tables. Either capitalize the first letter of all major words in a title (i.e., *Figure 6: Stress Fractures in Titanium-Aluminum Alloys*) or capitalize only the first letter of the first word (i.e., *Figure 6: Stress fractures in titanium-aluminum alloys*).
- Indicate the source of borrowed data.

time (minutes)	temperature (°C)
0	25
3	27
6	29
9	31
12	32
15	32

Figure 7.18 Example Table with Context on Left

temperature (°C)	time (minutes)
25	0
27	3
29	6
31	9
32	12
32	15

Figure 7.19 Example Table with Context on Right

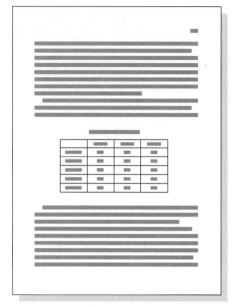 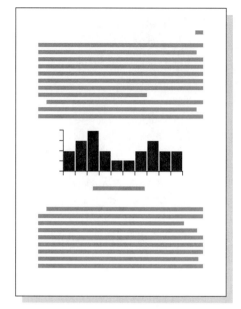

Place Label *Above* Table   Place Label *Below* Figure

Figure 7.20 Placement of Labels for Figures and Tables

When creating illustrations also remember the following:

- Mechanically produce illustrations if possible and otherwise draw them in black ink.
- Avoid out-of-focus pictures or closeups of complex devices. Where you must present complex diagrams, ensure that they are clearly and concisely labeled.
- Draw diagrams to scale or, if it is not possible to do so, indicate that to the reader.

When creating graphs also keep the following in mind:

- Use the horizontal *(x)* axis of a graph for the independent variable and the vertical *(y)* axis for the dependent variable. Include the zero baselines on your axes wherever possible. If necessary, show that the data is discontinuous with a break in the axis.
- Use simple axis labels, but include all the information needed to understand the graph (i.e., label the axes of graphs). Place scale units beside major tick marks only.
- Ensure that axes are heavier than grid lines and that curve lines are heavier than the axes. If possible, eliminate grid lines and when you use them, ensure they do not pass through bars or columns representing data.
- Where possible, use color to distinguish elements within the graphic, to identify parallel items, or to emphasize specific items (but ensure that you don't use too many different colors—a maximum of four or five is a good rule of thumb). When working in black or white, or preparing a document that will be photocopied, use patterns or shades of gray that are easily distinguished from one another.

You should be particularly careful to label graphs completely. Unlabeled or incompletely labeled graphs do not help readers, who are forced to reread the text in order to puzzle out the meaning of the graph. Readers are less likely to view a report with favor if their time is wasted. Figure 7.21 provides an example of a properly labeled graph.

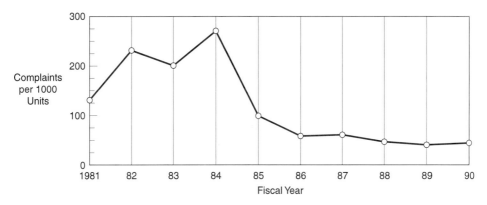

Figure 1: Complaints about XH17 Units Before and After Introduction of
Quality Control Program in 1985

Figure 7.21 Example of a Graph

When creating tables keep the following points in mind:

- Avoid presenting too many numbers (i.e., where possible, simplify larger spread-sheets and complex tables or present the same information with a graph).

- In complex tables, use horizontal and vertical lines to help readers follow the relationship between the data and the variables.

- Place the independent variable on the left and the dependent variable on the right.

Figure 7.22 provides an example of a table. In particular, note that the columns are labeled and the information source is provided.

7.4.3.2 Lists

As well as presenting information in figures and tables, technical reports frequently use lists to summarize information. Lists are a good way to make a document more concise

Table 1: Mohs Hardness Scale for Minerals			
Hardness	**Common name**	**Chemical formula**	**Sp. gravity**
1	Talc	$Mg_3Si_4O_{10}(OH)_2$	2.7–2.8
2	Gypsum	$CaSO_4 \cdot 2H_2O$	2.3–2.4
3	Calcite	$CaCO_3$	2.7
4	Fluorite	CaF_2	3.0–3.2
5	Apatite	$Ca_5(PO_4)_3(F,Cl,OH)$	3.1–3.2
6	Feldspar[†]	$KAlSi_3O_8$	2.5–2.6
7	Quartz	SiO_2	2.7
8	Topaz	$Al_2SiO_4(F,OH)_2$	3.4–3.6
9	Corundum	Al_2O_3	3.9–4.1
10	Diamond	C	3.5

Source: Chesterson, C.W. and K.E. Lowe, *The Audubon Society Field Guide to North American Rocks and Minerals*, 1978, (New York: Alfred A. Knopf).
[†]Figures are for Orthoclase Feldspar. The hardness of Plagioclase Feldspar ($NaAlSi_3O_3$) is the same, but the specific gravity is greater (2.62–2.76).

Figure 7.22 Example of a Table

simply because they eliminate excess words. When creating lists in reports, you should keep the following points in mind:

- Ensure that all items in a list have something in common.
- Create grammatically parallel lists (i.e., ensure all items begin with verbs or with nouns).
- Use bullets (or another symbol) when the items are of equal importance; use numbers to indicate step-by-step instructions, priority, or order of importance.
- Use hanging indents when formatting lists so that the bullets or numbers stand out clearly from the text.
- If possible, restrict lists to five or fewer points. If you have more points, attempt to break the list into sublists.

Lists can be punctuated in several ways: by placing periods or semicolons after complete sentences (with a period after the final item when the list is punctuated by semicolons), by using commas after sentence fragments and a closing period after the last item in the list, or by using no punctuation whatsoever. Whichever method you choose, ensure that you are consistent.

We also suggest that you introduce your lists with a complete sentence rather than with a fragment as the following example illustrates.

Original	**Revised**
In order to improve our facilities, we must	In order to improve our facilities, wc must *undertake the following*:
• repair our existing PCs	• repair our existing PCs
• purchase 11 more hard drives	• purchase 11 more hard drives
• increase system security	• increase system security
• expand our operating hours	• expand our operating hours

7.4.3.3 Equations

Technical documents often rely on equations rather than words to communicate information, and for good reason.[7] Mathematics is a language that is much clearer, more precise, and more concise in expressing certain types of concepts than is English. Rather than writing "the resistance of N series resistors is obtained by adding up the resistances of each resistor," we can write, "the final resistance is

$$x = \sum_{i=l}^{n} y_i,$$

where y_i is the resistance of the ith resistor." Further, we can then continue to manipulate the equation (for example, to describe what happens when another resistor is placed in parallel).

Note that when incorrectly used, equations can be as confusing to the reader as poorly written text. Consequently, we suggest you spend some time learning how to format equations in order to make them as clear as possible. We also recommend that you spend some time learning how to use the equation editor that is packaged with your word processor to automate many of the formatting tasks associated with equations that word processors do not easily accommodate.

---

[7] We would like to thank Professor Jacques Vaisey (Simon Fraser University, Burnaby, BC) for his assistance with the section about equations.

Apply the following guidelines when writing equations:

- Use as concise a notation as possible (for example, $\Sigma_{i=1}^{N} x_i$ is easier to understand than $x_1 + x_2 + x_3 + \cdots + x_N$).

- Ensure that you deal properly with superscripts and subscripts and space the elements of the equation appropriately because poorly spaced equations are difficult to read.

- To distinguish mathematical variables from regular text, use an italic font both for equations and when referring to variables in the text. (The exception is function names such as "sin," "cos," "log," "max," etc., which are set in the normal font.) Use boldface for vectors and matrices to distinguish them from other variables.

- Use Greek letters (e.g., α, β, χ, δ, ϵ, etc.) when appropriate because the extra symbol set makes it easier to recognize the concept that the symbol represents (unlike Roman letters—e.g., *a, b, c, d, e*, etc.—which we generally try to form into words). You may also want to use Greek letters to represent an entire *class* of symbols in your document so that readers know something about an equation as soon as they see the symbol.

Also consider the following formatting issues:

- Display equations on separate lines when the expressions become even slightly complicated.

- Number equations that are cross-referenced elsewhere in the text. (Numbers are not needed if the equations are not cross referenced or if references are obvious.) Center the equations and place numbers in parentheses at the right margin as follows:

$$x = mg \sin(\theta). \tag{1}$$

- Write equations so they can be read as elements of your sentences and use standard punctuation (i.e., put a period after an equation at the end of a sentence or use a comma if the sentence continues to explain the symbols).

- Generally reference equations in the text using the form "we can determine the force acting on the ball using (1)." An alternative is to use the expression "Eqn. (1)," but this is less common except at the beginnings of sentences.

- Where possible, avoid starting a sentence with a mathematical variable or an equation number.

7.4.4   Referencing Information Sources

Almost every document that you write is based, in one fashion or another, on the work of other people. You can acknowledge the contributions of others in one of three ways: by including an acknowledgments section in your document, by using footnotes to reference the contribution, or by using a formal set of referencing conventions.

7.4.4.1   Acknowledgments

When other individuals contribute broadly to a writing project, either by providing information, suggestions, or financing, or by helping to edit and revise the document, you are obliged to thank them for their contributions. In most cases, general contributions are recognized in the acknowledgments section of a document. These sorts of acknowledgments should be relatively short and should not be overenthusiastic. Acknowledgments that are excessive are not only embarrassing to the person mentioned, but those sorts of acknowl-edgments may also lead some readers to question how much *you* actually contributed to the document. Simply name the contributor and their contribution. For example, if you were writing a report about hydraulic erosion, you might say something like "I would like to thank John Smith for contributing his expertise in the area of hydraulic erosion to this report."

In other circumstances, individuals may have only contributed to the document in a re-stricted fashion; perhaps they provided a figure or suggested an avenue of investigation. In that case, a footnote acknowledging their specific contribution may be more appropriate.

7.4.4.2 Footnotes

Footnotes are generally used for one of two purposes: either to acknowledge a specific contribution to a document (a concept or a figure, for example) or to expand on an idea presented in the text of the document. In the case of a specific contribution, you generally use a footnote if the contribution is unpublished (i.e., a verbal suggestion, a photograph, a drawing on the back of a napkin, etc.). If the material is from a published source (includ-ing internal company documents), then use referencing conventions (i.e., a citation in the text and an entry in a reference list) to acknowledge the contribution. The exception oc-curs when you need to include only one or two acknowledgments. In that case, using foot-notes to acknowledge contributions is much easier than citing sources and preparing a reference list.

Another type of footnote, in which you expand on an idea in the text of the document, is becoming increasingly rare in technical documents, in part because it diverts the reader's attention from the main discussion. In many cases, if the extra information is suf-ficiently important that you really want to include it, then you should try to incorporate it into the text. In particular, a very lengthy footnote should be either incorporated into the text or made into an appendix. Nevertheless, footnotes are sometimes the best alternative.

When footnoting, you should consider several issues of format:

- Set the footnote in a slightly smaller font (i.e., 8- or 9-point font size) to clearly dis-tinguish it from the text.

- Use a separator (a horizontal line of about 1½ to 2 inches) to separate the footnote from the text.

- If you have more than 2 or 3 footnotes, use numbers for the footnotes; if you have only a few notes, you can use symbols such as * or †.

- Avoid using footnotes if your document uses footers; if you must use a footnote in these circumstances, ensure that you provide adequate blank space (¾ to 1 inch) between the footnote and the footer.

- If possible, avoid placing more than one footnote on a page.

These format issues are relatively easy to deal with if you use the footnote feature avail-able on your word processor.

7.4.4.3 Referencing Conventions

Professional ethics and the laws of copyright require that you acknowledge your reliance on other people's ideas whether you quote their words or use your own. The conventions for citing sources differ from journal to journal, editor to editor, publisher to publisher, and discipline to discipline. You will largely be expected to follow the conventions of a particular field, but even in the case of referencing conventions, rhetorical situation can be a determining factor. For example, in electronics-based engineering fields, the traditional referencing conventions are those sanctioned by the Institute of Electrical and Electronics Engineers (IEEE), but the IEEE approach is not always the most appropriate choice.

In this convention, references are acknowledged (or cited) by numbers, and the full reference appears in a numbered list at the end of the document. The numbers correspond to the first time a source is cited in the body of the document. IEEE citations appear as follows:

> Johnson [7] proposed the basic circuit shown in Fig. 2.
>
> Designing a robust controller requires a nominal model of the controlled plant [1].
>
> [10, 11, 12] propose alternative models.

As the last two examples indicate, readers often receive no information on a reference unless they flip to the reference list to discover the authority being cited. Because the IEEE convention was designed simply to save space in journals, it is not the best choice whenever an author's name or the date of publication could provide useful information for the reader.

For this reason, unless an instructor, employer, or publisher requests otherwise, the more user-friendly author-date system is preferable. The author-date system also has the advantage of being used in a wide range of disciplines, making it an obvious choice for anyone working in a cross-functional team. In this system, sources are cited by author and date and placed in parentheses within the text, and an alphabetically ordered reference list is provided at the end of the report. The following section includes a general guide for using the author-date system.

7.5 A GUIDE TO REFERENCING CONVENTIONS

In school, referencing conventions are important for two reasons. First, they provide a mechanism for giving credit to others for their ideas and their words. Second, they are part of the scholarly apparatus of a field that professors hope to teach their students. Just as engineering professors, in their roles as teachers, are training the next generation of engineers, history professors are training future historians, economics professors are training future economists, and so on. When you take a course in any field, one of the expectations is that you will follow the conventions of that field—as if you too were training to become a member of that discipline.

The first reason dictates the ethical need for references; the second reason indicates why details matter. We will turn our attention to the details in just a moment, but first we should note that the ethical need for references is at no time greater than the present. The Internet and its supporting technologies make unauthorized and unacknowledged use of intellectual property increasingly easy and common. Everyone, especially those seeking professional status, is obliged to do his or her part to ensure that copyright is respected and that intellectual property is acknowledged. Strong personal ethics and a good grasp of referencing conventions will help you meet these obligations.

In terms of the details, many instructors will provide you with style sheets or tell you what convention they are following. University bookstores and the reference section of libraries have books on a range of referencing and bibliographic conventions that cover the subject in great detail. What we offer in this guide is a general overview of the most common conventions and enough detail to help you cite and reference common sources of information. Note, however, that referencing conventions change, sometimes quite radically (most of the books on this subject have gone through many editions), and conventions are not yet established for every possible source.

As well as referring to reference guides, you can also learn about a particular convention by studying relevant published works, noting how authors cite sources and how they construct the items in their reference lists. In general, you will find that these conventions fall into two groups: those that use numbers for citations and those that cite the author's last name and the date of publication. We focus on the author-date system because it is more reader friendly and more widely employed. We also provide advice for placing citations in sentences, quoting directly from sources, and formatting reference lists.

7.5.1 Citing Sources by Author and Date

As its name suggests, the author-date system of citing sources involves acknowledging the source of information or the author of a quotation by providing an author's last name and the year of publication in the text. For example, to cite from a single author, place the author's last name and the year of publication in parentheses:

Recent research (Black 2000) supports this explanation.

Several variations of this basic pattern are described below.

7.5.1.1 Two or More Publications by the Same Author in the Same Year

If an author has two or more publications in the same year, distinguish between them by appending lowercase letters (*a, b, c*, etc.) to the year of publication and separate publications with semicolons:

(Chan 1999a; 1999b).

Also use these letters after the dates in the reference list.

7.5.1.2 More than One Author

When a publication has two authors, include both names:

(Smith and Jones 1994).

When a publication has three authors, include all three names the first time you refer to it, but for subsequent references use the first author followed by *et al.*:

(Smith, Trinh, and Matsui 2001) for the first reference

(Smith *et al.* 2001) for subsequent references

When a paper has four or more authors, cite it using *et al.* in the first citation: (McManus *et al.* 1993). Readers can check the reference list to discover the names of the other authors.

7.5.1.3 More than One Source Cited for the Same Author

If you provide two or more references together, separate them with semicolons:

(Smith and Jones 1994; Smith *et al.* 2001).

If you list several sources by the same author, list the name once and then separate the various dates with semicolons:

(Carver 1986; 1992a; 1992b; 1995).

7.5.1.4   Author Unknown

When the author is unknown, but the source is published or sponsored by an association, corporation, government agency, or other group, the name of the group serves as the author's name (both in the citation and in the reference list). If neither an author nor sponsoring group is indicated, use the name of the publication and, where possible, place the reference in the sentence rather than within parentheses:

These estimates are based on data provided by the National Research Council (1995).

If the organization was not named in the sentence, then the reference would appear as follows:

(National Research Council 1995)

If you use the abbreviation for an organization in the text, use it in the citation as well:

(NRC 1995).

7.5.1.5   Additional Information Often Included in Parentheses

Wherever possible, you should assist your reader in finding the information you have cited. To do so, place a volume, page, section, or equation number after the date, separated from the main citation by a comma:

(Chan and Jones 1994, 55) to refer to a single page

(Singh *et al.* 1985, 14–17) to refer to more than one page

(Brown 1996, sec. 11.5) if an entire section is relevant

(Peterson 2000, eq. 10) to refer to an equation

7.5.1.6   Omitting Information from Parentheses

Omit from the parentheses any information already given in the text. For example, if the author's name appears in the sentence, omit it from the citation. If both the name and date appear in the sentence, you can omit the parentheses or use them only for page, section, or equation numbers:

Saif and Zhou first proposed this method in 1984 (45–51).

Note that the convention is to refer to authors by their last names only. You may refer to authors by their given and last names the first time you refer to them, but thereafter use only the last name.

7.5.1.7   Citing Unconventional Sources

For sources such as interviews, personal letters, or mail exchanges, give the full name of the person you communicated with and the nature and date of the communication:

(John Brown, letter to the author, July 2000)

(Bill Smith, telephone interview, 10 Sept 1999)

(Susan Eaglets, mail exchange, Aug–Sept 1998)

The same principle applies to e-mail messages:

(Aaron Bates, e-mail to the author, 4 Nov 2000)

Note that none of these sources will appear in your reference list. You can create similar citations for films, videos, recordings, television programs, and the like, although such entries may be better suited to an author-date citation and an entry in the reference list.

Do not worry whether you are following *the* convention because one may not exist. You can always create your own convention for citing an unusual source, keeping two general principles in mind. First, create a citation that follows the same general pattern as conventional citations. Second, if the item is not included in your reference list, provide all additional information readers require to locate the source in the citation.

If you are *not* including a source in a reference list, identify the author, artist, director, or other appropriate equivalent to an author if one is available. If possible, provide a date of publication or equivalent. Also provide the name of the piece or program or another appropriate equivalent to a title. And include the name of a company, studio, station, or other equivalent to a publisher.

> (Buckner and Whittlesey, Directors, 1988, *Do Scientists Cheat?*, Videotape, Boston, MA, WGBH Educational Foundation)

If you included this source in your reference list, then the citation would appear as follows:

> (Buckner and Whittlesey 1988)

If the author or equivalent is unknown, then fill the author slot in the author-date citation with the organization, title, or whatever else appears first in your reference list. If you use a title, use no more than the first few words in your citation. Similarly, shorten the names of organizations if they are long, but ensure you begin with the same word used to position the entry alphabetically in the reference list.

7.5.1.8   Citing On-line Sources

Increasingly, research takes us online. While information acquired from *Usenet Newsgroups* and the *World Wide Web* (WWW) home pages may have changed or disappeared by the time someone reads your paper or report, proper citations and references are nevertheless important. When you find information you might use, copy all available information, including author, date, title, name of the newsgroup or home page, *Universal Resource Locator* (URL), page or paragraph numbers if provided, the date you accessed the site, and anything else that might help readers not only find an available source, but also locate information contained in a deleted source. With sufficient information, a member of a newsgroup or the owner of a home page may be able to provide the information a reader wants to locate.

If you are not providing a reference list, include all useful information in your citation.

> (XYZ Inc., *Thermistor Price List*, accessed 22 May 1999 <http://www.xyz.nz/~thermistors/pricelist/>

Note that your citation may also be as simple as the URL for a web site:

> The *Lightning!* exhibition at the Museum of Science featured the world's largest Van de Graaf Generator (http://www.mos.org/what's_happening).

We discuss electronic sources in more detail in the section on reference lists (pages 283–284).

7.5.2   Placing Citations within a Sentence

Whatever citation system you use, be sure to pay attention to the placement of citations. Wherever possible, place a citation at the end of a sentence (or just before a punctuation mark), and avoid placing one between a subject and verb, particularly if the citation is lengthy. If possible, revise a sentence to ensure the citation is at the end.

Sometimes, however, this sort of restructuring is awkward and you may need to place a citation within a sentence. Note, for example, how changing the location in the following sentence creates ambiguity:

> Researchers (Lightfoot and Jackson 1998) have reported findings that support this alternative explanation.

If you place the citation at the end of the sentence, Lightfoot and Jackson would appear to have provided the alternative explanation, rather than to have provided support for it. Examine another example:

> Existing methods (Brown 1986; Black 1987) fail to account for this potential problem.

Again, both Brown and Black describe existing methods, without necessarily being aware of the problem under discussion. If you put the citation at the end of the sentence, you would suggest that they were concerned with this problem.

7.5.3   Quoting

Most of the information you gain from others should be synthesized and related in your own words. However, sometimes an author's words are so to the point or so well put that quoting them is appropriate. Keep in mind that if you borrow even a phrase from someone else's writing, you are obliged to credit the author by putting those words in quotation marks and by providing a citation and a reference.

When you quote a few words or a relatively short passage of no more than four lines of text, enclose it in double quotation marks. The punctuation for the sentence will come *after* the citation:

> One of the requirements is that "Each bid must be accompanied by a certified check or cash to the amount of nine thousand dollars" (Brantwurst 1989, 46).

If you quote a longer passage, do not enclose it in quotation marks, but offset it from the rest of your text by beginning on a new line and indenting the entire quotation:

Eric Brantwurst notes the following requirement:

> *Each bid must be accompanied by a certified check or cash in the amount of nine thousand dollars. All certified checks must be drawn on some responsible bank doing business in the city of Vancouver, and shall be made payable to the City of Vancouver. (1989, 45)*

Note that if your document is double spaced, a long quotation is usually single spaced. Also note that the punctuation mark precedes the citation when the quotation is indented.

If you quote a passage that itself contains a quotation, indicate the quotation within the quotation in one of two forms, depending on whether your quotation is short or long. If your quotation is short, your source's quotation will be reformatted in single quotation marks:

> *Dagwood Brunster recalls an extraordinary engineering feat: "Sam Williams, the chief engineer, shrieked at all of the layabouts who were drinking coffee laced with rum, 'Get off your duffs, or I'll recall your engineering licenses, and throw you overboard to boot!'" (1999, 47).*

If your quotation is long (and indented), your source's quotation will remain as is, in double quotation marks:

> Dagwood Brunster recalls an extraordinary engineering feat:
>
> *Sam Williams, the chief engineer, shrieked at all of the layabouts who were drinking coffee laced with rum, "Get off your duffs, or I'll recall your engineering licenses, and throw you overboard to boot. Move it before you lose it!" Shortly, they were all back at work, sweating profusely under the warm Arctic summer sun. (1999, 47–48)*

In the latter example, note the quotation comes from two consecutive pages. The citation (1999, 47–48) indicates that your quotation consists of a passage on page 47 that is continued on page 48.

Another issue to consider when quoting is that the sentence structure of the quotation must agree with that of your text. You must therefore check the tense of the quotation and whether nouns are plural or singular, adapting your sentence to match the grammar and structure of the quotation. If you must change a verb in the quotation to match the structure of your sentence or change a pronoun to a noun for clarity, enclose the substitution in square brackets: *[were]* to replace *are* or *[the President's]* to replace *her*:

> As Humphreys explains, "Major investors [were] pleased with [the President's] successor."

Finally, except for changes noted in square brackets, quotations should be repeated exactly as written. If the quotation contains a grammatical, spelling, or other error, you can indicate that you are aware of the mistake in the quoted text by placing *[sic]* after the error. But before you use this convention, assess its affect *[sic]* on your readers. If used too often or for a minor offense, readers may view you as overly critical.

7.5.4   Preparing an Alphabetical Reference List

If you cite more than a few sources in a document, you should provide a *Reference List* (which is generally referred to as *References* in the sciences and applied sciences and as a *Bibliography* in the arts and humanities). The following descriptions and examples cover the types of references you are most likely to include in your list. Becoming familiar with the following conventions will help ensure that you know what information to record while you have access to a particular reference source. You will save time and effort if you add the item to your reference list when you initially review the material. Assume that the source will be unavailable the next time you look for it.

Note that the author-date system requires an alphabetical reference list. Also note that the conventions for the following examples are common in the sciences and applied sciences, but are not used consistently. Variations in reference list conventions are discussed later in this section on pages 285–286. Even though your particular field may require a different set of conventions, take careful note of the indentation, punctuation, capitalization, and the use of quotation marks and italics in the following examples. These details are important aspects of referencing conventions, and if you are familiar with the details of one approach, you will more easily adapt to another approach.

7.5.4.1   Two Works by the Same Author(s)

One of the oddities of alphabetical reference lists is the appearance of a line where you would expect an author's name. The line (or five unspaced hyphens) followed by a period at the beginning of an entry indicates that the same author (or authors) wrote this item as wrote the previously listed item. Following this convention, the author of the 1997 entry in the following example is R.E. Burnett.

> Burnett, R.E. 1993. Conflict in collaborative decision making. *Professional Communication: The Social Perspective*, eds. N. R. Blyler and C. Thralls. Newbury Park, CA: Sage Publications, 144–162.
>
> ———. 1997. Collaboration in workplace communication. Chapter 5 in *Technical Communication*, 4th ed. Belmont, CA: Wadsworth, 85–114.

For multiple authors, the names must be listed in exactly the same order before you can use this convention.

Note that the sample references also indicate that the 1993 Burnett entry is an article in a collection of articles edited by two other people and that the second entry is a chapter in a book written by Burnett. Different sources must be referenced somewhat differently. In the following sections, we supply a consistent set of conventions for referencing various types of entries. Use it if no other approach is specified.

7.5.4.2   Journal Articles

A basic reference-list format for journal articles follows. Pay particular attention to the punctuation, especially to the use of periods to separate the parts of the entries. This feature is used consistently for various types of entries.

> **Author(s)**. [last name first followed by initials for first author] **Year. Title**. [Capitalize only the first word of the title] *Name of the Journal*. [italicized] **Volume: number, page numbers**.

Journal entries in an actual reference list could appear as follows:

> Harney, M. 2000. Is technical writing an engineering discipline? *IEEE Trans. Prof. Commun*. 43:2, 210–212.
>
> Markus, L. 1994. Electronic mail as the medium of managerial choice. *Organization Sci*. 5:4, 504.
>
> Raymond, J. and C. Yee. 1990. The collaborative process and professional ethics. *IEEE Trans. on Prof. Commun*. 37:3, 77–81.

Note that journal titles and conference names are often abbreviated to save space. If you are unsure of the appropriate abbreviation, use full titles in your reference list. Also note that the volume for a journal is the collection of issues published in a year and the number is a specific issue that would be mailed to a subscriber. As the final example demonstrates, the last name precedes the initials only for the first author. Second and subsequent authors are listed with their initial or given names first. Finally, note the general format we are using that places the first line of an entry at the left-hand margin and indents all additional lines.

7.5.4.3   Articles in Collections Other than Journals

Study the following reference-list format for articles in collections:

> **Author(s). Year. Title of article**. *Title of Publication,* **edition number** and/or **names of editors. Publisher: place of publication** [if appropriate], **page numbers**.

The following examples include two we have already encountered. We have provided an example of an article in an edited collection, a chapter in a book by the same author, a paper in the published proceedings of a conference, and a paper in an edited collection of papers from a conference, respectively:

Burnett, R.E. 1993. Conflict in collaborative decision making. *Professional Communication: The Social Perspective*, eds. N.R. Blyler and C. Thralls. Newbury Park, CA: Sage Publications, 144–162.

———. 1997. Collaboration in workplace communication. Chapter 5 in *Technical Communication*, 4th ed. Belmont, CA: Wadsworth, 85–114.

Rittenbruch, M., H. Kahler, and A.B. Cremers. 1998. Supporting cooperation in a virtual organization. *Proceedings of the Nineteenth International Conference on Information Systems*. 30–38.

Robey, D, M. C. Boudreau, and V.C. Storey. 1998. Looking before we leap. *Electric Commerce: Papers from the Third International Conference on Management of Networked Organizations*, eds. G. St. Amant and M. Amani. 275–290.

Give yourself a pat on the back if you noticed that the last two entries are missing the publication information. In these cases, the conference organizers may be responsible for publication. All the papers from a conference are often published and distributed to participants; in other cases, a select number of papers are chosen by an editorial team and edited versions of these conference papers are published. These proceedings may, therefore, not have a publisher, *per se*, and so this information may have been legitimately omitted from these entries.

The place the conference was held is often provided in lieu of publishing information as demonstrated in the following example:

M.G. Barchilon, Sept. 1998. Technology's impact on online résumés. *Proc. Int. Professional Communication Conf.* Quebec City, Que. Canada.

Including this information is generally a good idea, but do you notice anything missing from the above example? *Proc.* is the abbreviation for Proceedings, which is a published collection of conference papers. Because the paper titled "Technology's impact on online résumés" is contained in published proceedings, page numbers should have been provided.

As this example suggests, missing information sometimes results when researchers, like students, forget to write everything down and then cannot relocate the source when compiling the reference list. If you have actually consulted a source and noted all the relevant reference information, we can imagine few reasons for not including page numbers. But if essential information such as the date of publication is not provided for a source, you may want to notify your readers that the omission is not an oversight by using a notation such as *n.d.* to clarify that no date is given.

7.5.4.4   Unpublished Conference Papers

Much cutting-edge information is shared at conferences and may be contained in papers that are not yet published. Researchers and scholars often share copies of these unpublished papers. While later versions of these papers are often published, the only references available may be to works that are never published or are not yet being prepared for publication. The following format applies to unpublished conference papers:

Author(s). Date. Title. Presented at Conference name, Place.

An actual entry from a reference list follows:

Divsalar, D. and J. K. Omura. June 1979. Performance of mismatched Viterbi receiver on satellite channels. Presented at Int. Conf. Commun., Boston, MA.

Note that the month as well as the year and the location of the conference are included for unpublished conference papers.

7.5.4.5   Unpublished Theses or Dissertations

The following format applies to unpublished theses or dissertations:

Author. Date. *Title*. Degree. University, Place.

An example follows:

> Newland, P.A. 1990. *Understanding Designers Knowledge Acquisition Processes: A Potential for Enhancing Information Transfer.* Ph.D. dissertation. Portsmouth Univ., Portsmouth, UK.

7.5.4.6   Books

Books are listed using the following conventions:

Authors. Date. *Title*. Volume [if applicable], **edition** [if other than first (i.e., 2nd ed.)]. **Place of publication: publisher**.

Examples of book entries follow:

> Berners-Lee, T. and M. Fischetti. 1999. *Weaving the Web: The Original Design and Ultimate Destiny of the World Wide Web by Its Inventor.* New York: Harper Collins.
>
> Vicente, K.J. 1999. *Cognitive Work Analysis: Toward Safe, Productive, and Healthy Computer Based Work.* Mahwah, NJ: Lawrence Erlbaum.

7.5.4.7   On-line Sources

The conventions for referencing on-line sources are still evolving, but for sources from the Web, the general pattern is to include as much of the information as you would for text-based sources as possible and then to include the date you accessed the site and the URL enclosed in angle brackets.

Web Articles   An example of a source that appears only on a web site follows:

> Beverly, C. 1993. The ethics of technology in education. 30 Aug. 2000 <http://rgfn.epcc.edu/programs/trainer/ethics.html>.

As her web site indicates, Carrie Beverly wrote "The Ethics of Technology in Education" in 1993. The second date is the date we accessed her site. Note that the lack of a period following the access date clarifies that 30 Aug. 2000 *is* the access date and thus belongs in the same part of the reference as the URL.

Published Articles Accessed on the Web   Often sources you access on the web are also available in published sources. The following reference indicates that the article "Problems and Cases" was published in the Fall 1996 edition of the journal *Professional Ethics*, but accessed on the web.

> Whitbeck, C. Fall 1996. Problems and cases: New Directions in Ethics 1980–1996. *Professional Ethics.* 30 Aug. 2000 <http://www.onlineethics.org/essays/education/index.html>.

On-line Publications   You can also access journals and newspapers that are published on the Web. Sample references follow:

> Greenleaf, G. June, 1996. A proposed privacy code for Asia-Pacific cyberlaw. *Journal of Computer-Mediated Communication, ed. A. W. Branscomb.* 2: 1.30 Aug 2000 <http://www.ascusc.org/jcmc/vol2/issue1>.
>
> Kaplan, C. S. 21 July 2000. Norwegian teenager appears at hacker trial he sparked. *The New York Times on the Web.* 30 Aug 2000 <www.nytimes.com/library/tech>.

On-line Discussion Groups   Information obtained from on-line discussion groups can be referenced in the following manner:

> Adamowski, T. Writer's resource. IEEE PCS Online Discussion Forum. 14 Dec. 1999 <http://ieeepcs.org/wwwboard/>.

7.5.4.8   Sample Reference List

A sample reference list is provided in Figure 7.23. Single space the individual items, but double space between them. Also note that while underlining is sometimes used to indicate italics, underlining is a holdover from the time when typewriters were common and specialized fonts were rare. It represents a convention for telling the typesetter to set the underlined text in italics when the document was destined for formal publication. Underlining is no longer appropriate because word processors support italicized fonts.

References

Cooper, R.G. 1995a. Developing new products on time, in time, part 1. *Research Technology Management.* 38:5, 49–53.

——. 1995b. Developing new products on time, in time, part 2. *Research Technology Management.* 38:5, 53–57.

Davenport, T. and L. Prusak. 1998. *Working Knowledge.* Boston, MA: Harvard Business School Press.

Engineering Accreditation Commission. 1996. *ABET Engineering Criteria 2000,* 2nd ed. Baltimore, MD: Accreditation Board for Engineering and Technology, Inc.

Fleder, R.M. and B. Solomon. 1998. Inventory of learning styles. 10 Nov 1999 <http://www2.ncsu.edu/unity/uer/f/felder/ILSdir/ILS-a.htm>

Hackos, J.A. 1994. *Managing Your Document Projects.* New York, NY: John Wiley & Sons.

Markus, L. 1994. Electronic mail as the medium of managerial choice. *Organization Sci.* 5:4, 504.

Talbott, L.S., ed. 1999. Technology and human responsibility. 17 Nov 1999 <http://www.oreilly.com/,stevet/netfuture>

United States Environmental Protection Agency, Office of Pollution Prevention and Toxics (MC 7409). 1996. *Valuing Potential Environmental Liabilities for Managerial Decision-Making: A Review of Available Techniques.* Rep. no. EDA 742-R-96-003. Washington, DC.

Figure 7.23 Sample Reference List Following Author-Date System

Note that this list includes two works by the same author, R.G. Cooper, and two works without specific authors. The citation for the fourth entry could be (Engineering) or (Engineering Accreditation) and the citation for the last entry could be (United States) or (U.S. Environmental Protection).

7.5.5   Variations in Reference Lists

As we mentioned earlier, referencing conventions are far from standardized and vary greatly among disciplines, fields, publications, and companies. However, certain trends are worth noting, including general stylistic differences between pure, applied, and social sciences on the one hand and the arts and humanities on the other. These differences are presented in Table 7.2. For example, compare the two versions of the following entry:

> Wise, Penelope. "Money Today: Two Cents for a Dollar." *No Profit Review* 2 (1987): 123–42.
> Wise, P. 1987. Money today. *No Profit Rev.* 2: 123–42.

Of course, the shorter entry represents the science-oriented style you will most often use. However, knowing something about the other style is useful when writing papers for humanities courses and when working on interdisciplinary projects. Also note that the version of the science style above is a middle-of-the-road option. You could use a somewhat longer or an even shorter entry:

> Wise, P. 1987. Money today: Two cents for a dollar. *No Profit Rev.* 2: 123–42.

> Wise, P. 1987. *No Profit Rev.* 2: 123–42.

What variation of the style you use may be a matter of the style adopted by a company, publisher, or instructor, but when you have a choice, the style should be governed by your readers' needs and expectations. The longer version above provides more information for the reader by providing a subtitle that indicates the content of the article. The shorter version provides just enough information for the reader to find the source, which may be all some readers require. Whatever variation you choose, be consistent throughout a document.

At the beginning of this section on referencing conventions, we mentioned the IEEE style. While the author-date system is more common across fields, systems using numbers as references are standard in some fields such as electrical and electronics engineering.

Table 7.2   Differences in Reference Lists

Arts/Humanities Style	Science/Social Science Style
Author's given name is spelled out.	Only uses initials for given names.
Date of publication is at the end.	Date of publication is after author's name.
Main words in titles are capitalized.	Only first word and proper nouns in titles are capitalized.
Full titles are provided for books and articles.	Subtitles are usually omitted.
Titles of articles are placed in quotation marks.	Titles of articles are not placed in quotation marks.
Titles of publications are generally written out in full.	Titles of publications are usually abbreviated.

The variation in this general approach is between numbering sources according to the order in which they are cited (IEEE format) and numbers assigned to sources based on alphabetical order.

An example reference list using the IEEE style follows. Because the list is not in alphabetical order, first authors' initials precede their last names:

[1] B. Oakley, II, "HyperCard courseware for introduction to circuit analysis," in *Proc. ASEE Annu. Meet.*, 1991, pp. 496–500.

[2] *Microsoft Video for Windows*, Microsoft Corp., 232-100-901, 1994.

[3] K. L. Conway, "Putting technology in its place: the classroom," *Institute for Academic Technology*, Spring 1991, p. 5.

[4] P. R. Keller and M. M. Keller, *Visual Cues*, Los Alamitos, CA: IEEE Computer Society Press, 1993.

Did you examine the style of entries closely enough to notice that like the humanities style, IEEE format uses quotation marks for articles and puts dates at the end? If so, you may also have noticed that only the first word of a title is capitalized as expected for a science-based style. Also note that, unlike either generalized style, commas separate items and page numbers are indicated by *p.* for a single page and *pp.* for two or more pages.

We draw your attention to IEEE conventions to emphasize a point. Referencing conventions vary widely. Even within the IEEE, conventions vary slightly from one society to the next. Whenever your situation changes or you are asked to publish, take a careful look at examples of appropriate referencing conventions.

For a little practice, study the following reference list representing the American Society for Engineering Education style and note the similarities to and differences from the IEEE example above:

1. Tonso, K. L., "Becoming Engineers While Working Collaboratively: Knowledge and Gender in a Non-Traditional Engineering Course," part of Margaret Eisenhart's Final Report to the Spencer Foundation entitled "The Construction of Scientific Knowledge Outside School," 1993.

2. Lunsford, A., Ede L., "Why Write . . . Together: A Research Update," *Rhetoric Review*, 5, 1986, pp. 71–76.

3. Gere, A. R., *Writing Groups: History, Theory, and Implications*, Southern Illinois University Press, Carbondale, Illinois, 1987, pp. 55–76.

4. "Learning Together Makes a Difference," *The Teaching Professor*, June/July, 1994, p. 5.

As you have surely noted, the general format is different, but did you also notice that the last name comes first for all authors, that all the major words in a title are capitalized, and that subtitles are provided? Did you note the shared use of commas to separate entries and *p.* and *pp.* before page numbers? An eye for detail is extremely useful when the time comes to prepare a reference list, but note that style sheets are often available, providing both instructions and examples to follow.

Finally, keep in mind that what may seem to be needlessly fastidious detail has a purpose: following the conventions of a given group right down to the way it cites sources and prepares reference lists helps establish your identity as a member of that group. Conventions of style, format, and organization change from group to group, and the ability to appreciate these differences is part of your education and preparation for an increasingly interdisciplinary profession.

7.6   ANALYZING FORMAT

If you have no sense of what proper format looks like, you should begin by looking through samples of well-formatted documents. The sections of this chapter about applying format provided samples of appropriate formats as well as general information about specific elements of format. While a student, your instructors may be able to provide you with samples of well-formatted documents that students have submitted in the past. Ask them if they can provide such samples. As an employee, ask your co-workers what formats have been used in the past. They will probably provide you with appropriate examples of report formats. If these sample reports are on a computer, you can often copy the overall formatting along with any attached styles or templates.

When you examine these sorts of sample documents, you should consider a range of specifics, and you should ask yourself some general questions about the format used. Although the following list of questions is not exhaustive, if you carefully answer the questions provided, the answers you generate should allow you to appropriately format documents for most situations.

7.6.1   Fonts and White Space

- What fonts are used in the document? What size is the font for the text of the document? Are italics or bolding used? Where?

- How wide are the margins for the document? What sort of line spacing is used (e.g., single spacing, one and one-half spacing, or double spacing)?

- How are new paragraphs indicated (e.g., are paragraphs indented or is a blank line inserted between paragraphs)?

- Does the document use one wide column or two or more narrow ones? How much space exists between columns?

7.6.2   Organizational Issues

- Where are page numbers placed on the document? Are the pages numbered in the same fashion throughout the document? Or are the prefatory pages, the body of the document, and the appendices numbered using different conventions?

- Does the document use headers and footers? What information do they "contain? How much space separates the headers and footers from the text? What font is used?

- Does the document contain section and subsection headings? How many levels of headings are used? How can you distinguish between heading levels (e.g., by capitalization, font size, or a numbering system)? How much space is left above and below the headings? Does the font used for the heading differ from the font used for the text? How long are the headings?

7.6.3   Document Structure

- Does the document contain prefatory pages? If so, how is the title page set up? What font is used? What information is emphasized? Does the document have an abstract or executive summary? How long is it? Is there an acknowledgments section?

- Does the document have a table of contents? How many heading levels does it have? Do dot leaders connect the section titles with the page numbers? Does the document have lists of figures and tables? How are they set up?

- How is the body of the document organized? Is it divided into sections? Do these sections follow a familiar pattern? Are there any sections you are unfamiliar with?

- Does the document have any appendices? What sort of information do they contain? How are the appendices identified? How are the pages numbered?

- Does the document have a glossary? How are glossary entries indicated in the text? Is the glossary in the prefatory pages or in the appendices?

7.6.4   Graphical Elements

- Does the document use figures, tables, or equations? How many? What sort of information do they contain? How are they labeled? Where are they placed in relation to being mentioned in the text? Are they numbered? What font is used? How much space separates a figure or table from its label? How much space separates the figure, table, or equation from the surrounding text?

- Does the document include any lists? Are hanging indents used? How are separate items indicated (e.g., bulleted or numbered)? How are they spaced?

- Does the document contain any listings of computer code? Where? What font is used? How is it formatted?

7.6.5   Referencing Conventions

- Does the document make use of quotations? Where? How often? What length? How are they formatted? How are they referenced?

- Does the document have any footnotes? How are they set up? Are they at the bottom of the page or at the end of the document? Are symbols or numbers used to indicate a footnote? What font size is used? What sort of information do they contain?

- Does the document have any references? What conventions are used for citing and referencing sources? How up-to-date are the references? What types of sources are represented (i.e., books, articles, conference papers, on-line sources)?

7.7   FORMAT CHECKLIST

Use this format checklist as you write various types of documents. We suggest that you review the points that it raises *before* you format a document. You can then determine which areas require particular attention in order for you to appropriately format your document.

7.7.1   Mastering Format

Have you examined some representative samples of well-formatted documents?   _____

Have you considered your format from the perspective of your readers?   _____

Have you mastered the operation of the word processor and graphics
programs that you use?   _____

Have you set up styles and templates for documents you must write frequently? _____

Do your documents look like the work of a professional? _____

7.7.2　Applying Format

Have you printed a copy of your document and revised it for format? _____

7.7.2.1　White Space

Have you considered whether the amount of white space in your document is appropriate for its audience and purpose? _____

Have you used appropriate sizes and styles of fonts in order to increase readability? _____

Have you used appropriate paragraph conventions, length, and spacing for your audience and purpose? _____

Have you used margins that are wide enough? _____

7.7.2.2　Organization

Have you sectioned your report in an appropriate manner? _____

Are the conventions and language you are using for headings clear? _____

Have you numbered your pages appropriately? _____

Have you used headers and footers if needed? _____

Have you included a Table of Contents and a List of Figures and Tables if required? _____

7.7.2.3　Graphical Elements

Have you ensured that your figures and tables are accurately and completely labeled? _____

Have you used lists in appropriate ways? _____

Are equations formatted properly? _____

7.7.2.4　Referencing Information Sources

Have you acknowledged anyone who helped with the document? _____

Have you used footnotes in an appropriate manner? _____

Have you cited sources appropriately? _____

If necessary, have you prepared a list of references? _____

7.8 EXERCISES

Ideally, you should apply the concepts, principles, and advice in this chapter to writing that you already do for school or work. We provide the Format Checklist earlier in this chapter for that purpose. The following exercises also reinforce the issues we raise.

1. Mastering Style Sheets and Templates

Prior to writing your next report for school or work, set up a style sheet with the goal of becoming familiar with how styles are organized and employed by the word processing package that you are using. Define styles and set up shortcut-keys for the table of contents, lists of figures, headings, headers and footers, paragraphs, labels, lists, footnotes, quotations, and references. Then try globally changing some elements of the style sheet to get a sense of how these changes affect the document. If the format of this report is one that you may need to use frequently, turn it into a template for subsequent reports.

2. Experimenting with Fonts

Using a very simple one-page sample of text, format the text using several different fonts and font sizes (you can be quite creative here). Also try experimenting with justification, underlining, bolding, italicization, and capitalization. Print out the different samples and consider what effects they might have on a reader.

3. Improving the Format of a Document

Swap drafts of a medium-length document (10 to 15 pages would be ideal) with a friend or colleague. Read the document specifically for the issues of format outlined in this chapter. What format changes would you suggest to make your colleague's document more persuasive, easier to understand, or easier to locate information in? What changes are suggested for your document?

4. Working with a Graphics Program

Working with some data that you have collected or invented, create a chart using an application such as Excel. Experiment with the format of the figure using the various options presented by the application. After you have created a graphic that you are satisfied with, review the principles outlined in pages 268–272 of this chapter. Would you make any further changes? Why or why not?

5. Working with Equation Editors

Find a journal article in your field that contains several complex equations. Try using your equation editor to duplicate the equations. What difficulties do you encounter?

6. Assessing Form on the Web

Search the web for sites dealing with a topic of interest. Access at least six sites, asking questions such as the following: How long did you wait for files to load? How easily could you navigate a site? Was important information presented on the first screen? Was information structured so that it was easily viewed on your computer screen? Were graphics used effectively? Was sound or animation used effectively? List the positive and negative features of each site. Next identify which sites you would recommend to a friend and which you would not. Do the sites you would recommend share particular features? What advice would you offer to improve the sites you rejected?

7. Learning about Referencing Conventions

Find a journal article in your field that contains a reference list and uses citations in the body of the article. Also find an article with a reference list from the humanities (English, History, Geography, etc.). or the social sciences (Education, Sociology, Psychology, etc.). Compare the details of the referencing conventions in your field with those of the other discipline. What differences and similarities do you find?

8. Changing Referencing Conventions

Find a journal article that uses numbers for citations. (Any of the dozens of IEEE journals in your university library will serve the purpose.) On a copy of the article, note the changes required to produce author-date citations. Then rewrite the reference list as an alphabetical list that follows the conventions provided in the reference guide on pages 280–286.

9. Examining Format in Detail

Obtain a sample of a well-formatted report from an instructor or a supervisor. Analyze the report using the heuristic provided on pages 287–288 of this chapter. What sorts of things do you note? What sort of improvements might you make? If the sample requires little if any improvement, try to model your next report on it.

Chapter **8**

Document Strategies
and Sample Documents

8.1 SUMMARY

One of the most important writing tasks you will undertake as an engineering student is mastering the various genres, or types of documents, that typify your particular field of engineering. This mastery communicates to your readers that you not only understand how to communicate like an engineer, but more importantly, that you also understand how to think like an engineer. Because requirements and conventions vary among fields and from one rhetorical situation to another, we cannot provide templates for you to copy. Instead, we discuss the common features and conventions of significant genres. But note that being able to follow conventions is only part of the equation. You must also respond to specific rhetorical situations.

As we have repeatedly pointed out throughout this book, communicating effectively is not a matter of filling in the blanks in a form. Effective documents result from your having a clear sense of purpose and from attending not just to the content of your message, but also to the needs and concerns of your readers, to readers' reasons for attending to your message, and to the contexts in which your message is received. Responding to these rhetorical concerns allows you to shape documents that are both appropriate to your discipline *and* unique responses to one-of-a-kind situations.

This chapter contains a number of sample documents that we use to point to the common features of a genre and to remind you of ways to tailor your communication to address a specific rhetorical situation. We focus on genres common to all fields of engineering that you must master for school and for work: e-mail, memos, letters, résumés, formal technical reports, and formal proposals.

1. **Introduction to Genre**

   This section introduces the concept of genre and outlines why mastering the genres of any given discipline is critical to your success. The degree to which you understand the documentation practices of your field of engineering will be a major determining factor in your success as an engineer. Further, because the forms that typify the genres of engineering are continually evolving, you must also be able to apply rhetorical principles to develop and to master new approaches.

2. E-mail and Netiquette

E-mail has largely replaced the traditional memos that used to circulate in the office environment. While this change has many benefits insofar as it allows us to send and receive information in a timely manner, it also has some drawbacks in that e-mails are too often sent without adequate consideration of audience, purpose, and tone. We provide general principles that you should follow in order to ensure that your e-mails have the intended effect.

3. Memos and Letters

Memos and letters are written less frequently in the era of e-mail, but they remain important genres for informal reports and proposals that are too long to be included in an e-mail message. We provide examples and discuss how to use rhetorical strategies to create a memo or letter when no samples are available. Later sections also provide examples of letters of transmittal and cover letters for job applications.

4. Résumés

Producing a good résumé is critical to your success in obtaining employment. We discuss the importance of organization, format, and editing, and discuss and provide examples of standard paper-copy résumés, scannable résumés, and cover letters addressed to specific employers.

5. Formal Technical Reports

Because the emphasis of technical reports written both at university and in industry are usually informative, this section addresses general issues relating to informative reports. We also provide an example of and commentary on an award-winning technical report, which includes an abstract and letter of transmittal.

6. Proposals

Throughout your career, you can expect to write numerous proposals in which your central concern will be to persuade your readers to adopt your recommendations or to follow a particular course of action. This section discusses proposals in general as well as a specific example written by a student project team. We also include a sample executive summary written by a working engineer.

Unlike the other chapters, we do not end this one—and the book—with a checklist, heuristic, and exercises. Instead, we encourage you to apply the rhetorical strategies provided throughout this book to your own writing and to use the heuristics in the previous chapters to analyze samples of the kinds of documents typical of your field of engineering.

8.2   INTRODUCTION TO GENRE

Genre (sounds like *John-ra*) may be defined as the rhetorical forms, styles, and types of content that characterize documents written for a particular discipline and communicative situation. For example, the essay is a very common genre used to analyze literature within the discipline of English while technical reports are much more common within the discipline of engineering as a means of extending technical knowledge. We can use these two types of documents to demonstrate differences among genres.

Note that essays and technical reports differ in terms of format. In essays, for example, we expect title pages but not the tables of contents, lists of figures, or appendices common in technical reports. Nor do we expect to find the reliance on headings and subheadings so characteristic of technical reports. Even similar sections may have different names so that while we expect to find a *Bibliography* at the end of an essay, we look for *References* in a technical report.

Essays and technical reports also differ in terms of style. A technical report may use dropped lists including many sentence fragments, contain relatively short sentences and paragraphs, and make little, if any, use of metaphors and similes. An essay, on the other hand, is unlikely to contain dropped lists and is more likely to contain more complex sentence structures and longer paragraphs and to make more frequent use of similes and metaphors.

The content of these genres also differs. For example, communicating personal anecdotes and using humor are rare in technical reports while including this kind of content is much easier in the essay genre. Explaining ideas through the use of figures, equations, and tables is a characteristic of technical reports not shared by essays. Further, although essays may draw or present conclusions, rarely do they provide a list of recommendations, which is a common feature of some kinds of reports.

In other words, the genre of the essay, which typifies writing in the humanities, and the genre of the technical report, which typifies writing engineering and the sciences, differ significantly. We would not expect to submit an essay in most engineering contexts any more than we would expect to submit a technical report within the humanities.

To a significant degree, your mastery of the forms, styles, and content typical of the documents common to a particular discipline represents your membership within that community. If you fail to master the genres that are typical of engineering, you may mark yourself as an outsider to the discipline and impact your future success.

8.2.1   Why Genre Matters

To illustrate how much genre matters, we provide the following example. When one of the authors of this textbook returned to university to obtain his Master's degree, he shifted from the discipline of English to that of education. While in English, he had mastered the essay form required for that discipline. Although the first few papers submitted in education received reasonable grades, he was surprised to note that most of the comments from the instructors dealt less with issues of content and more with issues of form and convention.

For example, the referencing conventions with which he was most familiar when in English were those of the Modern Language Association (MLA). So he used those conventions in the first papers he wrote when in education. But the conventions that were preferred within education were those of the American Psychological Association (APA). In order to deal with the comments expressed by the instructors, he learned to use the APA conventions (and his grades on the papers rose by 5 to 10 percent).

In this case, mastering the different conventions simply required obtaining a handbook put out by the APA and applying the conventions. But he also had to learn new ways of organizing information and of expressing ideas. Perhaps most importantly, he learned that conventions, organization, style, and even ways of thinking differ from English to education.

Why, you might ask, did the education instructors worry about something that might seem as relatively insignificant as referencing conventions? In part, by pointing out the differences, the instructors were trying to indicate that this writer was an outsider to the discipline of education. By indicating that certain changes were needed, the instructors were helping him to adapt to a new community.

Your engineering professors play a similar role in your education, introducing you to the genres of engineering. In the workplace, you must also learn to adapt to the document styles of particular organizations. If you are fortunate, when you move from school to industry someone will help you learn the genres of your workplace.

8.2.2 Learning New Genres

Whenever you change writing environments, you need time to master new genres. As engineering students and as practicing engineers, you must learn the ways of thinking and of organizing, expressing, and presenting information that signal membership in your particular engineering field. You must, therefore, become aware of the documentation practices and conventions typical to your field both in college and in the workplace.

In part, you can achieve this goal by reading samples of the various documents that you will be asked to produce. In industry, for example, you might ask a colleague or supervisor to provide you with sample documents. They may also provide you with templates commonly used when producing documents using word processors. In college, you can ask your instructors for any guidelines and/or samples they may have. You may also find it helpful to read some of the major journals common to your field. Your familiarity with how references, equations, figures, tables, and abstracts are handled in these professional journals, and your ability to emulate those conventions, help establish your membership in the professional community of engineers.

Keep in mind that within any field, the documents you may be required to write can vary greatly. As engineers you can expect to write short (1 to 4 pages) progress reports, site visit reports, application notes, and incident reports, medium-length (5 to 20 page) proposals, laboratory reports, case studies, and investigative reports, and lengthy business plans, functional and design specifications, and user's and technical manuals. Moreover, these documents may vary in terms of how formal they are, ranging from informal recommendations or observations provided in e-mail, memo, or letter formats to highly formalized specifications documents that may make great use of technical appendices and undergo a series of systematic revisions as a project progresses.

Beyond variations in length and formality, engineering documents also vary in terms of whether they are primarily informative or persuasive. For example, a laboratory report written primarily for a college instructor is generally more informative than persuasive. On the other hand, a proposal prepared for a client will emphasize persuasion.

Documents may also vary in terms of the medium employed. Consider, for example, the recent trend to supply user's manuals as on-line help files. A text-based user's manual and an on-line help file differ in terms of format, organization, and (rather unfortunately, we believe) the depth of explanation provided.

To further complicate matters, genres are not static; they evolve through time. And the computer revolution is accelerating the rate of change. Consequently, you must be aware of the potential for change. You must be flexible and accommodate changing norms and evolving practices.

As we pointed out in Chapter 6 (see pages 199–202), in the past, many scientific and engineering journals relied almost exclusively on the passive voice and third person (*A measurement was made*) as a means of appearing objective. Increasingly, journals accept, and even expect, the active voice and first person (*I/We measured*). Industry seems a bit slower to adopt this change, but active voice and first person are being used more frequently.

As we explained in Chapter 7 (see pages 249–250), format conventions are also changing. For example, the rule for placing the heading either above or below a table is evolving, and résumés are increasingly submitted on-line through the Internet rather than being printed and mailed to prospective employers.

In order to develop the flexibility needed to adapt to changing genres, we suggest that you familiarize yourself with some general principles of style and format. Your familiarity with the various stylistic and formatting options discussed in Chapters 6 and 7 will help you analyze the specific conventions currently being followed in college and industry. Experience with this sort of analysis should then help you shift from one convention to another and propose alternatives to conventions as the need arises. Given that formats and styles vary significantly even within the engineering profession and that the forms of documents are continually evolving, learning how to analyze genres is highly advisable.

8.2.3   Limitations of Templates

Many technical writers are most comfortable if someone gives them a template to follow, but keep in mind that generic document templates have limited applicability, may soon be outdated, and cannot account for the constraints of specific rhetorical situations. To help you develop the ability to analyze documents and create your own evolving models, we offer the heuristics at the ends of the chapters on style and format (see pages 240–241 and 287–288). These heuristics provide a series of questions that will direct your attention to the sorts of issues to consider when analyzing reports.

In this chapter, we discuss selected genres and provide annotated examples of e-mail messages, informal memo and letter reports and proposals, paper copy and scannable résumés with an accompanying cover letter, a formal technical report with a letter of transmittal and an abstract, and an executive summary and formal proposal. We offer both general advice for these genres and specific advice for adapting them to particular rhetorical situations.

We do not offer templates, so please do not attempt to copy these examples. Rather, consider how each example relates to its specific purpose, readers, and situation. To use these examples as models, you must compare their rhetorical situation to your own. To what degree is an example suited to your purposes? How would you change the organization, format, style, and content to address your purpose, your readers' needs and expectations, and your context?

Some points cannot be repeated too often: every document you write must reflect your specific rhetorical situation. It must address your precise purpose, accommodate the concerns, needs and fears of your actual readers, and respond to the constraints and opportunities of your particular context.

The discussion and commentary in the following sections relies on these basic principles and on various rhetorical strategies discussed in previous chapters. Our goal is twofold. On the one hand, we want to help you to master the genres of engineering and to present yourself through your writing as a *bona fide* member of the community of engineers. On the other hand, we want you to master adapting a genre to the constraints of a particular situation and to be able to create your own genre when no conventional one suits your needs.

8.3   E-MAIL AND NETIQUETTE

Much of the documentation that you write as an engineer could well be comprised of electronic mail (generally abbreviated *e-mail*). To a significant degree, e-mail has supplanted formal memos, telephone conversations, even face-to-face meetings, and it is rapidly becoming the most common way of communicating in business and industry. While e-mail is an efficient way to communicate, it is also subject to limitations. In particular, you must recognize when e-mail is appropriate and when it is not. For example, when writing about emotionally charged subjects (or when the person you are writing to may become angry, defensive, or otherwise upset about a subject), or when you are angry, a face-to-face conversation is generally more appropriate than e-mail. Similarly, discussing confidential matters or very complex issues is often better done in other ways than over e-mail.

In general, e-mail messages are equivalent to brief informal memos used to communicate information or to ask questions. In particular, e-mail messages should be *brief*. Ideally, keep your e-mail messages under 200 words. In some circumstances, such as writing an e-mail report, longer messages will be required. However, in most cases, short messages are likely to get your point across more clearly and are also more likely to get answered. Few people are interested in reading messages much longer than what fits on their computer screens. In fact, some people do not read lengthy messages or stop reading after the first few hundred words.

You should also exercise caution in terms of your tone. For example, a message from a junior employee that addresses the president of a major company in a familiar manner or a formal tone in an e-mail note to a colleague would be inappropriate. In other words, you should just as carefully consider your audience and purpose when writing e-mail as when writing formal reports.

Also, just because the medium is electronic, do not assume the messages you send are short lived. Many people archive their e-mail and system administrators can retrieve long-deleted messages. An ill-conceived e-mail could come back to haunt you in the future.

Similarly, poor spelling and grammar in e-mail messages could lead some readers to question your competence. Although problems with spelling and grammar are generally ignored in such forums as the various *Internet Newsgroups* (indeed, commenting on these sorts of issues is generally considered bad Internet manners), they are most often frowned on in school and industry. Sending an e-mail memo filled with spelling errors to an instructor, client, or supervisor is ill advised. Always take the time to ask yourself who will be reading your message and what effect tone, style, grammar, and spelling may have.

As well as the need to proofread and check the spelling in e-mail messages, keep the following points in mind:

- Identify yourself and your topic.
- Keep messages short and use a format and style that promotes readability.
- Ensure your messages are sent to the intended recipients.
- Do not use e-mail to send confidential information or personal information you would not want others reading.
- Always be polite and never send a message written while you were angry without first calming down and rereading it for tone and rhetorical effectiveness.

8.3.1 Identify Yourself and Your Topic

Set up your e-mail system so that your messages arrive at their destinations with the appropriate information appended and ensure that each message you send provides sufficient context for the reader.

- Where possible, identify yourself on the *From*: line using your full name rather than just your e-mail address. For example, use *"John Smith" js@imaginary.edu* rather than just *js@imaginary.edu*. Recipients are more likely to respond to you if they can easily identify who you are. In addition, knowing whom a message is from helps the recipient put the message in context.
- Include a subject line in your message and ensure that you use a meaningful phrase for the subject. A note without a subject or a note that has a general subject such as *Help* is less likely to get a response than one that is more specific: *How do I use Word 9.0 Master Documents?*

- At the end of your messages, include an alternative way to be contacted (i.e., phone number, FAX, postal address) along with your name. You can provide this information in a signature field that can be turned off for more personal e-mails. Providing contact information is especially important if you are asking for an answer that is likely to be quite complex. Often less time is required to explain something complex over the phone or in person than to type out the message.

8.3.2 Format and Style

Take care that the style and format of your e-mails promote readability.

- Avoid typing messages entirely in uppercase. As mentioned in Chapter 7, uppercase is more difficult to read than lowercase. And in some contexts (i.e., Internet Newsgroups), uppercase messages are considered the equivalent of shouting. Of course, capitalizing a few words is a good way to draw attention to an important point.

- Keep the format simple, using tabs for indentation and the asterisk for bullets. Note that only letters, numbers, and standard keyboard symbols are likely to appear as you intend them on the viewer's computer screen.

- Set your lines to a standard length. Few things are more frustrating than trying to read a message that seems to run off the right side of the screen for 23 feet. A good limit is 60–70 characters per line.

- Include sufficient material from the message to which you are replying so the reader is reminded of the context for your reply. Indicate the original text by starting lines of quoted text with a symbol such as > (most e-mail programs use this default). This practice helps readers easily distinguish between their own words and the new material you have added. But include only as much of the original message as needed to establish the context, and add your comments as early in the message as possible. Think about how frustrated a reader might feel after scrolling through a long message just to discover the words "I agree" added to the second-last paragraph.

8.3.3 Length and Attachments

E-mail messages should be as brief as possible. Because of the volume of e-mail many people receive everyday and the extra eyestrain involved in reading off a computer screen, you should assume most of your readers will not appreciate receiving long e-mail messages. Also, the printed format of e-mail messages is not particularly reader friendly. Consequently, documents of more than a few paragraphs that will be printed and may be read in hard copy should be word processed and sent as attachments. The basic points to remember relating to the length of e-mail messages and the use of attachments follows:

- If an e-mail message is of necessity more than a screen long, consider using headings to break up the text.

- Wherever possible, information that requires more than a few short paragraphs to express should be word processed and sent as an attachment—perhaps as a memo report or proposal. (We discuss these genres in the next section, pages 302–304.) The e-mail message will then serve to introduce the document and remind the reader of its origins or purpose, thus performing the same function as letters of transmittal, which are discussed later in this chapter (see pages 318–320).

8.3.4   E-mail Security

Beware of the ease with which e-mail messages can be sent to unintended readers and of the possibility that old e-mail messages will come back to haunt the sender.

- Ensure that personal messages are sent to the correct individual and not to a mailing list. Many e-mail users have been embarrassed by accidentally sending personal or confidential messages to an entire mail list or organization. Before you send an e-mail message, one of the last things you should do is check that you are sending it only to the intended recipient(s).

- Do not send confidential or personal material via e-mail unless you encrypt it because most e-mail systems are insecure. We suggest that you should treat any e-mail messages you send as a public document simply because e-mail servers typically maintain copies of e-mail and many people archive the e-mail they receive. An e-mail message sent several years ago could potentially turn up when you least expect it.

8.3.5   E-mail Tone

Take time to consider the tone of your message and how readers will likely respond.

- Do not publicly *flame* others. You will almost certainly regret sending a message that you wrote when you were angry. Before sending *any* message, ask yourself whether it is appropriate and consider the possible conscquences of the message. If you are angry, save the message and go for a walk or take a coffee break. After you have cooled down, reconsider whether you really want to send the message. Also consider asking a colleague to read the e-mail for his or her reaction to the tone of the message.

- Never send material with any potential to be offensive (i.e., racist or sexist jokes and comments). Remember that people's tastes and sensibilities vary widely. That joke you initially thought was so funny will definitely not seem very funny if you find yourself accused of harassment.

- Be polite. Although this point may seem simple common sense, people often pay little attention to manners when writing e-mail. If you are asking for something, say *please*. Many people are less likely to respond to a message that sounds like a demand than to one phrased as a request. And if you receive help, say *thank you*. People are more likely to offer future help if they feel they have been acknowledged for past help.

- Where appropriate, you can use *smiley faces* (sometimes called *emoticons*) to indicate the nuances of your message. Smiley faces are a series of ASCII characters that look like faces turned on their side. Use them sparingly because overuse destroys their effect. But used effectively, smileys can help to clarify the tone of your message. Of the dozens of different smiley faces, the following are the four most common ones:

    :-) smiling face that indicates amusement;
    :-(unhappy face that indicates disappointment or sadness;
    ;-) winking face that indicates irony or humor;
    ;-> mischievous face that indicates a comment is provocative.

    If interested, you can search the Web for a more complete list.

8.3.6   Examples of E-mail

The following is a fictitious example of an informal e-mail. Note that the e-mail responds to a query from a supervisor about the status of an investigation into a problem with a pollution control system at a power plant. Also note that the supervisor and engineer are on a first-name basis and seem to have a relatively friendly relationship (enough to play minor pranks on each other). The lettered circles on the e-mail draw your attention to a range of points discussed below.

```
Date: Friday, 07 April 2000 16:53:18
To: "Mike Markelson" <MARKELSO@selene.engr.cor.com>
From: "John Dunstone" <DUNSTONE@selene.engr.cor.com>        Ⓐ
Subject: Site Visit at KPL, Rockcliffe Installation

>John,        Ⓑ
>Once you get back from Rockcliffe, please send me an e-mail
>letting me knoew the status of the pump. Thanks. By the way,
>enjoy the motel Fred recommended? <grin> I don't expect you in
>tommorow cuz I figure you'll want to sleep in. Mike

Hi Mike,

I just got back from Rockcliffe and thought I'd send you this before head-
ing off home. As Joe, the maintenance supervisor, said in his phone call
there's a significant vibration problem with the X2-8000 pump running the
main scrubber.

This is the second pump that's developed this problem in the past 8
months. Joe suspects that the main bearings in the pump are not up to spec
and the manufacturer needs to use a better alloy. I'm not so sure that's
the case though.

I talked to the millwright who tore down the last pump, and she suggested
that the wear patterns on the bearings indicated that the main shaft is
out of balance. By the way, thanks for suggesting that I talk to her.        Ⓔ
Eileen kept the bearings from the last pump, and as she noted the wear
patterns are rather suspicious. Apparently, she's been telling this to Joe
for the last couple of months, but he didn't seem too receptive -- trying
to save a few bucks, I guess.

So I talked to Joe again, and suggested that the best way to resolve the
issue would be for them to ship the pump to us after they've replaced it
so we can do a more thorough analysis. He wasn't too happy with taking the
scrubber off-line to do the replacement (mainly because of the red tape
with the EPA), but finally agreed to do it.

They're going to ship the pump to us in about 3 weeks and would like a re-
port back a week after that if possible. I looked at our schedule, and I
think we can fit it in. If, in fact, the main shaft is improperly balanced,
it will need to be redesigned and we should be able to submit a competi-
tive bid on that. If it's just the bearings, we'll send the pump back rec-
ommending they go after the manufacturer. But I don't think that's the
problem, so maybe we should recommend KPL consider replacing Joe with
Eileen. ;-> She sure seems to know a lot more about the pumps than him.

Now on that other matter, the motel recommended by Fred was a HORRIBLE
DIVE. I swear he could sleep on a mattress filled with rocks. I told him        Ⓖ
that and he just grinned. You put him up to that, didn't you? <GRRR>

Regards,        Ⓗ

John D.
```

The lettered circles Ⓒ, Ⓓ, Ⓕ appear in the left margin alongside the respective portions of the e-mail text.

A Note that full names are listed on both the *From*: and *To*: lines. This practice allows the recipient to quickly identify who sent the message and also enables the postmaster to redirect the message should it go astray. Also note that the *Subject*: line provides a full description of the topic (rather than something more general such as *Site Visit*).

B The salutations on the e-mail (*John* and *Hi Mike*) indicate that the two people are on a first-name basis. Of course, this salutation would vary with context; in a more formal situation, an e-mail message might start with something like *Dear Mr. Johnson*.

C Note that some material is quoted from an earlier e-mail, allowing the reader to immediately situate the information in a context. The quoted material is also set off from the new material with the symbol > at the left-hand margin.

D The e-mail is generally informative, short, and well organized. If more information were required, the sender would probably write a site visit report.

E Note the politeness of "*By the way, thanks for. . . .*" Although being polite seems an obvious strategy, all too often, people neglect to attend to this issue.

F The use of the mischievous face (;->) indicates that the preceding comment is intended to be provocative.

G Note the change in both the tone and the subject in the final paragraph. The use of uppercase, HORRIBLE DIVE, emphasizes just how terrible the accommodation really was. Similarly, the use of a growl, <GRRR>, indicates a certain amount of anger (although given the tenor of the rest of the message, we can probably assume that the author of the e-mail accepted the prank with a fair degree of good humor).

H Note that the signature is simply the first name and the first initial of the last name. Because this e-mail is internal to the company, this abbreviated signature is sufficient as the recipient can easily contact the sender. But if the sender were to send an e-mail message outside the company, he would almost certainly use a more formal closing that provides alternative contact information:

Regards,

John Dunstone, Senior Engineer
Selene Engineering Corporation
123 Moon Drive
New Denver, Colorado 61811-1125
Phone: (123) 456-7890
FAX: (123) 456-0987

The following example includes a series of three messages. In this scenario, the Chair of a department asks the head of a departmental committee to explain an unusual circumstance. Note that the original message was copied to an e-mail list that includes all the members of the committee (ece-ucc@ctu.edu). Any reply meant for the entire committee cannot be addressed using the *Reply* function (that would send the message to swu@ctu.edu, but no one else). To ensure that all intended recipients receive a message, it must be addressed to the original mail list, which can be accomplished by using the *Reply All* function. You must also use *Reply All* when a message is addressed to several individuals and you want your reply to be sent to everyone who received the original message.

A The informality of this e-mail reflects the context in which it is written. It occurs in a university department where everyone knows each other well and where faculty are elected by their peers to serve five-year terms as department Chair. Any member of the Undergraduate Curriculum Committee could be a previous Chair of the department. Consequently, the sender chooses to omit a formal signature line. In this context, the formality of a signature that proclaims Dr. J.B. French as Chair of the Department of Electrical and Computer Engineering at Chicago Technical University is unnecessary and perhaps even inappropriate if maintaining collegiality is an issue.

```
Date: Friday, 15 September 2000 14:22:39
To: swu@ctu.edu
CC: ece-ucc@ctu.edu                        Ⓐ
From: "J.B. French" <jbf@ctu.edu>
Subject: Freshman Performance

Sharon,

I note that the number of students achieving grade point averages above
3.0 was significantly lower last year than in previous years. Can your
committee offer any explanations for this drop in performance?

JB
```

```
Date: Friday, 15 September 2000 15:10:42
To: "J.B. French" <jbf@ctu.edu>           Ⓑ
CC: ece-ucc@ctu.edu
From: "Sharon Wu" <swu@ctu.edu>
Subject: Re: Freshman Performance

The committee also noted a general drop in grades for the 1999/2000   Ⓒ
freshman year, specifically in the spring term, and has investigated the
situation. While the class average was about what we'd expect in the first
term, it was significantly lower than average in the second term. At the
same time, grades for the freshman design course were the highest ever.
Frank Simpson noted that in all the years he has taught the course, he was
most impressed with this group. He described the projects as the most
ambitious he has encountered.

The committee suspects that students became so involved in the design
course that other courses suffered. The students we talked to confirmed
that they spent the majority of their time last term on this course. Frank
assures me that special care will be taken in the future to ensure that
projects are not overly ambitious and students will be cautioned of the
need to give sufficient time to their other courses.

I would like to raise this general issue of time spent on project courses
at the next department meeting. Is that possible?

Regards,           Ⓔ
Sharon
```

Ⓑ   Note that the response is cc'd to the whole committee and not just sent to J.B. French.

Ⓒ   Note that the first sentence provides sufficient context to remind the sender of his request. Consequently none of the first message is repeated in the response.

Ⓓ   The message is divided into paragraphs to promote readability.

Ⓔ   While this e-mail contains no salutation to correspond to "Sharon" in the original message, the addition of "Regards," to the first-name signature maintains a sense of informal collegiality.

Ⓕ   As the third message indicates, academics routinely send colleagues drafts of articles they are preparing for publication for peer review and also share information with fellow experts and across disciplines. Because these documents are most often sent electronically, copies can be shared with much greater ease than when paper copies were exchanged. Before you pass on a draft copy of someone else's work, obtain the author's permission. They may have a revised document that they would rather share or they may have reasons for not wanting others to read their work.

```
Date: Monday, 18 September 2000 09:45:16
To: "Sharon Wu" <swu@ctu.edu>
CC: ece-ucc@ctu.edu
From: jbf@ctu.edu
Subject: Re: Freshman Performance

Attachment: Draft.doc (18.2KB)        F

Thank you to the Undergraduate Curriculum Committee. I'm pleased to see
your committee is on the ball. Your explanation is certainly plausible and   G
reminds me of an article a friend sent me a couple of years ago. I'm
attaching a copy for your information.

Sharon, how much time do you want for the discussion of time spent on
project courses? I'll give you the last 15 minutes of the next meeting if   H
that's sufficient.

Thanks again,    I
JB
```

G Note that once again none of the previous message is included in the response. This time, only someone who had read the previous messages would understand the context. If this message were part of an exchange that would be archived and referred to some time later (perhaps by a different department Chair and different UCC Chair), then the writer would be well advised to append at least part of the previous message. In this case, the context is sufficient.

H Note that while the message begins by thanking the entire committee, the second paragraph is clearly addressed to the Chair of the committee, providing an answer to the question she asked at the end of the previous e-mail. If the answer to her question were long and of no particular interest to the others receiving the e-mail, then a separate response would be appropriate.

I Being polite and saying "thank you" are not just good e-mail etiquette, but also ways to promote good working relationships.

8.4   MEMOS AND LETTERS

Memos are internal documents that remain within an organization. They are used largely for short reports and proposals and in three specific circumstances: *congratulatory* memos recognizing outstanding work, *disciplinary* memos documenting problems with performance, and *documentary* memos used to permanently record important information in a paper format. For these three occasions, printed memos are typically placed in personnel or project files. With the exception of reports and proposals, most other routine written office communication occurs via e-mail.

Letters are external documents that can be used for any relatively short correspondence with someone outside your organization. Consequently, whether a report is formatted as a memo or a letter depends on whether the document is internal or addressed to someone outside your organization, such as a client. The only essential differences in format are the obvious ones at the beginning and the end, as demonstrated in the examples in the following section. Letter formats vary, most often in terms of whether address, date, and signature are offset to the right-hand side of the page or at the left-hand margin, whether a subject line is included, and whether they will be printed on company letterhead.

Both memos and letters may be a few lines or several pages long. They may also contain headings, dropped lists, figures, and all the other features found in the body sections of technical reports and proposals. While content and organization can vary widely depending on the purpose, audience, and context, a few general principles apply:

- Begin with an introduction that accomplishes the following:
  - Clarifies the purpose of your report or proposal and indicates the scope and/or contents of the document
  - Provides sufficient context so that readers understand why they should pay attention to this report or proposal
  - Provides sufficient background to ensure readers comprehend the rest of the document

- Include one or more body sections that accomplish the following:
  - Provides the information required or the facts necessary to support your case
  - Informs readers using language appropriate to their level of technical expertise and explains any technical terms readers may be unfamiliar with
  - Discusses the information or facts provided and, where appropriate, indicates what actions should be taken and explains why that is the case

- Include a conclusion that accomplishes one or more of the following measures:
  - Summarizes the main points
  - Lists recommendations
  - Makes a request

You may well find the above list self-evident. The real challenge is not in knowing the categories of content to include, but in deciding what content is appropriate to a particular audience and what approach to take in order to persuade your readers. We have discussed these issues at some length in previous chapters and at this point provide only a couple of short examples to demonstrate the format and basic features of memos and letters.

8.4.1   Example Memo

The following informal report is sent from an employee to a supervisor in order to inform management of a potential problem. While the report can be considered informative, it also has a persuasive edge.

Ⓐ This format is standard for the opening section of a memo.

Ⓑ Note that even though this memo is businesslike, it is also informal. The level of formality could vary depending on the company's management style and how well the individuals know each other.

Ⓒ This memo comes straight to the point, providing sufficient context and listing the writer's concerns.

Ⓓ Although this memo does not contain a heading for *Introduction* or *Conclusion*, it nevertheless falls into three distinct parts with the body section divided by headings. Note that each point is made as briefly as possible because the purpose of this memo is simply to alert management to a potential problem. Excessive detail would not help Sadie make her point.

Ⓔ The memo concludes with a restatement of the recommendation made in the introduction, followed by two possible courses of action. Because the memo is addressed to a superior, the writer does not presume what should happen next, but merely offers her assistance.

Ⓕ Memos are not always signed, but adding a signature or initials is sometimes useful and may be required for legal purposes.

TO: George Montgomery, Regional Manager

FROM: Sadie Foster, Engineering Department

DATE: March 29, 2000

SUBJECT: Location of Seymour Meadows Access Road

George, I heard from Peter Wong that TimCo favors locating the access road to the Seymour Meadows site on the west side of the lake because of the shorter haul. I am familiar with the terrain in that area and strongly recommend the original plan of constructing the road east of the lake. Doing so will result in lower construction and maintenance costs and perhaps even help protect the watershed.

Construction Costs

The terrain to the west of Seymour lake is much more extreme than the terrain to the east. Although less than half the distance, I estimate construction costs will be at least 50% higher. If we ran into conditions similar to those encountered on the Black Mountain road, you could double or triple that estimate.

Maintenance Costs

A road to the west of the lake would also be much more expensive to maintain due to rockslides and washouts. In wet years, TimCo's operation could be hampered by lengthy road closures. Such disruptions are much less likely if the road is located as originally planned.

Ecological Costs

A road to the west of Lake Seymour would also open up the Salmon River Watershed to recreational use. An environmental assessment would be required to determine the road's impact on the watershed.

Because we assumed from the start that the road would be located east of the lake, we have conducted no studies to confirm my estimate of costs, but we have sufficient experience working in similar terrain that I am confident in recommending the original route for the road.

If you require additional information or want me to attend a meeting with TimCo representatives, please let me know.

Sadie

8.4.2   Example Letter

The following letter responds to the memo in the previous example. In this case, a letter is appropriate because the project manager is writing to a client. Note that the letter merely relates concerns and puts no pressure on the client to accept Sadie's assessment. We can assume that George Montgomery believes his client will need some time to consider the matter and will respond more positively if the process begins with an informal, face-to-face meeting. Such decisions are context dependent. In another situation, a fully developed written explanation of the problem could be more appropriate.

A

April 4, 2000

Barbara Wright, Project Manager
TimCo Industries Inc.
456 Turing Lane
Bellevue, WA 98008
B

SUBJECT: Location of Access Road to Seymour Meadows Site   **C**

Barb
Dear Ms Wright: **D**

I have been informed that you have had preliminary discussions with Peter Wong concerning
changing the route of the Seymour Meadows access road from the east side to the west side of
Lake Seymour. Sadie Foster, one of the planners in our engineering department, has expressed   **E**
serious reservations about this change in plans. She estimates much higher construction and
maintenance costs and points to the probability of road closures due to rockslides and washouts.

I suggest that before any decision is made, you meet with Sadie to discuss her concerns. Since
we are already scheduled to meet at 11 am on April 15th, I suggest that you stay for lunch and   **F**
that Sadie join us.

Please contact my secretary at 908-7654 to confirm the lunch meeting and to let her know if
you are bringing anyone else to lunch. I look forward to both of these meetings.

Sincerely,

George Montgomery   **G**

George Montgomery
Regional Manager

A This letter has been formatted for printing on company letterhead. Because the company address is included in the letterhead, it is not repeated above the date.

B A letter should include the full name and position of the person it is addressed to as well as
their full address.

C Subject lines are useful in allowing a recipient to locate the purpose of a letter at a glance. If a
subject line is not included, then the writer must ensure that the first sentence of the letter clarifies the subject.

D The context for this letter requires a certain degree of formality, and so addressing the project manager as Ms. Wright is appropriate. However, to acknowledge that the writer is on friendly terms
with the recipient, he crosses out the formal salutation, and writes in her first name.

E Note that the letter comes straight to the point. Brevity is always a virtue when writing letters. Also note that only the first two of Sadie's reasons are related here. The environmental protection issue is less well defined and could raise objections, so it is not included in this preliminary correspondence.

F Once the problem is identified, a next step is proposed and the letter is concluded on the assumption that this arrangement will be accepted. In a different situation, the writer might have been less presumptuous. In fact, in another situation, the regional manager may have phoned the client rather than writing to her.

G Note that "Sincerely" is an appropriate closing for almost any situation and that authors' names, and usually their position, are printed below their signatures.

Other examples of letters appear in later sections of this chapter. See pages 316–318 for a sample job application cover letter and pages 318–320 for a sample letter of transmittal accompanying a formal report.

8.5   RÉSUMÉS AND COVER LETTERS

When you apply for a position, your résumé and cover letter make a crucial first impression on potential employers, in terms of both their appearance and content. Therefore, both must be well organized, well written, and professionally presented. They must also be completely free of errors because a single spelling mistake can affect your chances of being considered for an interview. For instance, you place your competence in doubt if your cover letter contains a spelling mistake or if your résumé says you are familiar with *DOS*, but you have written *Dos*.

The best résumés and cover letters are those that most effectively present the writer as a suitable candidate for a particular position. While you may have a draft cover letter prepared ahead of time, you must tailor a revised version to the specific company and position you are applying for. Whenever possible, you should also write or revise a résumé with a specific employer and position in mind. Remember that your primary purpose is to persuade readers that you have the skills, abilities, and personal qualities they seek. The more you know about the company your reader represents and the position to be filled, the more successfully you can present yourself as someone worth interviewing.

While all-purpose résumés are rarely as effective as ones written or revised with a specific job in mind, they are nevertheless sometimes necessary. You might, for instance, write such a résumé for a general mailing to companies that hire engineering students for summer jobs. For this type of résumé, you must work with a fairly broad, general sense of audience and purpose.

You should also be prepared to produce two different forms of résumés: *paper copy* and *scannable* résumés. Many employers who receive large numbers of résumés require applicants to provide them in a format that allows a computer program to scan for key words. We discuss these scannable résumés and cover letters later in this section. But first, we address issues relating to paper copy résumés.

8.5.1   Paper Copy Résumés

Because your page layout strongly influences a reader's initial impression of your work, you should pay particular attention to the visual impact of your résumés. You should also strive to keep your résumés relatively short. Of course, what is considered short varies depending on how much experience you have and the preferences of reviewers or companies. Besides striving for a relatively short, professional-looking document, when preparing a paper copy résumé, you should bear the following in mind.

- Use white space effectively by providing adequate margins (about one inch on the sides and one to one and one-half inches at the top and bottom), by leaving sufficient space between columns ($\frac{3}{4}$ to 1 inch), and by clearly separating one section from another. For résumés of more than one page, you should also ensure the first page is well balanced and substantial without appearing crowded. Remember that you want your printed résumés to appear organized and uncluttered.

- Use a font size and a font style that is easy to read. We recommend using a serifed font of at least 11 points for most of the text of the résumé. For the headings, you may choose to use a larger, sans-serif font. (See Chapter 7, pages 256–259 for a discussion of fonts.)

- Design your résumés with busy readers in mind, arranging information so that it is easy to find. For example, isolate dates by listing them down one side of the page rather than embedding them in descriptions of employment history, education, etc. Also, when you must begin a new page in the middle of a section, ensure the page break falls between items, such as between job descriptions in the employment section.

- Because many people may handle a résumé, we suggest you put your first initial and last name followed by the page number at the top right-hand corner of additional pages (i.e., *H. Brown, 2*).

- Use the best-quality paper possible for your résumé and cover letter. Always print originals on a laser printer so that the print is crisp, clean, and dark, and if you require photocopies, consider spending a few extra cents per copy to have them copied on good-quality paper on a well-serviced machine.

Finally, to ensure that your résumé is error free, edit it extremely carefully. Triple check everything, including the spelling of your name, your address and phone number, and the dates in your education and employment sections. When you are sure that your résumé is error free, ask at least one other person to read for errors and to comment on layout and content.

8.5.2 Examples of Paper Copy Résumés

The following fictitious résumés represent an engineering student with two years of engineering education. Appropriately, the various sections in each résumé are organized chronologically and highlight the skills an employer might be looking for in a student. We present a two-page and single-page version for comparison and to allow us to discuss formatting features of two-page résumés. In this particular scenario, the one-page version is likely more appropriate. An individual with more experience could require a second page.

Someone with more employment experience or who is seeking a higher-level position might also organize a résumé differently. For example, such an individual might include a statement of career goals or expertise at the beginning of her résumé followed by her employment history. Education and awards would accordingly be listed later in the résumé. Someone far enough along in his career may use an entirely different organizational strategy and arrange his résumé according to job functions, providing a section on various categories of experience such as management and international project experience. As always, context helps determine how to organize and present information in a document.

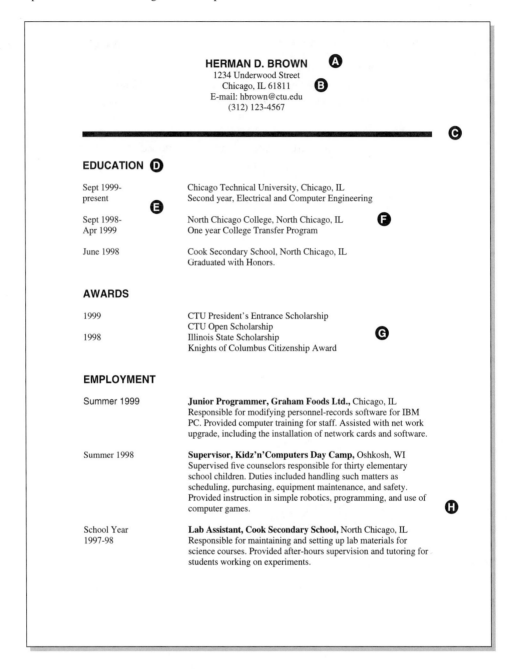

When you have read the following two-page résumé, take a few minutes to identify any information you would leave out in order to produce a one-page version.

A Note that in order to emphasize the person's name, it is presented in bold uppercase using a larger font than the text of the résumé.

B The contact information is centered and includes the mailing address, e-mail address, and phone number (including area code). The writer might also list a home page if one is set up. Alternative ways of presenting the name and address would be to line them up on the left side of the page rather than center them or to place the name on the left side of the page with the contact details flush right.

Ⓘ   H. Brown, 2

Summer 1997

Counselor, Kidz'n'Computers Day Camp, Oshkosh, WI
Provided instruction and leadership to small groups of elementary
school children. Taught simple electronic circuitry and helped
participants construct simple robots.   **Ⓙ**

SKILLS

Software

Ⓚ

- Competent in BASIC, TurboPascal, and C11.
- Familiar with UNIX, DOS, OS/2, and Windows operating
  environments.
- Experienced with Microsoft Word, Word Perfect, Excel,
  CorelDRAW!, and MathCAD.

Hardware

- Competent with basic lab equipment such as oscilloscopes,
  DMMs, power supplies, and function generators.   **Ⓛ**
- Familiar with multilayer PCB design and fabrication.
- Experienced with setting up and maintaining networked computer
  systems.

Other

- Excellent communication and teaching skills.
- Familiar with basic darkroom procedures
- Fluent in Spanish.
- Typing: 60 wpm.

VOLUNTEER AND EXTRA-CURRICULAR ACTIVITIES

1997	• Fund raiser, Big Brothers Recreation Fund
1995, 1996	• Metro Chicago Debate Competitions
	• President of Photography Club **Ⓜ**
1994-present	• Big Brothers recreation assistant
1993	• CTU Math Enrichment Program
1992-1995	• Junior Achievement Program

INTERESTS

Electronics:
- Built a number of devices including a radio security alarm from
  schematics.

Debate:
- Competed on team that placed 2nd in regional high school   **Ⓝ**
  championship.

Drawing & Design:
- Worked on school yearbook design and designed the North
  Chicago Junior football team logo.

Sports:
- Football, skiing, and swimming.

REFERENCES

Available upon request.   **Ⓞ**

Ⓒ   Note that a black line (generally called a *rule*) has been used to separate the name and address from the remainder of the information in the résumé. In some cases, lines are also used to separate the various sections. Although the occasional use of such lines can be an effective way to make your résumé visually appealing, when overused, a résumé appears cluttered.

Ⓓ   The section headings are set in a bold sans-serif font in order to distinguish them from the rest of the résumé. In lengthy résumés, the headings are sometimes numbered as in technical reports.

Ⓔ   The dates of various activities are often listed in a column on the left-hand side of the résumé. Note that, by convention, they are ordered in reverse chronological order, starting with the most recent activity and ending with the oldest activity. Also note that when months are listed, they are generally abbreviated into three- or four-letter contractions (*Jan, Feb, Mar, Apr, May, Jun* or *June, Jul* or *July, Aug, Sep* or *Sept, Oct, Nov, Dec*).

F For those without significant work experience, the *Education* section is generally presented first. However, a person with sufficient or impressive work experience may choose to put the employment section first. In this example, employment experience could have been placed before education. The choice of placement would depend on a number of factors, including relevance of work experience, assessed strength of experience compared with other applicants, qualifications highlighted in a job ad, and so on.

Because Herman has not yet finished his first degree, he may legitimately indicate high school graduation but should include nothing further back. Also note that he lists the college transfer program attended prior to starting university. The dates attended and details of all degrees or diplomas are listed along with the full name, city, and state for each institution. Additional information can be included under each education entry: relevant courses that are not part of a major, courses that indicate special skills or abilities, or brief descriptions of special accomplishments or awards. For example, Herman could have included the title of a special project followed by a brief description as part of an education entry.

G Note that in this example the awards are presented in a separate section in order to draw attention to them. This strategy is particularly useful if the person has many awards or some that are especially prestigious. An applicant with only one or two awards could choose to list them as part of the relevant entry in the *Education* section.

H In the *Employment* section, the job title and the employer's name and location are clearly listed. Note that the position and the company name are bolded in the example in order to emphasize them. A short description of the duties and responsibilities of the position is also included unless the duties are self-evident from the position (i.e., Cashier or Short Order Cook). Descriptions are generally written in an abbreviated form to eliminate "*I*" subjects. For example, "*I was responsible for . . .*" becomes "*Responsible for. . . .*"

Note that the titles of sections should be tailored to the individual situation. For example, a student without work experience could provide a section entitled *Work-related Experience* and list volunteer work, projects undertaken for courses, and the like. In other cases, the *Employment* section can be expanded to include relevant project courses. If you want to include such projects, change the heading *Employment* to *Employment & Related Experience* and include a description of the project and your role in it.

Also note that if you have spent more than five or six years in the work force, you may find it most appropriate to create a section for early, nontechnical work:

Other employment: Lifeguard (summers 1995–98)
Ski instructor (winter weekends 1994–97)
Short-order cook (part-time 1996)

I Including an abbreviated version of your name along with the page number helps ensure that your résumé does not get mixed up with other résumés should one of the pages get separated from the rest.

J Note that the *Employment* section in this example was too lengthy to list on one page, so one of the listings appears on the second page. The section is appropriately separated between entries rather than half-way through an item.

K In the *Skills* section, note that the various items have been organized into subsections. Depending on your background and the particular job being applied for, a variety of subtitles could be used.

L In this résumé, the various skills are presented in a bulleted list of related items. For example, under the *Software* subheading, the first bullet introduces a list of programming languages, the second bullet introduces a list of operating environments, and the third bullet introduces a list of applications. Also note that the *Skills* section could have been located earlier in the résumé. Someone with more impressive skills than work experience could locate this section following *Education* and *Awards*, for example.

Ⓜ At one time, employers required such personal details as age, height, and marital status, but such information is now considered discriminatory. Instead, sections such as *Interests* and *Volunteer and Extracurricular Activities* are used to provide information about citizenship, teamwork, and leadership activities and to demonstrate that an applicant is well rounded. In the example, the information provided about the person's fund-raising activities with Big Brothers and as president of the photography club lets prospective employers know that the applicant is active in the community and has leadership skills. Keep in mind that most employers are looking for more than just someone with good technical skills; they also seek people who show the potential for leadership within the company, an aptitude for teamwork, and the ability to relate to the larger community.

Ⓝ Note that you can use the *Interests* section not only to list interests, but also to provide examples of activities that demonstrate your interests. Doing so is highly recommended because demonstrating your interests is much more effective than simply listing them. This section provides a good opportunity for you to express your individuality and distinguish yourself from other applicants. For example, one of our brothers-in-law gained employment with a medical imaging company largely because he stood out in the interviewer's mind because of a shared interest in romance languages. This section is also useful if you have a noticeable gap in your education and employment record. You may, for example, account for the missing time by including the subheading *Travel* in this section.

Ⓞ At the end of a résumé, you usually indicate that references are available on request, or you list the names, titles, addresses, and telephone numbers of two to four people who have agreed to provide you with a reference. If you indicate that references are available on request and you are successful in obtaining an interview, ensure that you take a list of referees with you to the interview.

Never list anyone as a referee who has not already agreed to be one. Doing so is both unprofessional and risky. Imagine the impression you would make if an employer phoned up one of your Physics professors from a few years ago and the person did not remember you, or worse still, confused you with another student who never handed in assignments on time. You should always give a potential referee the opportunity to decline your request because most people will refuse to provide a reference if they cannot provide a good one. Potential referees may also want to spend time with you to refresh their memories prior to providing references.

One of the challenges of preparing a résumé is packing a great deal of information into relatively little space. You may frequently be limited to only a page. The following example suggests one way to condense Herman's two-page résumé to fit on a single page. Compare our solutions to those you anticipated.

Ⓐ The text font, which was 12 point in the two-page version of this résumé, is reduced to 11 point to allow for more information on the page.

Ⓑ Note that the right-hand column has been shifted further to the left and that all indentation is eliminated at the left-hand margin. Top, bottom, and side margins can also be narrowed slightly to help squeeze a few more characters on a line or an extra line or two on the page.

Ⓒ Note that Herman's one-page résumé eliminates all education and awards prior to entering CTU. The two remaining awards are impressive and suggest that this student received others before entering university.

Ⓓ The descriptions for each of the entries in the *Employment* section have been severely edited but still provide a good sense of the type of work involved. Details can be provided at an interview or, if relevant, added to the cover letter.

Ⓔ The *Skills* section eliminates such phrases as *familiar with* and *competent in* to save space and uses bolding to draw attention to specific skills. Note that typing is no longer listed in the *Other* category while leadership ability has been added.

HERMAN D. BROWN
1234 Underwood Street
Chicago, IL 61811
E-mail: hbrown@ctu.edu
(312) 123-4567

EDUCATION

Sept 1999-present

Chicago Technical University, Chicago, IL
Second year Electrical and Computer Engineering
Awarded CTU President's Entrance and CTU Open Scholarships.

EMPLOYMENT

Summer 1999

Junior Programmer, Graham Foods Ltd., Chicago, IL
Modified software, provided computer training for staff, and assisted with
upgrading network hardware and software.

Summer 1998

Supervisor, Kidz'n'Computers Day Camp, Oshkosh, WI
Supervised counselors and handled scheduling, purchasing, equipment
maintenance, and safety and taught robotics and programming.

1997-98

Lab Assistant, Cook Secondary School, North Chicago, IL
Set up labs for science courses and supervised and tutored students.

Summer 1997

Counselor, Kidz'n'Computers Day Camp, Oshkosh, WI
Provided leadership and taught children to build electronic circuits and robots.

SKILLS

Software

• **BASIC, TurboPascal, and C11.**
• **UNIX, DOS, OS/2, and Windows** operating environments.
• **Microsoft Word, Word Perfect, Excel, CorelDRAW!,** and MathcCAD.

Hardware

• Oscilloscopes, DMMs, power supplies, and function generators.
• **Multilayer PCB design and fabrication.**
• Set up and maintain **networked computer systems.**

Other

Excellent **communication skills, leadership ability,** and **fluent Spanish.**

ACTIVITIES AND INTERESTS

Electronics   Built a number of devices including a radio security alarm from schematics
Design   Designed North Chicago Junior football team logo and high school yearbook
Debate   Competed on team that placed 2nd in Metro Chicago Debate Competitions
Photography   President of CTU Photography Club and skilled in dark room procedures
Big Brothers   Fundraiser and Recreation Assistant since 1994
Sports   Football, skiing, and swimming

REFERENCES   Available upon request.

F *Activities and Interests* have been combined to save space. Note that the only information missing from the shorter résumé is the Junior Achievement Program and the Math Enrichment Program. If Herman had a strong interest in entrepreneurship, he could have added the heading *Entrepreneurship* to the Activities and Interests section and added a one-line description of his activities while in the Junior Achievement Program. The math program is too dated to be worth including. In fact, this item should also be omitted from the two-page version—a reminder that you should periodically review your resume to eliminate dated material as well as to add new items.

8.5.3 Scannable Résumés

Most large and many medium-sized companies use optical character recognition (OCR) programs to scan paper copy résumés and to store résumés sent via e-mail. Some companies also provide online résumé-building forms to assist applicants. Recruiters can then search the resulting résumé databases for applicants with specific kinds of experience, attributes, and skill sets.

The challenge when preparing scannable or electronic résumés is to create a document or file that is computer friendly. In this case, you need not be reader friendly (although expect a recruiter to look at your résumé if the computer identifies you as a potential candidate). Computer programs do not care what your résumé looks like, they simply find a recruiter's list of key words and identify those résumés that provide the best match. Consequently, to assure as many matches as possible between the words you use and those the recruiter inputs, you must also represent your qualifications for a job in the key words a recruiter is most likely to use.

When you prepare a scannable résumé, you must also use a format that ensures the OCR program will accurately represent the information on your paper copy. Many of the OCR programs currently in use cannot decipher the characters and features you would normally include in your paper copy résumés. You should, therefore, produce a format that is stripped to the bone. If you submit this type of résumé electronically, provide an ASCII or plain text file.

When you prepare a scannable or electronic résumé, you need not worry about length. Although you must be concise in presenting information, you must nevertheless list as many facts, skills, and attributes as possible. The more information and key words you provide, the greater the chance a computer program will match you with a position. Once that match is made, you begin to deal with people and a more reader-friendly paper copy of your résumé should be available, either before or at an interview.

While you are dealing with a computer program, use descriptive nouns and noun phrases wherever possible. For example, the program will search for nouns such as *surveyor, programmer*, and *manager* rather than for verbs such as surveyed, programmed, and managed. Also rely heavily on technical language and use the current industry buzz words and common technical acronyms. But avoid abbreviations, except for degrees (i.e., B.A.Sc., M.S. or Ph.D. are acceptable). If you aren't sure how common an acronym is, then provide both the full form and the acronym. (Remember, a machine is reading your résumé.)

To ensure you are using the same key words as recruiters are inputting, study job ads and job descriptions for the kind of position you are applying for and make lists of the words used that match your qualifications and abilities. Be sure to pay attention to nouns and noun phrases that could describe your personal traits and work ethic such as *skill in time management, dependable, leadership, sense of responsibility*, and the like. Include as many relevant words and phrases from your list as possible in your résumé.

Also include as many facts as possible. For example, rather than writing that you have experience managing a software-development group, write that you have *six years'* experience as manager of a software-development group. Provide similar technical detail. For example, if you use digital simulation tools, note the ones you use (i.e., *Cypress Semiconductor's WARP VHDL simulator* and *SimuCAD's SILOS III Verilog simulator*).

Other issues relating to content are as follows:

- Place your full name on a line by itself as the first item on each page.
- State your employment objective using key words that match a particular job description or that are commonly used for the type of position you are applying for.
- Provide a section for your relevant employment history or experience.
- List degrees and degree majors.
- Provide sections that list skills, qualifications, accomplishments, licenses, and certifications as well as personal information relating to your abilities and work ethic.

To ensure your résumé is correctly interpreted, you must also attend to the following formatting details:

- Put contact information on different lines and if you list more than one phone number, put each on a separate line. Also use the standard address format:
  Herman D. Brown
  1234 Underwood Street
  Chicago, IL 61811
- Left justify the entire document, avoid columns, and use at least one-inch margins.
- Use standard fonts and reduce formatting features to a minimum. Font size should be 10 to 12 point and the font should be a standard one, such as Courier, Arial, New Century Schoolbook, or Times New Roman (no less than 11 point). Avoid tabs, italics, underlining, brackets, bullets, dashes, horizontal or vertical lines, shading, symbols, and pictures.
- Use white space to indicate where one section ends and another begins. You may also use bolding and capitals for headings, but make sure letters do not touch each other.
- If the résumé is printed, use white paper and a laser printer to produce crisp, dark type on one side of the page only. Do not fold or staple. If the résumé is to be e-mailed, save the file in ASCII or plain text format.

The example in the following section presents a scannable version for the examples from the previous section.

8.5.4   Example Scannable Résumé

Compare the following example of a scannable résumé to the example of a one-page résumé in the previous section on paper copy résumés, noting the differences between a document written for a person to read and one written for a machine to scan.

A Note that the format is very simple with no indenting and with sections indicated by capital letters and bold and separated by a blank line. The font for this résumé is 12 point Arial.

B The name, address, and other contact information are on separate lines and the address is presented in a standard format. Note that if two phone numbers were included, each one would appear on a separate line.

C Because this résumé has been prepared for a job the applicant learned about at a job opportunities fair, it includes a very specific objective that repeats key words a recruiter used to describe the job.

D Note that the *Skills* section relies heavily on nouns to accommodate a key-word search (i.e., *fabrication* rather than *fabricated*), but also contains information more suited to a human reader, such as the last sentence describing specific design experience. Even though scannable résumés are produced to be read by computers, recruiters may also read them in order to decide whether those applicants the computer identifies should indeed be interviewed.

E To help ensure the computer scans your résumé accurately, use only left justification, keeping the right-hand margin ragged. Check for tabs or extra spaces at the ends of lines that could create problems for the computer program scanning your résumé.

F Listing personal strengths is important because a computer program cannot make judgments based on job descriptions or the overall impact of your skills, activities, and experience. In this case, some of the items listed in this section would also be useful additions to the paper copy résumé.

G The active verbs appropriate to paper copy résumés have been replaced with nouns to facilitate a key word search. But careful attention is still paid to style, grammar, and spelling so that a recruiter reading this document will be left with an appropriate impression of the applicant's written English.

Herman D. Brown
12234 Underwood Street
Chicago, IL 61811 **B**
312-123-4567
hbrown@ctu.edu

C **OBJECTIVE**
Seeking work term position that involves set up, upgrade, or redesign of computer networks.

D **SKILLS**
Software and programming with BASIC, TurboPascal, C11, UNIX, DOS, OS/2, Windows operating environments, Microsoft Word, Word Perfect, Excel, CorelDRAW!, and MathCAD.

Hardware experience with set up and maintenance of computer networks, upgrade of network hardware and software, electronic circuits, multilayer PCB design and fabrication, **E** oscilloscopes, DMMs, power supplies, and function generators. Built radio security alarm and constructed other electronic devices from schematics.

Other skills include fluent Spanish, drawing and design, darkroom procedures. Designed logo for North Chicago Junior football team and was design editor for high school year book.

F **PERSONAL STRENGTHS**
Communication skills including oral and written communication, organizational skills, leadership ability, self starter, enthusiastic, strong work ethic, detail minded, goal oriented, industrious, hard worker, team player, effective teacher and trainer, creative, committed, and active in team and individual sports.

G **WORK EXPERIENCE**
Junior programmer, Graham Foods Limited, Chicago, Illinois.
Duties included software modification, computer training for staff, helping upgrade hardware and software for the computer network.
May to August 1999.

Supervisor, Kidz'n'Computers Day Camp, Oshkosh, Wisconsin.
Supervisor for computer camp counselors. Duties included event and program scheduling, purchasing, equipment maintenance, safety, teaching robotics and programming, and introducing computer games to elementary school children.
June to August 1998.

Lab Assistant, Cook Secondary School, North Chicago, Illinois.
Two years experience with set up and maintenance of science laboratory. Independent supervision and tutoring for students working on science experiments.
1997 and 1998.

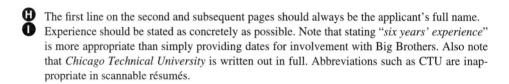

Herman D. Brown

Counselor, Kidz'n'Computers Day Camp, Oshkosh, Wisconsin.
Provided instruction and leadership to elementary school children. Taught electronic
circuitry and helped participants construct robots.
June to August 1997.

OTHER ACTIVITIES
Fund raiser and six years experience as recreation assistant for Big Brothers.
Current President of Photography Club at Chicago Technical University.
Play football, ski, and swim.

REFERENCES
Available upon request.

 The first line on the second and subsequent pages should always be the applicant's full name.

I Experience should be stated as concretely as possible. Note that stating *"six years' experience"* is more appropriate than simply providing dates for involvement with Big Brothers. Also note that *Chicago Technical University* is written out in full. Abbreviations such as CTU are inappropriate in scannable résumés.

8.5.5   Example Cover Letter

A paper copy of a résumé should be accompanied by a cover letter that identifies the position you are applying for. The letter may be only a few lines long, but can be up to a full page and is often used to explain why you should be considered for the position, to highlight specific aspects of your education, training, and experience, and to provide information not included in your résumé.

Word for word, a cover letter can be one of the most time-consuming kinds of writing you will undertake. In relatively few words (never more than a page), you must convince a prospective employer or recruiter that you are a person they should want to hire and must demonstrate your enthusiasm for the job and the organization that would hire you. Your content must interest readers; your style must be active and concise; your format must be professional; and your grammar and spelling must be without fault.

The example cover letter is one that might be written to accompany the one-page paper copy résumé for Herman Brown. In this case, Herman has been asked to apply for a specific position, and thus the letter is more detailed than might otherwise be the case.

A Note that you include your contact information at the top of your cover letter just as you would with a résumé. In addition, you include the date on which the letter was written. The example letter places this information on the right-hand side (*modified block format*) in order to create a more balanced look to the letter. Alternatively, you could use *full block format* in which the address, date, and signature are all placed flush left. See the letter of transmittal in the next section for an example of full block format.

B Include the recipient's full name, position, and address. If you do not know the person's full name or his or her position, you can generally phone the company and ask, or you can try finding it with a web search.

1234 Underwood Street
Chicago, IL 61811
E-mail: hbrown@ctu.edu
(312) 123-4567   **Ⓐ**

October 2, 1999

Joanne Greene
Director of Personnel   **Ⓑ**
SuperSoft Systems
4321 Adanac Street
Chicago, IL 61810

Dear Ms. Greene:

As you suggested during last week's Employment Opportunity Fair at Chicago Technical
University, I have enclosed a copy of my résumé for your consideration. I am especially
interested in the possibility that you mentioned of setting up the new computer network for   **Ⓒ**
SuperSoft. As you will note, I have prior experience assisting with setting up and maintaining a
computer network at Graham Foods Ltd.

During my time at Graham Foods, I was entirely responsible for modifying the company's
personnel-records software. Although that job was expected to take the entire Summer, I was
able to complete the task in only two months, and so I was asked to train the staff in the use of
the new system and to assist the systems administrator (Ms. Marsha Thurgood) with upgrading
the computer network. Because she went on holidays during the final month of the Summer, I
was provided the opportunity to complete the upgrade on my own.

I am particularly excited about the possibility of redesigning your computer network to eliminate
the various problems you have been encountering. As you mentioned, part of that job will
require that I integrate the new network with *NT 4.0* and *Microsoft Office 2000*. I feel my   **Ⓓ**
considerable experience with that operating environment, my familiarity with a range of
Microsoft products, and my prior training with networks provide me with the skills needed to
successfully complete the task.

I am a self-starter and would enjoy working independently in the informal and creative
environment of SuperSoft. The experience I would gain working with the programmers at
SuperSoft would prove of considerable benefit as I continue my education. Your programmers
would also gain from the implementation of a seamless networking environment.

Please refer to my résumé for additional details about my work experience and qualifications. I
am available for an interview at your convenience, and I can be contacted at 123-4567 in the   **Ⓔ**
evenings or via e-mail at hbrown@ctu.edu. I look forward to hearing from you. Thank you for
your consideration.

Sincerely,

Herman Brown   **Ⓕ**

Herman D. Brown

Enclosure: Résumé   **Ⓖ**

Ⓒ A job application letter may contain a subject line that identifies the position being applied for
(i.e., Subject: Junior Programmer position advertised in the *New York Times*, competition number 32-P55). This subject line generally appears before the salutation (Dear Somebody). If no
subject line is provided, the first paragraph should identify the position you are applying for
(including any competition number), explain how you learned about the opening, and state
why you want to be considered for it. You may also want to draw attention to a major qualification or highlight of your experience or education that makes you particularly well suited for
the position. Save details for the next section, however.

Ⓓ The central section generally includes one or more paragraphs that draw attention to and elaborate on particularly relevant aspects of your education, employment, or extra-curricular activities. But be careful not to turn the letter into a restatement of all the information in your résumé. Be selective and provide more detailed information than that given in the résumé. You can also use this central section to indicate your knowledge of and enthusiasm regarding the company in question.

Ⓔ The concluding paragraph should refer the reader to your résumé for additional information and politely, but confidently, request an interview at the reader's convenience. Make sure you include your phone number and e-mail address. Finally, do not forget to thank readers for their consideration.

Ⓕ Be sure that you sign the letter. Although a seemingly minor detail, employers receive many unsigned letters from potential applicants and often interpret this omission as a lack of attention to detail on the part of the applicant.

Ⓖ Include this enclosure line whenever you are sending something, such as a résumé or transcripts, along with a letter. This practice helps ensure that whoever is opening the mail does not inadvertently leave something inside the envelope and that whoever reads your application knows when something is missing.

8.6 FORMAL REPORTS AND LETTERS OF TRANSMITTAL

Reports are perhaps the most common documents that you will write both as engineering students and as engineers. Consequently, your success—both in school and in the workplace—will partly depend on your ability to produce effective reports. Unfortunately, learning to produce the various types of reports required is not always a simple matter. However, certain conventions provide guidelines for organization and content. The sections of a formal report are conventionally tenfold (in the following order):

1. Title Page
2. Copyright Page or Revision History Page
3. Abstract or Executive Summary
4. Acknowledgments
5. Table of Contents
6. List of Figures and Tables
7. Glossary (sometimes included with the appendices)
8. Body of the Document (typically divided into numbered sections)
9. References
10. Technical Appendices

Many reports are also accompanied by letters of transmittal, addressed to whomever requested the report. The purpose, format, and structure of these letters and of the typical sections of a report are discussed in some detail in the following section.

8.6.1 Example Report with Letter of Transmittal

We would like to express our thanks to Eric Hennessey, who has given us permission to use his report, *Seeing the Light: Multi-Wavelength Optical Networks*, as an example in this section and to adapt the report to meet the stylistic and format requirements of this textbook.[1] The central purpose for Eric's report is to provide information to his supervisors about the work he performed for the Optical Networks Division of Nortel Networks Corporation during cooperative education (co-op) employment in 2000. After completing

---

[1] We would like to thank Nortel Networks Ltd. for permission to reprint the following figures included in this report: Figures 1, 2, 4, and 5, copyright © Nortel Networks Ltd., 1998, all rights reserved; Figure 3, copyright © Nortel Networks Ltd., 2000, all rights reserved.

the co-op, Eric submitted another version of the report to the Simon Fraser University School of Engineering Science to fulfill the work-report requirement for co-op. This second version of the report provides more background explanation than the first version because it is written for an audience unfamiliar with his work.

In addition to being very well written, we have chosen to use his report for several reasons. First, the report provides a good example of the various format issues that must be considered when writing reports. Second, the report is quite interesting, examining as it does a contemporary engineering topic. We hope that as well as considering the style, format, and organizational issues in the report, you will also read the report for its content.

First, we turn your attention to the letter of transmittal.

12345 Genre Drive
Surrey, B.C. V3X 1E1
(604) 345-6789
henness@sfu.ca

December 1, 2000   **A**

Dr. John Jones, Director
School of Engineering Science
Simon Fraser University
Burnaby, B.C. V5A 1S6

Re: ENSC 196 Work Report, *Seeing the Light: Multi-Wavelength Optical Networks*

Dear Dr. Jones:

Please find attached the work report from my fall-semester work term with Nortel Networks in Ottawa, entitled *Seeing the Light: Multi-Wavelength Optical Networks*. Working in the optical link area of Nortel's optical networks division, I had the opportunity to observe and make allowances for the many physical effects manifest in the node-to-node transmission of multiple optical signals over a single fiber. My report details these experiences, discussing the primary effects, Nortel's correcting response to these non-idealities, and my specific role in ensuring consistent and reliable data transmission.

To demonstrate the consequences of these effects, I discuss the impact on system performance encountered with the mid-span addition of an optical add/drop multiplexer that extracts and introduces new signals entirely at the optical level. The use of such equipment destabilizes the system, potentially introducing unacceptable transmission errors. Sik Heng Foo and Mike Moyer both provided guidance for my work in this area.

My work on multi-wavelength optical transmission has perfectly supplemented my prior experience with optical networks at the optoelectronic circuit pack level and furthers my goal of a career in optics. If you have any questions regarding my report or my experiences, please contact me at the above phone number or e-mail address.

Sincerely

Eric Hennessey

Eric T. Hennessey

Enclosure: ENSC 196 Work Report, *Seeing the Light: Multi-Wavelength Optical Networks*

A As mentioned in the previous section, reports are typically accompanied by letters of transmittal addressed to whoever requested them. This letter should identify the report and provide enough information to allow the addressee to determine whether to read it or pass it on to someone else. The letter of transmittal should also include the following:

- The title of the report
- A clear statement of its subject and purpose
- A brief summary of major results, conclusions, or recommendations
- An acknowledgment of any assistance received in terms of advice, information, or special equipment or facilities
- A closing invitation that tells readers how they may contact you if they have any questions

B Note the descriptive title used here, *Seeing the Light: Multi-Wavelength Optical Networks*, rather than something more generic such as *Co-op Report*. A good title briefly summarizes the content of the report so that readers have a sense of what the report is about prior to reading it.

C As well as a descriptive title, the title page usually includes the version number and date (if applicable), the author's name and institutional affiliation (or course name and number), and the name and institutional affiliation of the person to whom the report is submitted. Note that the format Eric used to present this information is relatively plain and linear. The font is a sans-serif font (*Arial*) and only two font sizes are used (24 point for the title and 16 point for the remainder of the information). Nevertheless, the placement of the title itself is rather interesting, occurring as it does about $\frac{1}{3}$ of the way down the page.

One way to increase the impact of the visual elements of a document (and for this purpose, the title can be considered a visual element) is by placing them strategically on the page. Readers tend to look first at certain parts of a page when they initially glance at it or when reading it. As Figure 8.1 indicates, one of the first places that readers initially look at on a page is about $\frac{1}{3}$ of the way down the page, sometimes referred to as the *optical center*. By locating the title where he has (rather than in the middle of the page as do many writers), the author has subtly emphasized it.

Nor is this the only possibility that the author had for the placement of information on the page. As Figure 8.2 indicates, other information could have been located in one of the posi-

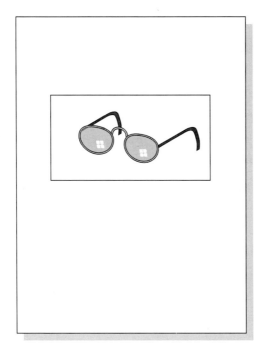

Figure 8.1 Optical Center of the Page

Seeing the Light: Ⓑ
Multi-Wavelength Optical Networks

Version 2.0 (January 2001)

Prepared by

Eric T. Hennessey

Simon Fraser University   Ⓒ
School of Engineering Science

Supervised by: Mike Moyer

Optical Link Budgets
Nortel Networks

tions of emphasis in the "Z" reading pattern. When readers first scan a page, they tend to let their eyes linger slightly longer at the beginning and the ending of the "Z", thereby emphasizing the information at those locations.

You might find it useful to experiment with different ways of presenting the information until you find one that is visually appealing (bearing in mind that usually the title and author are given the greatest prominence). In addition, you might consider how to visually emphasize graphics placed elsewhere in your reports (something that is not always possible, but worth keeping in mind).

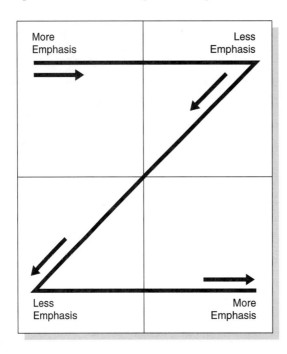

More
Emphasis

Less
Emphasis

Less
Emphasis

More
Emphasis

Figure 8.2 Positioning Graphics for Emphasis

D The *revision history* of a document is frequently included in the prefatory pages of project reports because these documents often undergo several iterations during the life of a project and may be revised by several different people. Project team members (or technical writers) need to know whether they are working with the most current document in order to avoid needless work (i.e., making changes that have already been made by someone else). In addition, they may need to know the nature of the changes made to previous documents, especially when reverting to an earlier design.

In this case, we chose to add the revision history page to this other type of report because we had worked with several different versions of the report. In other documents (particularly theses and user's manuals), the second page is generally used for a *copyright* notice, such as the following:

> © Copyright 1997 by Jane Rumplestiltskin. All rights reserved. This work may not be reproduced in whole or in part, by photocopy or other means, without the express written consent of the author.

This page can also be used to indicate that the report is confidential and that the material should not be circulated outside of the company. (Sometimes this confidentiality notice is placed on the title page and repeated in a header or footer throughout the document.) When this page is not included, then page *ii* is usually the abstract or executive summary.

E Note the convention used here for numbering the prefatory pages: lowercase Roman numerals (*ii, iii, iv*, etc.) centered at the bottom of the page. This convention, while traditional, is frequently not used in industry. Instead, reports are often numbered using Arabic numbers (*2, 3, 4*, etc.) placed in the upper right-hand corner of the document (or in the upper outside margins when printing on both sides of the page). Note that the title page rarely has a page number placed on it, even though it is counted as page *i* or *1*.

F An *abstract* should clearly indicate in less than a page the technical content and scope of the report, providing the main points without the supporting detail. In the abstract, you should generally avoid using acronyms (except for the most common ones), referring to other documents, or duplicating the language used in the introduction to the report. The abstract is not an introduction, but rather a succinct summary of the contents of the report that readers use to determine whether to read the entire document. Keep in mind that an abstract can be copied and placed in on-line databases or indexes. Researchers search by keyword to identify abstracts for

Revision History   **D**

Version 1.0 (December, 2000):   Original version of report prepared for Nortel Networks.

Version 1.1 (January, 2001):   Version of report prepared for Simon Fraser University, School of Engineering Science in order to fulfill co-operative education requirements. Changes included adding an equation and a general introduction.

Version 2.0 (January, 2001):   Version of report prepared for inclusion as an example in the text *Strategies for Engineering Communication* (Steve Whitmore and Susan Stevenson). The style and format of the report were revised to ensure that they agreed with the general recommendations of the text, In addition, a glossary was added.

ii   **E**

articles on a particular topic and then read the abstract to determine whether the full text is worth locating and reading. Abstracts should, therefore, be written as standalone documents.

The sample abstract, for example, was copied and placed in an index of abstracts in order for other students to determine whether they were interested in working for the same employer or whether they wanted to read the entire report to ascertain whether they were interested in working on this type of project.

An *executive summary* (sometimes just called the summary) is superficially similar to an abstract. The purpose for an executive summary, however, differs significantly from that of an abstract in that executive summaries are generally persuasive in nature while abstracts are informative. We provide more information about executive summaries and present an example in the section covering proposals later in this chapter (see pages 343–344).

Abstract

Many physical effects are manifest in the node-to-node transmission of multiple optical signals over a single fiber, all of which impair the reliability of data transmission. Some of these effects include linear chromatic dispersion, self- and cross-phase modulation, four wave mixing, stimulated scattering, and mid-span optical amplifier properties such as non-uniform signal amplification and the introduction of noise. Many strategies exist to compensate for these negative effects, and this report discusses several of the principal ones: dispersion compensation modules, fiber type, wavelength allocation plans, peak signal power control within the amplifiers, and transmitter launch power adjustment. To ensure reliable data transmission, some of these techniques must be specifically and procedurally employed by introducing additional mid-span equipment such as optical add/drop multiplexers, which extract and introduce new signals entirely at the optical level. An example of such a need involves adjusting the input signal power levels to compensate for amplifier-induced noise.

iii

 Often an acknowledgments section is included in the prefatory pages, but in this report, Eric chose to place his acknowledgments in the final paragraph of the conclusion because they were relatively brief. Had his acknowledgments been more extensive, he likely would have placed them on a separate page after the abstract. The acknowledgments section is used to thank individuals who have played a significant role in the preparation of the document. You may choose to thank people for their technical, writing, or clerical contributions, for funding supplied to a project, or even for personal support. Generally, acknowledgments are best kept short and to the point as in Eric's conclusion: "I thank Mike Moyer and Sik Heng Foo for their support in the writing of this report and for readily providing helpful technical assistance throughout the

Table of Contents   Ⓗ   Ⓘ

Ⓙ

iv

work term." A more effusive thank you could be embarrassing to the contributor or could suggest to the reader that Eric's contribution was minimal.

Ⓗ   A table of contents is generally provided for reports that are longer than 10 or 12 pages in order to enable readers to easily find various sections of a report. In addition, a table of contents can function as an overview of the structure of the report for readers, thereby increasing their comprehension of the technical material covered in the report.

In this particular report, the word processing program generated the table of contents automatically. That made it relatively easy to update when various sections were added and removed. However, the automatic table of contents does have some drawbacks of which you should be aware. First, most programs will automatically include a listing for the table of con-

List of Figures   Ⓚ

List of Tables   Ⓛ

v

tents in the table of contents (a rather illogical inclusion if you think about it). Consequently, you may find it necessary to edit the table of contents to eliminate this circular reference. Also note that the title page should not be listed.

Ⓞ Typically, the table of contents lists the headings down the left-hand margin and the page numbers down the right-hand margin. Often the headings are connected to the page numbers using *dot leaders*. These dots help the reader locate the page number that goes with a specific section. Although this feature is of little concern when the table of contents is relatively short, it becomes more important in lengthy tables of contents.

As we explained in Chapter 7, dot leaders are not simply a series of periods manually typed in, but rather they are related to the *tab* function and can be inserted automatically. In fact, if you try to use periods rather than dot leaders, you will quickly discover that not only do you waste a lot of time, but the page numbers will fail to line up properly. Also note that you use the left tab rather than the right tab to line up page numbers.

Another point that you should bear in mind with the automatic features for inserting a table of contents is that most word processors will take the default formats from the style sheet and apply them to the dot leaders and the page numbers. Consequently, you could end up with dot leaders and page numbers in varying font styles (plain, bold, italic, etc.) and sizes (12, 14, 16, etc.). This process results in a decidedly odd appearance for the table of contents. We recommend that you set up your own basic style for the table of contents. Keep in mind that the success of any report depends not only on its content but also on its appearance.

J The format of the table of contents should also clearly indicate the level of each heading. In the example, indentation is used to identify subsections. The numbering system, of course, also helps distinguish the level of the headings. The table of contents typically lists at least two levels of headings. In the vast majority of cases, you should need no more than three levels. An important exception to this guideline is that you may choose to list all heading levels for functional and design specifications to help readers locate specific information.

K If you have only one or two figures or tables in a relatively short (up to 10-page) document, you can generally omit the lists of figures and tables. In this case, however, the report has five figures and one table so including a list of figures is important. Note that lists of figures and tables are formatted in a fashion similar to the table of contents. Again the automatic features of the word processor can be used to produce these lists.

L Although figures and tables are generally listed on separate pages, if they are relatively short you can include both lists on the same page. In that event, the list of figures is physically separated from the list of tables by about an inch of space. Make sure that you do not simply mix together the figures and tables into a single list. In most cases, you exclude from these lists any figures or tables included in appendices. Graphics in appendices are typically referred to in the text by appendix and page (i.e., "See Appendix A, p. 18"). You can use the cross-referencing feature included in your word processor for these references.

M A glossary (sometimes called a list of acronyms or technical terms) defines any specialized terms or abbreviations used frequently throughout the document. Typically, a glossary is formatted by listing the technical terms on the left-hand side of the page (often in bold) and the definitions on the right-hand side. Providing a glossary is especially important when you are writing for a nonspecialist audience. Also note that when you are writing for a nonspecialist audience, you should further explain these terms in appropriate places in the body of the document.

N While page numbers are usually placed in the upper right-hand corner, they are sometimes located in the lower right-hand corner. (On reports printed on both sides of the page, numbers would, of course, alternate between left and right sides of the page.) Page numbers are sometimes omitted from page 1, but this practice is becoming increasingly rare.

Although not employed in the example report, headers are sometimes used to provide the company name, to indicate that a report is confidential, or in the case of a lengthy report, to identify subsections for the readers. Headers are usually located about one-half inch from the top of the page and about one-half inch above the text. The example proposal in the next section demonstrates how to use headers and footers.

O The headings for reports usually employ a different font than the body of the report. In this case, a sans-serif font, *Arial*, is used for the headings and a serifed font, *Times New Roman*, is used for the body of the document. The heading levels are also indicated by both a numbering system and decreasing font sizes (from 18 point for level one to 14 point for level two and 12 point for level three). Other methods you can use to distinguish between heading levels include the use of capital letters, italics, and indentation. As a final point, note that the numbering for this report is automatically supplied by the word processor, and we suggest you use this feature because it easily allows you to promote, demote, remove, add, or move sections without manually changing heading numbers.

Glossary

BER:	Bit Error Rate
DCM:	Dispersion Compensation Module
DSCM:	Dispersion Slope Compensation Module
DWDM:	Dense Wavelength Division Multiplexing
EDFA:	Erbium-Doped Fiber Amplifier
MOR:	Multi-wavelength Optical Repeater
NDSF:	Non-Dispersion Shifted Fiber
OADM:	Optical Add/Drop Multiplexer
OC:	Optical Carrier
OSA:	Optical Spectrum Analyzer
OSNR:	Optical Signal to Noise Ratio
STS:	Synchronous Transport Signal

vi

P The first paragraph of the example report provides general background about where the author was working and what he was writing about. Because this report was submitted to various university instructors who might be unfamiliar with both the company and the topic, this general background is necessary. However, in the initial draft of this report, which was solely intended for circulation inside the company, this section was omitted.

Q This section provides the technical background about Dense Wavelength Multiplexing necessary to understand the rest of the report. Note that the original report provided to the employer started with this section.

1

1. Introduction to Optical Networks

1.1. Nortel Networks

Nortel Networks has grown to become a global Internet and communications leader, with 75,000 employees worldwide generating over 30 billion dollars in revenues. Though Nortel has a broad focus, spanning optical, wireless, and local Internet, a significant portion of its revenue can be attributed to its optical networks division; sales of Nortel optical networks alone are expected to exceed 14 billion dollars in the year 2000. Optical networks, the primary means of fast and large-volume data transfer, are evolving at an incredible speed. They boast ever faster data transfer rates, with 400 gigabits per second (Gb/s) capacity commercially available, and continually increased reliability. Nortel currently boasts a bit-error rate (BER) of 10^{-15} on many of its networks.

1.2. Dense Wavelength Division Multiplexing

One attractive advance in network technology involves the transmission of multiple signals simultaneously over a single optical fiber, called dense wavelength division multiplexing (DWDM). DWDM technology enables an upgraded system using existing fiber plant to multiply its capacity many times over. Nortel is currently working towards 160-wavelength DWDM systems.

1.3. Complications of DWDM

Advances in DWDM, however, bring with them associated difficulties involving the fundamental properties of light and the method of its transmission. As optical signals simultaneously propagate down a fiber, they experience many physical effects, including the following:

- Linear chromatic dispersion
- Four wave mixing
- Self- and cross-phase modulation
- Stimulated scattering

Because these effects hinder the correct interpretation of signals at their destinations, allowances must be made and counteracting procedures performed. Additionally, a signal can be optically amplified a number of times between its point of transmit and its final destination. This amplification further modifies the signals, and the relationships among them, through the following amplifier properties:

- Non-uniform signal amplification
- Noise introduction

A variety of techniques must be used together to minimize these effects.

1.4. Mid-Span Access Equipment

Transmitting data reliably is complicated even more by the introduction of additional optical equipment between transmit and receive locations. This equipment performs useful tasks, such as routing with the use of optical add/drop multiplexers (OADMs) that extract and introduce new signals entirely at the

---

You should also note the convention employed for placing the various headings in relation to the text. The headings are closer to the text that they entitle (1 blank line) than they are to the preceding text (2 blank lines). We recommend that you follow this convention as it more clearly delineates sections than the practice of placing headings midway between sections of text.

Note that the references for where the information was obtained are provided in the text in the following manner: (*Willner 1997, 35*). This author-date convention helps readers identify the source for the information listed in the reference list. Note, however, that referencing conventions differ from discipline to discipline, and you should examine sample documents for your field and/or speak with your instructors or supervisors about which convention to use. Refer to Chapter 7 for an all-purpose convention (pages 274–286).

2

optical level. However, it also unintentionally modifies signal relationships such as power levels and noise quantities that are relevant to the quality of the signals. Special consideration must be given to the overall node-to-node transmission when this equipment is included. Later sections of this report describe techniques that are effective in circumventing such performance decreasing factors.

2. DWDM Optical Networks

2.1. DWDM Links Defined

Dense wavelength division multiplexing (DWDM) is a powerful approach to increasing the bandwidth of existing networks without laying more fiber. It involves transmitting multiple signals simultaneously over a single optical fiber. This approach is possible because a property of light is that multiple photons can occupy the same space, as illustrated by the unimpeded crossing of two flashlight beams (Willner 1997, **R** 35). Simultaneous transmission is accomplished by allocating certain signals to certain wavelengths of light, so that DWDM is analogous to the technique used to broadcast many different radio channels by placing them on carrier waves of different frequency. However, when many signals are concentrated in a narrow span of wavelengths, they must be spaced closely together, which leads to unwanted physical effects (see Section 3, Optical Layer Physical Effects).

An optical link is shown in Figure 1.   **S**

Service Terminating OC-192 Network Element   *Cascaded OC-192 Amplifiers*   Service Terminating OC-192 Network Element

Mixed Tributaries (OC-12/12c, STS/OC-48/48c)   Mixed Tributaries

Figure 1: An Optical Link[1]   T

An optical link begins when a network node transmits signals as light into a fiber span. These signals may be coming from many different sources and line-rates, called tributaries, with common sources including the Optical Carrier (OC) 12 and 48 levels (where the n in OC-n represents a capacity of (n * 51.84) megabits per second capacity). The electrical analog to the OC standard is the Synchronous Transport Signal (STS). Signals are combined into an OC-192 signal (10 gigabits per second per channel), and the light can travel long distances through optical fiber because it is optically regenerated at line amplifier sites as needed to preserve signal levels while not degrading the signal quality past a certain threshold. When the signal arrives at its destination, a network node receiver converts the light back into an electrical signal and further routes it. This conversion terminates the link.

As indicated in Figure 2, in DWDM, multiple transmitters operate at different wavelengths, sending optical signals into a multiplexer which combines onto one fiber the signals that were previously traveling separately. These signals then travel together and, for the most part, are amplified together. For reasons discussed in Sections 3 and 4, some wavelength groups are amplified separately. Often a link will be bi-directional, with many wavelengths travelling in different directions to reduce the interaction among signals. Some wavelengths may be routed away from the link and others added at special mid-span sites using optical add/drop multiplexers (OADMs). Once the signals are near their destination, they go

---
[1] All figures, except Figure 3, are from "S/DMS TransportNode: Advanced Optical Networking Solutions for Global High-Capacity Transport Applications", courtesy of Nortel Networks.   **U**

S   As the example report illustrates, prior to presenting a figure or table, you should mention it in the text. We recommend that you develop the habit of mentioning the figures and tables by number (*Table 1* or *Figure 3*) rather than by phrases (*the figure below* or *the table on the next page*). Although not that important an issue in the reports you prepare for school, when submitting articles for publication, failing to identify figures and tables by number could prove problematic. Because of space constraints, a figure that you expected to appear below where you mention it may end up appearing earlier or even several pages later.

We also recommend that you use your word processing software to automatically number figure captions. This practice will allow you to easily add, delete, or move figures without expending time and effort to manually renumber them. Similarly, we suggest that you cross

3

through a demultiplexer and are again separated into their individual fibers, after which they may be received. The Multi-Wavelength Optical Repeater (MOR) and MOR Plus series of optical regenerators support up to 16 and 32 wavelengths, respectively. A full MOR Plus DWDM link is shown in Figure 2.

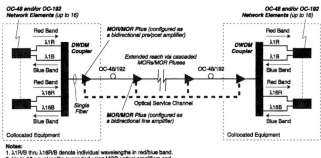

Figure 2: A DWDM Link

2.2. DWDM Systems

Two of Nortel's DWDM systems are the Multi-Wavelength Optical Repeater (MOR) Plus system capable of supporting 32 wavelengths and the OPTera 1600G series of amplifiers that commercially support 40 wavelengths (uni-directional), with plans for extensions up to 160 wavelengths. Both systems involve specific hardware used to reliably and consistently amplify multiple wavelengths, and follow specific standards, including which wavelength signals may be assigned to OADM sites and which may be added or dropped at these sites. For example, the MOR Plus standard allocates sixteen wavelengths between 1547.5 nm and 1561 nm to the red band, which propagates in the opposite direction to the sixteen wavelengths between 1527.5 nm and 1542.5 nm in the blue band. An extra wavelength is allocated for a spare and another for service.

Table 1 lists the wavelength allocation plans on the International Telecommunications Union Grid for the MOR and MOR Plus standards, which are valid on the fiber type known as non-dispersion shifted fiber (NDSF). Symbol X represents a valid channel, and OADM represents a valid channel that may be optically added or dropped, mid-span.

reference the mention of a figure (or table) in the text with the figure (or table) caption. This practice ensures that if figures are added, deleted, or reordered, the reference will automatically adjust to accommodate that change.

Ⓣ   Note that, following the traditional convention, the figure caption is placed below the figure, whereas an evolving convention places the caption above the figure. Also note the manner in which punctuation and capitalization are used in the figure caption. In the example report, the figure number is followed by a colon and all major words are capitalized. In addition, the caption is presented using a boldface font. Alternatively, you might use a period in place of the colon, and you might choose to capitalize only the first word following the colon:

   Figure 1. An optical link

4

Table 1: MOR/MOR Plus ITU Grid

ITU-T λ Grid 100-GHz Spacing (nm)	Number of Wavelengths in Application			
	Up to 32	Up to 24	Up to 16	Up to 8
1560.60	SPARE	SPARE	SPARE	SPARE
1559.79	OADM			
1558.98	OADM	OADM	OADM	
1558.17	X			
1557.36	OADM	OADM	OADM	OADM
1556.55	OADM			
1555.75	OADM	OADM	OADM	OADM
1554.94	OADM			
1554.13	X	X	X	
1553.33	X			
1552.52	X	X	X	X
1551.72	X	X		
1550.92	X	X	X	X
1550.12	OADM	OADM		
1549.32	OADM	OADM	OADM	
1548.51	X	X		
1547.72	X	X	X	
1541.30	SPARE	SPARE	SPARE	SPARE
1540.56	OADM			
1539.77	OADM	OADM	OADM	
1538.98	X			
1538.19	X	X	X	
1537.40	OADM			
1536.61	OADM	OADM	OADM	
1535.82	X			
1535.04	OADM	OADM	OADM	OADM
1534.25	OADM	OADM		
1533.47	OADM	OADM	OADM	OADM
1532.68	OADM	OADM		
1531.90	X	X	X	
1531.12	X	X		
1530.33	X	X	X	X
1529.55	X	X		
1528.77	X	X	X	X
	MOR Plus Applications Only		MOR and MOR Plus Applications	

2.3. Link Performance Measurement

To assess how well our system (link) is performing, and to know how we must modify it to enhance its performance, we need techniques of predicting how many errors will result over some length of time while transmitting our data. Two popular methods for predicting this bit-error rate (BER) involve the optical signal to noise ratio (OSNR) and system Q.

In other cases, figures are numbered according to the section of the report in which they are placed. Using this convention, Figure 1 in the example report could be numbered as follows:

Fig. 2.1-1: An optical link.

As you may also note, the above example abbreviates the word *Figure* to *Fig.*—another common option. Which particular convention you choose will often be a matter of which convention is widely used in your particular engineering field or what is standard within your company.

The author of this report used footnotes to indicate the sources of figures he adapted for the report. Rather than using footnotes, he could have simply made the reference part of the caption:

Figure 1: An Optical Link (Nortel Networks, 1998, 25)

He chose to use a footnote in this case because most of the figures came from the same source. This footnote uses a separator line and a smaller font to clearly distinguish the footnote from the text.

5

2.3.1. Optical Signal to Noise Ratio

OSNR is obtained by taking the optical signal power and dividing it by the noise power in a certain bandwidth, which can be easily observed through the use of an optical spectrum analyzer (OSA). Most demultiplexers include a monitor tap that the OSA can be connected to in order to obtain the OSNRs of all channels just before they reach their receiver sites. Although we observe a signal of reduced power, the noise floor is reduced proportionally, yielding the same OSNR. This information tells us how easily the receiver will be able to interpret the signal without errors. We also use data on the channel OSNRs relative to each other in a process called equalization, where input transmitter power is adjusted to obtain roughly equal OSNRs over all the channels.

2.3.2. System Quality

Another measurement technique involves obtaining system Q. Using this number in connection with an understanding of how an optical receiver makes optimizing decisions in the resolution of signals allows us to determine channel BER performance.

System Q is defined as in Equation 1

Ⓦ

$$Q = \frac{I_1 - I_0}{\sigma_1 + \sigma_0},$$
(1)

where I_1 is the signal level for the 1 bit and I_0 the signal level for the 0 bit, with σ_1 and σ_0 the respective root mean square noise variances in a 1 and a 0 bit respectively, at the receiver (Verreault, 1999, 599). For more information on system Q, see *Fiber-Optic Communications Systems* (Agrawal 1997, 170-172).

3. Optical Layer Physical Effects

3.1. Effect Descriptions

Optical layer physical effects such as fiber loss, linear chromatic dispersion, four wave mixing, self- and cross-phase modulation, and stimulated scattering all contribute to the degradation of optical signals as they travel through an optical fiber.

3.1.1. Fiber Loss

Every type of fiber has an associated amount of attenuation per length of fiber and is wavelength dependent. As the signal propagates down the fiber, it is progressively attenuated. As an example, Nortel Network's "S/DMS TransportNode: 200 GHz, MOR/MOR Plus, 2 to 16-lamda Optical Layer Applications Guide" lists a 0.21 to 0.25 dB/km loss over NDSF fiber. Optical amplifiers are necessary to amplify the attenuated signal if system reach is to be extended.

3.1.2. Chromatic Dispersion

In most current DWDM systems, chromatic linear dispersion is the largest factor in signal distortion (Verreault, 1999, 600). This effect occurs because silica fiber has the innate property that its index of

Ⓥ In the convention used here, the caption for the tables is placed above the table while the caption for a figure is placed below it. A different convention places all captions below the figures and tables while yet another convention places all labels above the figures and tables. Which particular convention you choose to use generally depends on the conventions typically used in the company or discipline for which you are writing. Perhaps the most important point here is being consistent irrespective of the convention you finally adopt.

Ⓦ Note that equations are centered from left to right and are punctuated as if they were part of the sentence. Also note that the equations are numbered using a convention that places the numbers inside parentheses at the right-hand margin. As with headings and captions, the automatic number feature of the word processor was used. For more information about using equations, see Chapter 7 (pages 272–273).

refraction is a function of wavelength. But the speed of a lightwave is dependent on the refractive index of its medium, meaning that different wavelengths propagate down a fiber at different speeds. A signal experiences no dispersion at a particular wavelength, λ_o, which is dependent on the type of fiber being used (see Section 3.2.1). A wavelength pulse is made from a number of spectral components that travel at different speeds down the fiber. Consequently, the transmission pulse spreads, with slower energy components lagging and faster components proceeding the main energy group. These variations create intersymbol interference and a higher BER.

3.1.3. Four Wave Mixing

Four wave mixing involves the transfer of power between wavelengths. The fiber's physical properties cause signals centered at optical frequencies ω_1, ..., ω_n to interact and create new frequencies at combinations and multiples of the parents. For example, we may find lower power signals at $\omega_i - 2\omega_j$ and $\omega_i - \omega_j + \omega_k$.

3.1.4. Self-Phase Modulation

Self-phase modulation occurs in high-speed optical systems due to fiber non-linearity. Specifically, the refractive index varies with light intensity. Because the rise and fall times of a bit are finite, the intensity variations induce a phase change (Willner 1997, 39), thereby creating new frequencies that then travel at different speeds down the fiber, resulting in either time compression or spreading of the bit.

3.1.5. Cross-Phase Modulation

Two pulses overlapping in a fiber cause a local power increase that changes the refractive index and causes another phase change in the same manner as that caused by self-phase modulation. The intensity at one wavelength modifies the refractive index at its own wavelength as well as the refractive index experienced by neighboring wavelengths. In this case, dispersion is beneficial because it spaces out the signals, reducing overlap.

3.1.6. Stimulated Scattering

The nonlinear effects of stimulated scattering have a variety of sources. Stimulated Rayleigh scattering is a loss mechanism involving light scattering from small-scale differences in the internal refractive index of the fiber. These differences are caused by microscopic density fluctuations introduced during the manufacturing process. Stimulated Brillioun scattering and stimulated Raman scattering both involve inelastic scattering of light with phenons, causing the wavelength of the scattered light to shift upwards (Agrawal 1997, 59-62). These effects reduce signal power and cause degradations.

3.2. Solutions

Some strategies that are employed to compensate for these negative physical effects include the use of specific fiber types, dispersion compensation modules, pre-distortion, and wavelength allocation plans, as described below.

7

3.2.1. Fiber type

Dispersion can be partially reduced by carefully selecting the fiber type to be used in the optical system. The following is a list of typical fiber types and their center frequency λ_o, as described in Nortel Network's "S/DMS TransportNode: 200 GHz, MOR/MOR Plus, 2 to 16-lamda Optical Layer Applications Guide" (1999):

- Non dispersion shifted fiber (NDSF): λ_o near 1310 nm, meaning positive dispersion for the typical wavelengths, in the 1550 nm range
- TrueWave Classic: $\lambda_o < 1530$ nm
- TrueWave RS: $\lambda_o < 1452$ nm
- LEAF: $\lambda_o < 1513$ nm
- E-LEAF: $\lambda_o < 1500$ nm

3.2.2. Dispersion (Slope) Compensation Modules

Because different wavelength channels travel at different speeds, we can observe their dispersion relative to one another to determine how much dispersion compensation is required. We can also determine polarity, which is positive if the shorter wavelengths travel faster than the longer ones, and negative if the shorter wavelengths travel slower than the longer ones (Willner 1997, 39).

One main method of dispersion compensation involves the use of dispersion compensation modules (DCMs), which are modules that introduce dispersions with opposite signs to the transmission fiber. As the wavelengths used become more numerous and over a greater range, the difference in dispersion experienced by each channel is no longer minor, and a dispersion slope compensation module (DSCM) must be used that provides wavelength dependent dispersion compensation.

Alternating between allowing the signals to disperse and correcting them as they travel along the fiber is useful because dispersion along the line helps to decrease nonlinear effects such as cross-phase modulation.

3.2.3. Pre-distortion

Most transmitters can be adjusted to pre-compensate signals with either a positive or negative chirp (which should be set to the opposite of the expected dispersion of the fiber). A chirp affects a pulse's spectral contents over a bit period such that the signal experiences compression rather than expansion over its propagation distance, which counteracts the fiber dispersion.

3.2.4. Wavelength Grids and Spacing

Standards are set for the wavelength distribution with three main points in mind.

1. We want as many wavelengths as possible, with as small a spacing between them as possible.
2. The larger the spacing between wavelengths, the more dispersion issues, but the less inter-channel cross talk, such as four-wave mixing.
3. Channels moving in opposite directions experience less inter-channel interactions.

See Section 2.2 for an example of four wavelength plans for the MOR/MOR Plus systems.

 Note the use of a bulleted list to outline typical fiber types and their center frequencies. The list is introduced by a grammatically complete sentence but is not punctuated as part of that sentence; however, in some reports, bulleted lists are punctuated. Choose the approach common in your field, or else pick the one you prefer and use it consistently.

Also, if you haven't already noticed, the convention being used for paragraphing involves inserting a blank space between paragraphs rather than indenting them. This blank-space convention is most commonly used in technical reports.

4. Amplifier Effects

4.1. Effect Descriptions

Similar to the optical layer physical effects treated above, mid-span optical amplifiers possess properties that degrade signal performance, including non-uniform signal amplification and the introduction of noise.

4.1.1. Non-uniform Signal Amplification

Doped fiber amplifiers are currently the most popular amplifier technology sold commercially (Verreault 1999, 600), with erbium-doped fiber amplifier (EDFA) technology leading the way. EDFA modules are capable of amplifying signals over the massive spectrum of 180 nm in the 1550 nm range. One drawback, however, is that the gain profile of the modules is not flat over this wavelength range, causing discrepancies in the amount of amplification each channel experiences. If these amplifiers are cascaded, this effect is significantly more pronounced.

4.1.2. Amplifier-Induced Noise

EDFA modules consist of silica glass fiber doped with erbium ions. The erbium ions are then excited to a higher and metastable energy state from which photons can cause them to fall, releasing a photon with the same phase and wavelength as the initial photon in the process and thus amplifying the signal (Willner, 1997, 33-34). This process is called stimulated emission. However, if not stimulated by a photon, excited erbium ions have only several micro- to milliseconds to live before decaying. This natural decay emits a photon with a random phase that adds noise to the system. Again, cascaded amplifiers greatly increase this effect.

4.1.3. OSNR Differences

We have achieved an equalized system when all of our channels have the same OSNR. We strive toward this goal, but it is hindered strongly by the two effects listed above. Because both contribute to channels at the receiver end of a link with unequal OSNRs, these effects must be countered.

4.2. Solutions

Tilt control, peak power control, grid amplification, and transmitter power adjustment can be used together to minimize many negative amplifier effects.

4.2.1. Tilt Control

Tilt control is an amplifier setting that helps to minimize the effects of EDFA non-uniform gain.

9

4.2.2. Peak Power Control

By putting the amplifiers in peak power mode and giving them a peak power goal, the amplifiers will attempt to ramp up their entire gain envelope to set the peak channel's power to that of the goal. If the total output hits an upper limit (also software-provisionable) before the peak power goal is reached, gain increases are stopped. This technique allows maximum amplification, while making sure a particular signal power (and thus OSNR) does not become too large in relation to the other signals – which would degrade them – and limits the extent of nonlinear effects experience by the channel.

4.2.3. Grid Amplification

In effect, we amplify each group of wavelengths independently of the other groups, which allows us additional control over amplification strategies such as the peak power setting. For example, in the MOR Plus system, we amplify the blue band in one direction and the red band in the other.

4.2.4. Equalization through Transmitter Power Adjustment

OSNR values can be adjusted by changing the link input (transmitter output) powers. A popular iterative approach to equalization involves repetitive calculation of average OSNR at the receiver site and adjusting all of the transmitters in steps toward this average. Slowly the OSNRs converge. Another method of transmitter adjustment involves gathering channel power information from each amplifier along the link. The following equation (Chraplyvy 1993, 428) indicates the output power of the ith transmitter should be set to an n span link:

$$P^i_{new} = P_{TOT} \left[\frac{P^i / (S/N)_i}{\sum_{j=1}^{n} P^j / (S/N)_j} \right], \qquad (2)$$

where $(S/N)_i$ is the OSNR of the ith channel. This equation allows us to perform transmitter adjustments remotely without any manual measurements. In large systems, this option is very attractive.

5. A MOR Plus with OADM System

5.1. System Diagram

Figure 3 shows a link that was setup in the lab to test the effects of mid-span OADM equipment.

The symbols S1...S6 represent six 100 km spans over E-LEAF fiber. NX represents a blue Pre-(MSA) amplifier and a red Post-(MSA) amplifier, and PX represents a red Pre-(MSA) amplifier and a blue Post-(MSA) amplifier. MLs are optical attenuators.

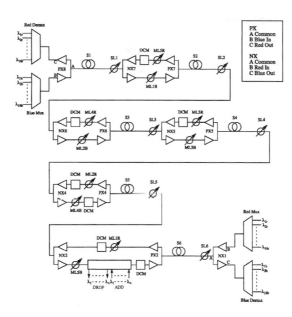

Figure 3: OADM System Diagram[2]

A typical one-wavelength (per direction) OADM site is shown in Figure 4. The above OADM site would look similar if expanded.

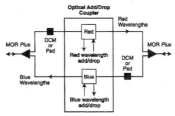

Figure 4: A Typical OADM Site

---

[2] Figure from "Level3 MOR+ OADM Equalization Test Plan," 2000, courtesy of Nortel Networks.

5.2. OADM Explained

The purpose of OADM is to route (re-direct) signals not just from node to node, but also in between nodes at the optical level. This routing saves time and effort because the signal does not need to be converted back into electricity, broken down, examined, re-built, routed, and converted back into light.

Figure 5 shows three typical applications of OADM equipment.

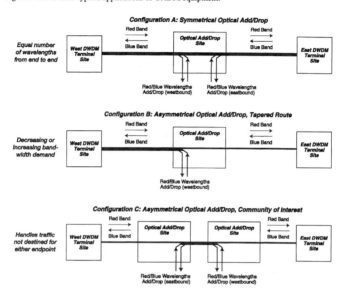

Figure 5: OADM Applications

6. OADM Performance Considerations

6.1. OADM Effects on Performance

OADM systems must be designed with concern for the physical and amplifier non-idealities discussed above. One of these considerations involves amplifier-induced noise.

12

6.1.1. Non-Equal Amplifier Noise

The channels being added have not necessarily passed through the same number of amplifiers as those channels from the west-to-east or east-to-west DWDM terminal sites, known as express channels. This discrepancy is a problem because, as noted in Section 4.1.2, amplifiers introduce noise onto the channels. If all channels pass through the same number of amplifiers, the amount of noise they have accumulated is constant, resulting in a roughly horizontal noise floor, neglecting the effects of tilt. Added channels may contain significantly different amounts of noise, yielding largely varying OSNRs, which put the system out of equalization.

6.2. OADM Performance Solutions

The problem with amplifier-induced noise encountered in OADM systems can be compensated by adjusting the added signal power.

6.2.1. Adjust Add Channel Power

A large signal OSNR existing on an added channel can be reduced by attenuating the signal at the time it is added to bring it into an acceptable range. Ideally, the signal power should be appropriately set at its transmitter site. When installing an OADM site in a link, this strategy must be kept in mind.

7. Conclusion

Optical networks are growing; they are becoming faster, larger, and more reliable. DWDM technology is a prime player in increasing network bandwidth. Increasing bandwidths and speeds, however, create an increased number of complications that must be taken into account in the design of optical network equipment and systems. Some performance-degrading effects to be aware of include linear chromatic dispersion, four wave mixing, self- and cross-phase modulation, stimulated scattering, and mid-span optical amplifier properties such as non-uniform signal amplification and the introduction of noise. To counter these effects, we can tailor systems with a combination of dispersion compensation modules, fiber type, wavelength allocation plans, peak signal power control within the amplifiers, and transmitter launch power adjustment. OADM systems, in particular, must account for the effects of amplifier-induced noise, which can be corrected through the dampening of signal power on the added channels.

As further advances are made in the field of optical networks, the effects discussed in this report will become increasingly significant. Fortunately, a strong base of knowledge and many effective counter techniques already exist to serve as guides to future solutions.

I thank Mike Moyer and Sik Heng Foo for their support in the writing of this report and for readily providing helpful technical assistance throughout the work term.

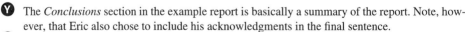

Y   The *Conclusions* section in the example report is basically a summary of the report. Note, however, that Eric also chose to include his acknowledgments in the final sentence.

Z   We draw your attention to the referencing conventions used in this report. In particular, note the general convention of providing author, date, title, publication details, and URL (in that order). Sufficient information is provided that should readers wish to find any of the sources, they can do so. A few other points to note are that the titles are in italics, not underlined, and that references are listed in alphabetical order. You will also note that five hyphens (———) are used to avoid repeating the author, *Nortel Networks* (in this case, the company is the author). Keep in mind, however, that referencing conventions differ from discipline to discipline and

13

8. References

Agrawal, G.P. 1997. *Fiber-Optic Communication Systems*. Toronto, ON: John Wiley & Sons, Inc.

Chraplyvy, A.R. *et al*. April, 1993. "End-to-End Equalization Experiments in Amplified WDM Lightwave Systems." *IEEE Photonics Technology Letters*, Vol. 4, No. 4, 428-429.

Foo, S.H. October, 2000. "Level3 MOR+ OADM Equalization Test Plan." Nortel Networks.

Nortel Networks. 2000. "OPTera Long Haul 1600 Optical Line System: 1600G Amplifier Optical Layer Applications Guide." Canada: Nortel Networks.

-----. July, 1999. "S/DMS TransportNode: 200 GHz, MOR/MOR Plus, 2 to 16-lamda Optical Layer Applications Guide." Canada: Nortel Networks, 1-7 to 1-20.

-----. November, 1998. "S/DMS TransportNode: Advanced Optical Networking Solutions for Global High-Capacity Transport Applications." Canada: Nortel Networks, http://www.nortelnetworks.com/products/01/sonet/collateral/56088_16Iss_2.pdf

-----. October, 1996. "Introduction to SONET Networking." Canada: Nortel Networks.

-----. No Date. "SONET 101." Canada: Nortel Networks, http://www.nortelnetworks.com/products/01/sonet/collateral/sonet_101.pdf

Verreault, M. *et al*. September, 1999. "BER Performance Equalization Strategies for DWDM Optical Networks," *The 15th Annual National Fiber Optic Engineers Conference Technical Proceedings*, 599-606.

Willner, A.E. April, 1997. "Mining the Optical Bandwidth for a Terabit per Second," *IEEE Spectrum*, 32-41.

from engineering field to engineering field, so you must identify the particular conventions required for any given writing situation. For further information about referencing, see Chapter 7 (pages 274–286).

Many technical reports end with appendices, which are particularly useful when writing for an audience comprised of both specialists and nonspecialists. In such a case, place most of the highly technical material in one or more appendices. Even when writing for an audience comprised solely of specialists, consider putting subsidiary information in an appendix (such as product monographs or specifications), particularly when including the information in the body of the document would prove disruptive to the flow of the document.

Generally, appendices are lettered rather than numbered (i.e., *Appendix A, Appendix B,* etc.). Each appendix is given a descriptive heading indicating its contents, such as *Appendix C: Test Data.* Tables and figures in appendices are sometimes identified using the appendix letter. For example, *Figure B.2* would be the second figure in Appendix B. Similarly, the pages in appendices are sometimes numbered using the appendix letter. For example, *C-3* would indicate the third page of Appendix C. This convention for numbering pages and figures and tables using the appendix letter seems to be decreasing in popularity; a simple Arabic numbering system (1, 2, 3, etc.) is more common in the contexts we know best.

8.7   PROPOSALS AND EXECUTIVE SUMMARIES

At school and at work, you can anticipate writing many proposals. Some of these proposals may be relatively simple internal one-page documents, but you may also be expected to write or assist with very lengthy and complex project proposals. In either case, you must keep in mind that the central purpose of a proposal is to *persuade* your audience to grant your request, to adopt your recommendations, or to award you the contract. The need to persuade is paramount whether you are asking permission to follow a particular course of research, bidding to provide a service or to design a system or device, seeking funding from an investor, or attempting to restructure a program or department.

When writing proposals, you must pay careful attention to the needs and desires of your audience. Precisely what do they want? What do they need? What are they concerned about? To the degree that you analyze and understand your audience, your proposal will be successful. Consequently, you must adopt a different stance than you might take with an informative report.

For example, if you are trying to persuade an investor to fund an innovative project, the investor will probably be less concerned with the technical details of the project than with the potential to make a profit from the investment. In this case, your treatment of the financial and economic issues involved will be key to successfully persuading the investor. In another case, you could be working with someone who requires a quick solution to a particular problem. Here, cost may be less important than clearly convincing your reader that you can meet the required deadline. In yet another case, the quality of a given device may be at issue, and you may want to emphasize the superiority of your technical design.

Most often, however, you will find yourself working with a client who has a combination of needs. For example, for a complex government or industrial project, many people will read over the proposal to assess whether you can successfully complete the project. Financial officers will assess its cost-effectiveness or potential profitability; senior engineers will assess the technical feasibility of the proposal; and senior company administrators will determine whether the plan supports overall corporate goals.

You should also keep in mind that preparing a proposal is often a competitive process. Sometimes, a client requests proposals from several different groups, hoping that one of them will offer an advantage the others do not. For important projects, governments often request proposals from dozens of engineering firms.

The sort of advantage a client might be looking for could be a lower cost, an earlier completion date, a higher-quality product, or a more marketable product. But, in general, you must convince your client of three things: that you understand their needs, that you possess the expertise (and, in some cases, the facilities) to design and deliver the device or service, and that the benefits outweigh the costs. In other words, you must persuade your client that your proposal is realistic.

In order to persuade a client that you understand their needs, you should ask several basic questions and address these in your proposal:

- When is the proposal due and can you meet that deadline?
- Can you actually complete the project? Do you possess the necessary expertise? Will additional information, material, or personnel be required? How soon would you need additional resources in order to complete the project?
- What does the client *say* they need? What does the client *really* need? (For example, what constraints in terms of time, size, cost, and safety are mentioned? Should other constraints be considered? Can the job be completed within these constraints? What is possible? Is the budget sufficient?)
- What are the alternatives? Does a similar system already exist?
- How long will this project take?
- How much will it cost?
- What are the possible technical solutions?
- What issues of safety and ethics are involved?
- What information that you need has the client left out?

In the body of the proposal, you should carefully and comprehensively define the nature of the problem to be addressed or solved. Generally, your proposal should be written from a *functional* perspective rather than a *design* perspective: *what* a device needs to do rather than *how* it accomplishes that task. In other words, if you were proposing to build a temperature controller, you would define what the device will do, why the device is needed, where it might be used, and who might use it. If your recommended approach is innovative, you should clearly explain the advantages of your approach over the alternatives. This explanation is particularly important if your approach is more expensive than others. Be sure to explain the additional benefits the client can expect for the increased cost.

You should also provide as much specific detail as possible about your plan for the project. Where do you propose to obtain the materials? What will the device or system do? How will you measure its performance? What are its limitations? Is a similar device or system already on the market? What are its limitations in terms of functionality and cost? What standards for quality control do you propose? What standards must be met (e.g., ISO, MIL, IEEE, etc.)? What issues of safety are involved? How will you deal with them? How feasible is the project given your constraints of time, funds, and expertise? What precisely are those constraints? Who is the user? What are their limitations in terms of knowledge, manual dexterity, and so on? What training will they require?

Once you have answers to these sorts of questions, you should be prepared to write your proposal.

8.7.1   Example Executive Summary

While the executive summary is presented at the beginning of a document, keep in mind that it is generally written last. The executive summary is a relatively brief one- or two-page summary of the document's substance. Although it is superficially similar to an abstract, the two should not be confused because they differ quite dramatically in terms of their intended audiences and purposes. Abstracts are typically designed for the technical specialist while executive summaries are expressly designed for the often-nontechnical readers who have final authority to approve the proposal and will not likely read the entire document. Consequently, they should be written in plain language for someone with a nontechnical education. Further, you should devote considerable effort to an executive summary because it *sells* your proposal to those who are in a position to grant your request and/or to approve expenditures.

The following example demonstrates how to structure the executive summary of a 25- to 30-page business plan for a high-tech company. The intended audience was the board of directors of Spectrum Signal Processing Inc., a world leader in digital signal processing (DSP). The summary was prepared as part of an assignment for a continuing-education program (*Management Skills in Advanced Technology*) for mid- to upper-level managers of high-tech companies. We would like to thank Glenn Mahoney for giving us permission to reproduce his summary. As you read through the summary and the commentary on it, note how much he achieves in a single page.

Ⓐ Note the use of a graphic down the left-hand side of the page. Although the meaning of this particular graphic may not be readily apparent to people unfamiliar with the company, the board of directors would recognize that it captures the "business" of the company and also illustrates what is being proposed in the business plan. As such, it helps to persuade the audience that the author has a fundamental understanding of the company and its core business.

Ⓑ Rather than simply calling this an *Executive Summary*, the author has chosen to provide a descriptive title that captures the central intent of the proposed business plan: *Leveraging Our Value Added*.

Ⓒ The author further demonstrates his understanding of the company by describing the three product areas the company focused on at the time.

Ⓓ Note that the author outlines, in brief, the logic of the business plan, explaining that higher-level (and more profitable) services and products can be generated by using the strength of the company's underlying DSP component business. He also mentions that the details of this approach are provided in the full business plan.

Ⓔ The author also briefly describes potential risks of the plan as well as how to reduce them. Note that by including these two sentences, the author has indicated to the board of directors that he is aware of the risks, but at the same time, he has not focused on them excessively in order to maintain a positive tone throughout the summary.

Ⓕ In this paragraph, the author has outlined the potential profitability of the plan, projecting that the company will increase revenues from 7.8 million dollars in 1995 to 92 million dollars in 2000. For many board members, this paragraph is probably the most persuasive because it focuses on some of their central values: growth, profitability, and risk management.

Ⓖ The final paragraph, with a few strategically selected words, captures the key reason for the board to adopt this business plan. Note that part of the impact lies in simple, direct statements.

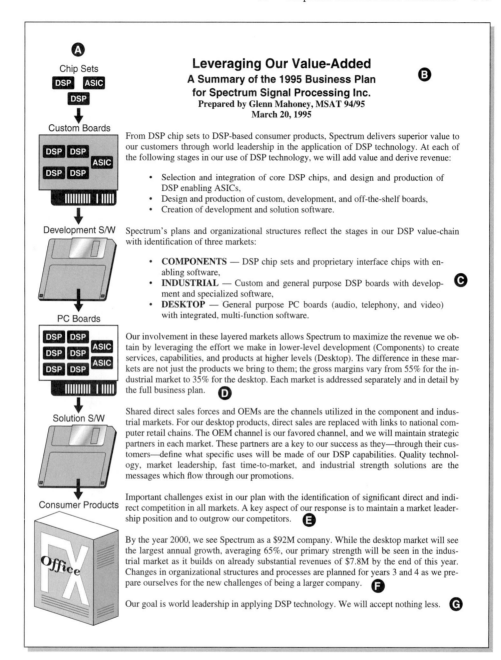

A Chip Sets

Leveraging Our Value-Added
A Summary of the 1995 Business Plan **B**
for Spectrum Signal Processing Inc.
Prepared by Glenn Mahoney, MSAT 94/95
March 20, 1995

From DSP chip sets to DSP-based consumer products, Spectrum delivers superior value to our customers through world leadership in the application of DSP technology. At each of the following stages in our use of DSP technology, we will add value and derive revenue:

- Selection and integration of core DSP chips, and design and production of DSP enabling ASICs,
- Design and production of custom, development, and off-the-shelf boards,
- Creation of development and solution software.

Spectrum's plans and organizational structures reflect the stages in our DSP value-chain with identification of three markets:

- **COMPONENTS** — DSP chip sets and proprietary interface chips with enabling software,
- **INDUSTRIAL** — Custom and general purpose DSP boards with development and specialized software, **C**
- **DESKTOP** — General purpose PC boards (audio, telephony, and video) with integrated, multi-function software.

Our involvement in these layered markets allows Spectrum to maximize the revenue we obtain by leveraging the effort we make in lower-level development (Components) to create services, capabilities, and products at higher levels (Desktop). The difference in these markets are not just the products we bring to them; the gross margins vary from 55% for the industrial market to 35% for the desktop. Each market is addressed separately and in detail by the full business plan. **D**

Shared direct sales forces and OEMs are the channels utilized in the component and industrial markets. For our desktop products, direct sales are replaced with links to national computer retail chains. The OEM channel is our favored channel, and we will maintain strategic partners in each market. These partners are a key to our success as they—through their customers—define what specific uses will be made of our DSP capabilities. Quality technology, market leadership, fast time-to-market, and industrial strength solutions are the messages which flow through our promotions.

Important challenges exist in our plan with the identification of significant direct and indirect competition in all markets. A key aspect of our response is to maintain a market leadership position and to outgrow our competitors. **E**

By the year 2000, we see Spectrum as a $92M company. While the desktop market will see the largest annual growth, averaging 65%, our primary strength will be seen in the industrial market as it builds on already substantial revenues of $7.8M by the end of this year. Changes in organizational structures and processes are planned for years 3 and 4 as we prepare ourselves for the new challenges of being a larger company. **F**

Our goal is world leadership in applying DSP technology. We will accept nothing less. **G**

(Labels in figure: Custom Boards, Development S/W, PC Boards, Solution S/W, Consumer Products, Office FX)

8.7.2 Example Proposal

We would like to thank Caroline Dayyani, Frederick Ghahramani, May Huang, and Shirley Wong for permission to use their proposal as an example in this text. This proposal was written for a third-year engineering project course integrated with an engineering documentation course. These courses provide an opportunity for the students to take part in a simulation of project development as it might occur in industry. Over a period of 13 weeks, teams of students in this course are expected to undertake a significant engineering project, proceeding from the proposal phase through analysis and design to the prototype stage.

The audience for the example proposal was the two faculty members who taught the courses. Although this proposal was prepared as an educational exercise, the instructors had a range of very practical goals for their students to achieve throughout the course: mastering engineering design and documentation practices, demonstrating teamwork, planning, and entrepreneurial skills, developing critical and creative reasoning processes, and acquiring various technical skills. A proposal would, therefore, be deemed successful only if it persuaded the two faculty members that a team's project would result in the students achieving these goals.

The example proposal meets this criterion and offers an excellent example of a persuasive proposal with the potential to result in a marketable device.

SMART SENSE
innovations

School of Engineering Science ● Burnaby, BC ● V5A 1S6
http://ssi.cjb.net ● ssi-ensc@sfu.ca

January 19, 1999

Dr. Andrew Rawicz
School of Engineering Science
Simon Fraser University
Burnaby, British Columbia
V5A 1S6

Re: ENSC 370 Project Proposal for a Mobile Paging Car Security System

Dear Dr. Rawicz:

The attached document, *Proposal for a Mobile Paging Car Security System*, outlines our project for ENSC 370 (Transducers and Embedded Systems). Our goal is to design and implement a programmable unit that will alert a car owner, through a mobile page, when the car alarm has been activated.

The purpose of this proposal is to provide an overview of our proposed product, an outline of the design considerations, our sources of information and funding, a tentative projected budget, and information on project scheduling and organization. This document also explores alternative forms of car security and this system's market potential.

Smart Sense Innovations consists of four motivated, innovative, and talented third-year engineering students: May Huang, Shirley Wong, Caroline Dayyani, and Frederick Ghahramani. If you have any questions or concerns about our proposal, please feel free to contact me by phone at (604) 123-4567 or by e-mail at ssi-ensc@sfu.ca.

Sincerely,

Shirley Wong

Shirley Wong
President and CEO
Smart Sense Innovations

Enclosure: *Proposal for a Mobile Paging Car Security System*

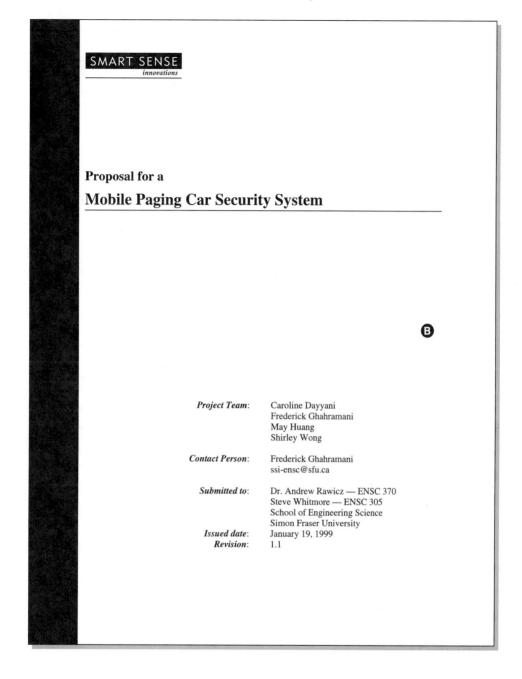

SMART SENSE
innovations

Proposal for a

Mobile Paging Car Security System

Project Team:	Caroline Dayyani
	Frederick Ghahramani
	May Huang
	Shirley Wong
Contact Person:	Frederick Ghahramani
	ssi-ensc@sfu.ca
Submitted to:	Dr. Andrew Rawicz — ENSC 370
	Steve Whitmore — ENSC 305
	School of Engineering Science
	Simon Fraser University
Issued date:	January 19, 1999
Revision:	1.1

A The letter of transmittal serves both an informative and persuasive function by briefly explaining the nature of the product, the scope of the proposal, and the individuals involved in the project. Also note that the team has provided contact information in the form of a phone number, an e-mail address, and a web site URL.

B The title page identifies the people on the project team as well as the intended recipients of the document. In addition, the title is clearly descriptive and the layout of the title page demonstrates team members' understanding that readers initially judge them on the appearance of their document.

 Proposal for a Mobile Paging Car Security System

Executive Summary

Jane has had the most stressful workday of her life. Her clients had her running all day, constantly paging her for support, all demanding immediate answers and attention. Finally, her workday comes to a close, and she longs to leave for a hot bath. Car keys in hand, Jane experiences the biggest shock of her life in the parking lot: some time in the last 14 hours, her car has been stolen and her day just got worse . . .

Jane's story is all too familiar these days. It seems we all know of someone who has been a victim of auto crime. Provincial statistics highlight a large increase in auto theft, and according to the Insurance Corporation of British Columbia (ICBC), more than 60,000 vehicles were stolen or vandalized in British Columbia in 1997 alone. This figure has increased more than 750% since 1989 and costs tax payers nearly $130 million every year.

Many preventative devices have been developed to deter, and even stop, potential auto thieves from stealing a car. Physical restraint systems have been designed to impede the movement of the steering wheel or to disable the forcefully entered vehicle. Alarm driven deterrent systems are popular, and in theory, alert passersby of illegal activity. However, false alarms have begun to annoy the public. A recent study suggests 68% of car drivers pay little or no attention to car alarms, and instead assume they are false alarms (RCMP, Burnaby Detachment sources). Moreover, these devices provide little information as to the exact timing of the theft or vandalism.

This document proposes developing a device that will interface with an already existing car security system. The device sends an alphanumeric mobile page to the car owner when the car is being vandalized or stolen. Such a device can be designed to interface with both existing preventative and deterrent systems. Using this device, the car owner will be provided with immediate feedback as to the state of their car's security, thereby empowering the car owner to take action instead of expecting passersby to be alerted by these alarms.

Smart Sense Innovations (SSI) consists of four third-year engineering science students with experience in analog/digital circuit design, telephony, and signal processing. SSI members are also well trained in a wide range of software design, from real time operating systems to microprocessor assembler programming.

We propose the engineering cycle for this project will encompass research, design, and construction. This cycle will span a 13-week period with April 1, 1999 as the scheduled completion date for an operational prototype. The entire project is tentatively budgeted at $800.00, which we expect to obtain from a variety of sources.

 The executive summary provides a comprehensive summary of the project, lists the experience of those doing the work, indicates the timeline and budget, and, most importantly, explains why the project matters. In particular, note how the brief story at the top of the page cleverly explains the students' motivations for undertaking this particular project. Also note the sense of design that comes through in the layout of the headers and footers.

Proposal for a Mobile Paging Car Security System

TABLE OF CONTENTS

 Proposal for a Mobile Paging Car Security System

1. INTRODUCTION

"Where there is wealth, there is crime"—This ancient Assyrian proverb couldn't be more true today. Car theft and vandalism is on the rise in British Columbia, costing tax payers upwards of $130 million every year, which is greater than the 1997 provincial deficit.

Car owners have some options in attempting to counter this rise in auto theft and vandalism. Many third-party after-market systems have been developed and marketed. These products range from physical preventative devices that lock the steering wheel or disable the car, to alarm driven deterrent systems that sound a siren alerting passersby of illegal activity.

The objective of our project is to develop a stand-alone module that will interface with an already existing car security system. The device will receive as input, the output from the car security system. The module will ultimately send a mobile alphanumeric page to a preprogrammed commercial pager number. The paging will be executed through the use of DTMF code generation and an embedded mobile telephone. Depending on the quality of the existing security system, our system will also be able to differentiate between the class of alarms (a "you are too close to the vehicle" alarm versus a "the window has been broken" alarm), subsequently sending 2 different types of pages.

A mobile paging car security system will give the car owner and law enforcement officials immediate warning when an auto theft or vandalism is occurring. This immediate surveillance feedback is promising, considering the statistics of the timing of auto theft and vandalism. According to ICBC, nearly 54% of auto thefts occur between 8 pm and 5 am, presumably after cars have been parked for the night. Another 28% occur between the hours of 9 am and 4 pm, presumably when commuters have parked for the day. Both time frames share the characteristic that the report of the theft or vandalism will occur several hours after the criminal act.

This delay between the realization of a crime and the occurrence of the crime, places law enforcement officials in a very difficult situation when attempting to investigate the theft. According to the Burnaby detachment of the RCMP "auto crime and property crime are recorded, but [are] rarely investigated unless we catch the thieves red handed, or have a special project to target a series of crimes." A device that alerts the car owner when a theft is occurring could aid law enforcement and finally catch all such thieves red handed.

This document is a proposal providing an overview of our product, outlining design considerations, sources of information and funding, and project scheduling. Alternate solutions and existing forms of this system are discussed and critiqued. Projected financial requirements and sources of funding are provided, as are project, Gantt, and milestone charts.

 The introduction expands on the reasons for undertaking this project and also provides facts and figures to support the authors' assertions. Providing this information persuades readers that team members have a clear understanding of the problem and have already undertaken some research.

innovations   Proposal for a Mobile Paging Car Security System

2. SYSTEM OVERVIEW

Figures 1 and 2 show the basic function of the mobile paging car security system. The sensing stage of the module is attached to the existing car security system. The state of the alarm system is monitored, and analyzed until it is determined that the alarm system is activated. At that point, the system utilizes the embedded mobile telephone to generate and output the alphanumeric page to the pre-programmed number.

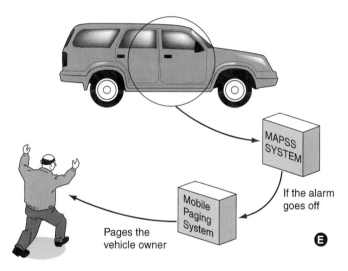

Figure 1: Conceptual Overview

ⓔ The conceptual overview is persuasive insofar as it clarifies for readers how the system might work. Note how the careful use of a graphic not only helps readers understand the system, but is also subtly persuasive.

Proposal for a Mobile Paging Car Security System

A simple programmable user interface is provided to allow pre-programming of the mobile pager number. Inside the processing unit, the output from the car security system is sampled on a consistent clock cycle frequency to detect an activation.

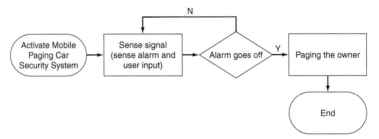

Figure 2: System Block Diagram

 Proposal for a Mobile Paging Car Security System

3. POSSIBLE DESIGN SOLUTIONS

Currently, with the help of new technology, there are many different ways to catch car burglars. Although current deterrent and restraining devices minimize car crime rates, these systems still have many disadvantages. One of the biggest problems in preventing crime is a need for a system which can be activated quickly in order to determine the actual time and place the crime occurred. Some of the devices currently used are listed below.

3.1. Car Alarm Systems

One of the problems with recent car alarms is that once activated, they send out an annoying sound which in most cases disturb other people until the owner of the car is informed. Moreover, we lack a fast way to inform the car owner about his/her car burglary. Also, in many cases people have difficulty determining whether their car alarm system has been activated or someone else's. Another problem is that even with new and expensive car alarms, the rate of auto thefts is still so high that some people do not trust car alarms at all.

3.2. Steering Wheel Locking Systems

Although using a steering wheel locking system (also known as the *car club*) is a less expensive way of preventing car thefts, it can easily be bypassed once the car door is broken. Another disadvantage is that the owner will not be informed, in any way, of when the crime took place. Since these devices are physically obstructive, they do slow down crime rates—but they don't stop the crime. For this reason, many people distrust using car clubs.

3.3. Existing Custom Pager Systems

Another option for car owners is to use custom pagers that will alert them when their car has been broken into. But these pagers don't have the range of commercial pagers. Moreover, the car owner has to buy another device on top of his/her existing pager. This system requires a person not only to pay more, but also to carry an extra device.

 By outlining several existing ways of deterring theft (as well as their limitations), the authors further indicate that they have analyzed the problem in some depth. Again, this detail has a persuasive function, in that evaluators are most likely to approve those projects they perceive to bc carefully thought through.

innovations   Proposal for a Mobile Paging Car Security System

4. PROPOSED DESIGN SOLUTIONS

Our proposed solution is to build a module that interfaces with an existing car alarm system and delivers a distress mobile page to the car owner once the alarm has been activated. Such a device would be very useful to law enforcement officials who would be given a chance to catch auto thieves or vandals *red handed*. This device would also be beneficial to the general public on two accounts. First, the financial burden that the crown insurance corporation (ICBC) must pass on to tax payers every year would be decreased. Second, this device would immediately alert the car owner of an alarm, thus eliminating the annoying phenomena of the neighbor who left his alarm on for 30 minutes.

A mobile paging car security system is not a novel idea. Already several car alarm developers have designed such a solution. Unfortunately, these solutions involve the purchase of a system-specific pager and do not offer as large a coverage (2 miles at best) as conventional paging networks. The simplification offered by our design is that any commercially available pager can be used as the alert device. As such, we are proposing an inexpensive and simple way of providing peace of mind to car owners who are at an increasing risk of being the victims of auto-theft and vandalism.

The main constraints in completing this project are the limited timeline and funding. We have only been allotted thirteen weeks to complete the project, and we must seek funding on our own initiative in that same time frame. Within these constraints, we will be able to build a simple module that will be customized only for a specific type of car security system and will send a distress mobile page to the car owner when the car alarm has been activated.

With more time and money, we would develop a more robust mobile interface unit. Such a unit could interface with any given security system or we could even develop our own security system. As well, a module that delivers a mobile page upon sensing a certain action could have a variety of other applications, such as implementations for home security or for the hearing impaired (in this last case, the device could be adapted to differentiate between the door bell, a phone ring, a fire alarm, etc.)

The number of applications for a robust mobile page interface unit is considerable. For our ENSC 370 project, however, we have decided to design the module specifically to combat auto theft.

G This section presents the advantages of the proposed solution to the problem and suggests alternative uses for the system. Because the intent of the course is for the teams to develop a potentially marketable device, pointing out the possibility of a broader market than might initially seem to be the case is quite persuasive. On a more subtle level, suggesting that the device could aid the hearing impaired is persuasive because the team members are aware that their primary readers are also interested in devices to assist the disabled.

Also note that this project team does not simply outline the advantages of the system, they also acknowledge its limitations, tight time frame, and the relatively large development funds required. These acknowledgments demonstrate that the authors have assessed their proposed project in a realistic fashion.

 Proposal for a Mobile Paging Car Security System

5. SOURCES OF INFORMATION

In researching and analyzing our problem, we will obtain information from a variety of sources: course textbooks, electronics periodicals, telecommunication publications, and manufacturers' component specification sheets for mobile phone and security systems.

The Internet will likely be a valuable resource for locating sources of funding, industrial contacts with similar projects, and other technical information. For example, the Statistics Canada website led us to the RCMP website, where we managed to find the phone number of a contact person who informed us of several grants that we could apply for. He also connected us with the ICBC information resources coordinator.

In addition, several faculty at SFU are currently involved in telecommunications research and will no doubt be invaluable resources for our project. As well, some undergraduate students in Engineering Science have previously worked on related projects and can be contacted for technical information. For example, Matt Stewart and Tim Norman conducted some student-level wireless research for their wireless fencing sensor.

Finally, perhaps the most intriguing and eccentric individual source of information is a team member's colleague from industry. At the age of 16, this individual was prosecuted for *hacking* several private corporation PBXs for long distance usage (a practice that earned the now reformed individual an instrumental design position at a telecommunications company two years later).

 Providing a section on sources of information is also persuasive, indicating that team members have considered what they *do not know* and have identified ways to gain the necessary knowledge.

Proposal for a Mobile Paging Car Security System

6. BUDGET AND FUNDING

6.1. Budget

Table 1 outlines a tentative budget for the mobile paging car security system. Many of the sub-components have been grouped with their functional equivalents. For example, a microphone will most probably be used and has been grouped with the "Mobile Phone and Accessories". Also, a pager must be activated for testing and has been grouped with the costs of the mobile phone. Most components have been overestimated by at least 15% to provide for contingencies.

Table 1: Tentative Budget

Equipment	Estimated Cost
Moblie Phone and Accessories	$400.00
Dialer	$200.00
Back up Power System	$100.00
User Interface	$50.00
Signal Splitter	$20.00
Cables	$15.00
Case	$15.00
Total Cost	**$800.00**

6.2. Funding

As with the design of any prototype, the initial engineering cycle will require more capital than the actual cost of the finalized product, especially given economies of scale once the original cycle has produced a prototype.

Due to the high cost of this project, many sources of funding are being considered. SSI is in the process of applying for the Engineering Science Student Endowment Fund, the Wighton Development Fund, and the ICBC Auto Crime Prevention Grant. As well, the Burnaby detachment of the RCMP has shown interest in contributing used equipment. We are also currently involved in lobbying two local government officials for nominal financial donations (the Burnaby MP and MLA), and a Burnaby consulting firm (AGF Robertson Inc.) has pledged to contribute financially to our project.

Our team members are willing to accept that we may not be able to generate enough original capital to sufficiently fund the entire project. If such circumstances arise, our team members are willing to share the remaining financial costs of the project equally. An accurate account of all financial transactions will be kept to ensure proper reimbursement to members. We also plan on entering our design in several competitions in the future, such as WECC (Western Engineering Conference and Competition) in January 2000 and miscellaneous other engineering competitions in hopes of reimbursing our members.

7

 The section on the budget demonstrates that team members have estimated their costs, and more importantly, have given considerable thought to potential sources of funding for their project. The inclusion of a 15 percent contingency figure indicates that they have also given some thought to dealing with cost overruns. Further, the two faculty members running the course were pleased that the authors recognized their own responsibility for potential cost overruns and funding shortfalls.

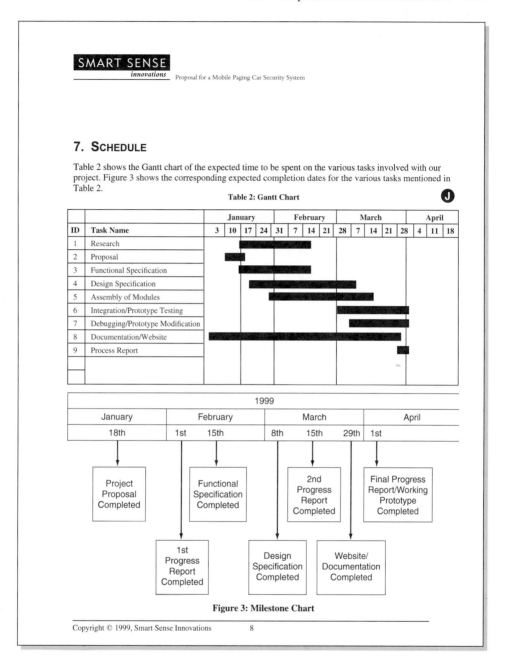

SMART SENSE
innovations   Proposal for a Mobile Paging Car Security System

7. SCHEDULE

Table 2 shows the Gantt chart of the expected time to be spent on the various tasks involved with our project. Figure 3 shows the corresponding expected completion dates for the various tasks mentioned in Table 2.

Table 2: Gantt Chart   J

ID	Task Name	3	10	17	24	31	7	14	21	28	7	14	21	28	4	11	18
		January					February			March					April		
1	Research																
2	Proposal																
3	Functional Specification																
4	Design Specification																
5	Assembly of Modules																
6	Integration/Prototype Testing																
7	Debugging/Prototype Modification																
8	Documentation/Website																
9	Process Report																

Figure 3: Milestone Chart

Copyright © 1999, Smart Sense Innovations   8

J The inclusion of the *Gantt Chart* and the *Milestone Chart* lets readers know that the team has devoted sufficient time to planning the project and is aware of project deadlines. These charts help persuade readers that the project can realistically be completed by the deadline.

Proposal for a Mobile Paging Car Security System

8. TEAM ORGANIZATION

Smart Sense Innovations consists of four talented and creative engineers: May Huang, Caroline Dayyani, Shirley Wong, and Frederick Ghahramani. All members are third-year engineering undergrad students, with differing program specialization interests. Our diversity of interests is important to note, for each member will contribute his/her specific expertise to achieve a common goal for this project. The members' specific skills are highlighted in the next section, Company Profile.

SSI's corporate structure is loosely organized in the following manner: each member is responsible for a specific field of operation in the corporation. But with a project this large, a great deal of the work will be shared and delegated among team members. Shirley Wong, President and Chief Executive Officer (CEO), is in charge of the overall progress of the project and is also responsible for resolving organizational conflicts. Caroline Dayyani, Chief Financial Officer (CFO), will manage the budget and resolve financial issues. May Huang, Vice President of Operations, is in charge of the technical operations of the project and is also responsible for acting as the "design elder" in suggesting alternative solutions to technical issues. Frederick Ghahramani, Vice President of Marketing, is responsible for generating capital through product marketing, shareholder recruitment, and web page design.

To ensure proper group dynamics and concise communication, the team has designated a meeting time once every week to discuss the progress of individual tasks. Instead of permanently appointing a member to be a scribe or chairperson for meetings, we have instead decided to have an open forum meeting structure with the only rule being "respect the opinions of others." At the beginning of the meeting, the team sets a timeline for the end of the meeting to guarantee that the dialogue remains on topic. Members who miss, or are late to meetings, will end up missing the doughnuts provided that week, so we don't foresee anyone missing a meeting too often.

The task assignment will be allotted based upon member's strengths and weaknesses. Upon functionally decomposing our solution into smaller modules, we plan on pairing up into groups of two to complete specific tasks. At this time, we foresee the need for three groups of two operating every week. This leads us to the structure where one person each week will be involved in two projects. The member performing a "double duty shift" will be alternated weekly to ensure that each member of the team receives a proper rest period and sufficient preparation time for other courses.

Many engineering students who have taken ENSC 370 in the past have noted the importance of team work and group dynamics in successfully completing their projects. With this in mind, we believe that SSI— with its open-minded yet focused teamwork ethic — will be successfully complete its project Moreover, SSI's members may still be friends after four months.

K One of the course instructors emphasizes team dynamics and communication and stresses how problems with student projects often arise from failures in how work is assigned and completed. The project team addresses his concerns in at least three ways: by each team member assuming a formal role in keeping with his or her individual expertise, by considering how to structure meetings, and by scheduling breaks for each team member to accommodate the demands of other courses.

Proposal for a Mobile Paging Car Security System

9. COMPANY PROFILE

Shirley Wong — Chief Executive Officer (CEO)

I am a third year Systems Engineering student at Simon Fraser University with a previous co-op term placement at PMC-Sierra. My skill set encompasses both software and hardware. Through course work and work experiences, I have programmed in C++ and 8086 assembler. I have implemented and designed test schemes for custom designed integrated circuits, designed TTL to PECL converters using MC10ELT20 translators, configured FPGAs as well as various other digital hardware. However, more important than my technical experience is my ability to communicate and work well with others.

Caroline Dayyani — Chief Finance Officer (CFO)

I am a third year Systems Engineering student at Simon Fraser University with previous co-op term experiences in the Opto-electronics Lab of Nortel Networks and Biomedical Departments of Vancouver General Hospital. I have programming experience in object-oriented design (C++) and Assembly language. I am familiar with the operations of most electronics equipment used in the lab such as oscilloscopes, power supplies, function generators, digital multi-meters, and FFT spectrum analyzers. I have experience with the design of PCB layout using PSpice and assembly of PCB boards by hand. As well, I am able to design and implement circuits at the logical switch and transistor levels. Further, I am familiar with the operation of different kinds of sensors, actuators, motors, generators, and feedback systems. Above all, I have good communication and team-work skills.

May Huang — Vice President of Operations (VP Operations)

I am a third year Electronics Engineering student at Simon Fraser University with experience in a variety of fields. In past years, I have taken courses in semi-conductor devices, analog and digital communications, and real-time and embedded systems. Some of the projects completed relating to these courses include a Plate Sorting System using pneumatic sensors and PLC, games implemented on the 68HC11 and displayed on an analog oscilloscope, and a Train Simulator and Emulator programmed using C and QNX.

Frederick Ghahramani — Vice President of Marketing (VP Marketing)

I am a third year Computer Engineering student at Simon Fraser University with two co-op work term experiences (Ballard and Nortel wireless). My strengths lie predominantly in software and digital processor design. More specifically, I have extensive experience with low level assembly and machine code debugging on a variety of processors (8086, HC11, PowerPC 604e). I've undertaken several self initiated telephony projects in the past and have experience with fundraising and marketing in general.

 The brief descriptions of the various team members are also persuasive in that they demonstrate that team members have the technical expertise to successfully complete the project.

 Proposal for a Mobile Paging Car Security System

10. CONCLUSION

Smart Sense Innovations is dedicated to applying technology to help reduce auto thefts and vandalism. The result of our goal is financially beneficial to car owners as well as tax payers. Along with the financial savings, comes the peace of mind in knowing that a reduction in auto crime can mean an increase in the number of safe neighborhoods.

Our proposed security system module would empower law enforcement officials in their pursuit of car thieves and provide car owners with instant feedback as to the state of their cars. Our approach is more cost effective than similarly existent systems through the use of already existing commercial paging networks. Our system is functionally superior to conventional car alarms due to the instant remote feedback.

The Gantt and milestone charts in the schedule section demonstrate that this project can and will be completed in the time frame allotted. We have highlighted our sources of information and research material. We have presented our potential financial sources, and have clearly defined our solution and proposed a strategy to achieve this objective.

We are confident that by April 1999, no matter how long Jane decides to stay in the office, she can rest assured that she'll never again have to face an empty car stall at the end of the day.

 The conclusion provides a final opportunity for the authors to persuade readers that they have the resources, expertise, and enthusiasm to complete the project. The ending is particularly effective in returning full-circle to the story presented in the executive summary, which leaves readers with a reminder of the team's motivation for undertaking the project.

 *innovations* Proposal for a Mobile Paging Car Security System

11. SOURCES AND REFERENCES

1) Apex Electronics (Kingsway, Burnaby)

2) BCTEL Mobility (Metrotown, Burnaby)

3) Radio Shack (Metrotown, Burnaby)

4) RP Electronics (Richmond)

5) Sergeant Cooke, Burnaby Detachment of the RCMP

6) Constable Scott Sheppard, Burnaby Detachment of the RCMP

7) Janice Knapp, ICBC Information Resources Coordinator, ICBC Public Affairs

8.8   WHAT'S LEFT OUT

In this chapter, we have not attempted to provide you with instructions for writing all the various kinds of documents engineers routinely write. Instead, we have provided basic guidelines and discussions of the most common types of documents. Using the rhetorical principles and strategies provided throughout this book, you should be able to problem-solve your way to a solution for just about any situation that requires a written document. The examples provided in this chapter serve two purposes: to demonstrate responses to specific rhetorical situations and to alert you to features common to a genre that you will normally want to follow in order to demonstrate your membership in a particular group. We have also not included an example of a hypertext document, but note that the discussion of hypertext documents in the previous chapter, pages 250–251, provides guidelines that can assist you in designing a document for presentation on the Web.

8.9   LAST WORDS

We conclude this chapter, and this book, not with another checklist, heuristic, and set of exercises, but rather with a reminder that a major aim of this text has been to prepare you to apply a range of strategies so that you can adapt to any number of communicative situations. The heuristics at the end of Chapter 6 (pages 240–241) and of Chapter 7 (pages 287–288) are designed to help you analyze documents, distinguish among genres, and craft your writing to present yourself as a *bona fide* member of the engineering community. As a final exercise, we encourage you to obtain samples of the documents that typify your particular field of engineering and then apply those heuristics in order to fully understand the documentation practices current in your field.

We both wish you great success in your careers as engineers. We hope that your work changes the world for the better and that you are successful in communicating to others why your work matters.

Index